Panicle Architecture of Rice and its Relationship with Grain Filling

Pravat K. Mohapatra • Binod Bihari Sahu

Panicle Architecture of Rice and its Relationship with Grain Filling

Springer

Pravat K. Mohapatra
School of Life Science
Sambalpur University, Jyoti Vihar
Sambalpur, India

Binod Bihari Sahu
Department of Life Sciences
National Institute of Technology Rourkela
Rourkela, India

ISBN 978-3-030-67895-1 ISBN 978-3-030-67897-5 (eBook)
https://doi.org/10.1007/978-3-030-67897-5

This Springer imprint is published by the registered company Springer Nature Switzerland AG
The registered company address is: Gewerbestrasse 11, 6330 Cham, Switzerland

Preface

The antiquity of rice cultivation on Earth is around 10,000 years old. Rice is the most important food crop regarding human nutrition and caloric intake. Ever since domestication, rice crop has been the most widely consumed staple food for over half of the human population on Earth. As of now, the crop has fed more people than any other cereal like wheat, maize, or barley. For densely populated Asian nations, it has been the most important food source. It is an economic crop for millions of poor marginal farmers and workers, who grow the crop in vast stretches of land areas under a wide spectrum of ecological niches differing prominently in land contour, soil moisture regime, ambient temperature, and incidence of pests, diseases, and other natural calamities. Being sturdy, rice survives in multiple habitats, unlike other cereals. Habitat-specific resilience garnered has sustained the life and livelihood of the hapless native farmers far removed from modern agriculture. Although rice contributes richly to the economy and food bowl of Asian nations and is complimented twice by United Nations as an expression of gratitude by human society, there has been inadequate effort to revise the yield potential for global food security and socio-economic stability.

Panicle of rice inflorescence is the topmost organ of the plant borne on the last internode. It is composed of primary ramifications (primary branches) which carry secondary branches themselves. Both the branches carry spikelets, one on each node atop a pedicel. The architecture of panicle is a complex trait inclusive of quantitative features like panicle size and the numbers of branches and spikelets/grains and their distribution pattern, which directly determine grain yield. The regulatory mechanisms underlying this complex trait are scarcely explored precluding genetic resources for rice breeders to break yield ceiling and improve rice production. Spikelets are the individual flower units of rice panicle, each capable of bearing a well-developed grain at maturity. However, the interplay of genetic and environmental factors underscores panicle grain number to the detriment of grain yield very often. The molecular and genetic mechanisms underlying the generation of a large number of spikelets in the ontogenic phase and degeneration of many of them later in the pre-and post-anthesis stage are crucial issues for the development of designer rice in achieving high yield. Panicle traits of rice are most intuitive; over the last two to three decades, several genes characterizing various events of panicle development have been identified and functionally illustrated. It looks optimistic now to explore the mechanisms related to panicle diversity, abortion of grain development in basal two florets of rice spikelets, and degeneration of spikelets numbering beyond the intrinsic sustainable capacity of the panicle. The book guides rice scientists in improving rice yield potential.

Sambalpur, India Pravat K. Mohapatra

Rourkela, India Binod Bihari Sahu

Contents

Chapter 1
Importance of Rice as Human Food

Abstract Rice is the most important grain crop in the world. The worldwide diffusion has established rice as the most important human food. Rice farmers feed and have fed more people than any other important cereal crop like wheat, barley, and maize. The Antiquity of rice reveals the domestication of rice as one of the most important developments akin to human civilization culture and food habit. Rice is unique for biodiversity numbering cultivars more than 140,000 diffused across an enormous range of geographic locations and ecologic environments of the southern and northern hemispheres of the globe. For success in providing food security, rice has been the unique crop honored by the United Nations twice as an expression of gratitude by mankind. The cultivated rice originated from a perennial stock. In the long process of evolution, the putative progenitor-like wild *Oryza ruffipogon* has advanced into an annual form *Oryza nivara* before the origin of annual cultigen *Oryza sativa* L.. Matching with the unabated growth of rice consumers, breeders have manipulated changes in rice phenotype for higher grain yield by accessing the rich diversity. Yield apart, the diversity has also assured the climate resiliency of the crop in the years to come. Progression of phenotype changes with the intent for revision of genetic yield ceiling is evident in the comparative assessment of rice cultivars used by the farmers, such as traditional, semi-dwarf, hybrid, new plant type, and super rice.

Keywords Rice antiquity · Food security · Origin · Biodiversity · Yield potential

Introduction

Rice is one of the most important grain crops in the world. It is grown extensively in many parts of the world at different times of the year under different climatic conditions. The grains are exclusively used for human consumption while other plant parts are utilized as animal feed and several domestic needs of the farmer. Rice is a staple food for nearly half of the seven billion people on the earth (Mohanty 2013). No other crop of any significance provides food calories to as many people on the earth. Of the 90% world's rice, China and India account for 30 and 22% respectively. The majority of the rice consumers belong to the low-income countries

P. K. Mohapatra, B. B. Sahu, *Panicle Architecture of Rice and its Relationship with Grain Filling*, https://doi.org/10.1007/978-3-030-67897-5_1

of tropical Asia and for them, rice accounts for more than 76% of caloric intake (Fitzgerald et al. 2009). Especially for the poor marginal farmers, rice is the principal food crop and also the major source of employment and income. For them, rice has been the single largest food source. The crop is the source for one-fourth of the global per capita energy. Since the last 12,000 years or so, the cultivation, domestication, dissemination, and diversification of the crop have been a facet of human history, particularly in the rice- eating nations of Asia and Africa, being intricately related to culture, tradition, and societal rituals and above all employment. Rice is grown traditionally on the farm and exclusively consumed on the farm. It is seldom traded outside for commercial purposes. In tropical Asia, climatic conditions are unstable and annual rainfall is limited to 2–3 months of the year; the seasonal availability of water sustains one crop per annum. Heavy monsoonal rains inundate crop fields and make the soil swampy. Soil submergence coupled with a high temperature of the wet season renders the field unsuitable for the cultivation of any grain crop other than rice. Additionally, the lack of infrastructure for the storage of run-off water precludes the possibility of large-scale crop cultivation in the lean season and makes it a viable economic option for the farmers. The success of the agrarian economy in South and South East Asia, therefore, depends on the success of rice cultivation. The seasonal rainfall pattern and other environmental factors are conducive for rice only and rice cultivation dominates the agro- ecosystem.

Rice Is Unique

Further, unlike other cereals, rice is versatile in being comparatively sturdy and resilient to various physicochemical stresses. Rice grows under a wide spectrum of ecological conditions, ecosystems, and cropping patterns (Huke and Huke 1990); the most important factor for adaptation being the soil water regime (Mahapatra et al. 1985) when extreme temperatures are not limiting growth. In particular, the depth of water level of the field dictates the phenology of the crop. Low-yielding ephemeral type cultivars are preferred for cultivation in upland terrains, where the absence of accumulated field water curtails growth duration and restricts biomass production. In contrast, plant morphology changes significantly where floodwaters gather. In this condition, crop duration increases because of the extension of the vegetative growth period essential for survival under challenging conditions. In the sequel, flower initiation is delayed, while the plant waits for the recession of floodwater in the period following the rainy season. Prevalence of shorter day length at this time makes the low land cultivars strongly photo-sensitive. The expression of phenotypic differences in response to variation of field water depth is evident in the wide diversity of rice cultivars adapted to different water regimes worldwide. The plant survives soil submergence due to the presence of an efficient air passage system from the roots to the leaves that functions as a snorkel for the supply of free oxygen to the submerged parts.

Rice in Different Parts of the Globe

Additionally, rice is the only crop that adapts and sustains growth in the typical nutrient cycling that occurs in the flooded fields. This uniqueness in rice has helped the dissemination of the crop into a large part of the globe ranging from 59° north latitude in the banks of Amur River, Russia to 40^0 south latitude of central Argentina. Apart from Asia, where it is grown abundantly, it is also becoming an important food crop in Latin America and Africa. There are over 140,000 rice varieties of *Oryza sativa*. In the rice gene bank of International Rice Research Institute, the Philippines over 80,000 samples of cultivated and wild rice have been conserved *ex-situ* although *in-situ* conservation has been proposed to be an important conservation strategy for the future (Jackson 1995). This large diversity includes landraces, high-yielding inbred cultivars, hybrids, and related wild relatives dwelling in various agro-ecosystems *viz.* altitude, temperature, soil moisture regime, and the type and finally developing varied morpho-physiological features. Apart from the phenotypic studies, the genetic diversity of rice has been assessed using new techniques like RAPD, ISSR, AFLP, STMS, and SNP markers. Rice cultivars are grown in widely contrasting habitats on every continent save Antarctica. Rice habitats range from the wettest areas of 5100 mm rains in Arakans of Mynamar to feeble rains of 100 mm in Al Hasa Oasis, Saudi Arabia, the average temperature of 33 °C in upper Sind, Pakistan to 17 °C in Otaru, Japan, sea level fields of coastal plains and river delta regions throughout Asia to high upland fields at the height of 2600 m in the mountains of Nepal; solar radiation potentials ranging from 25% under cloudy skies of south Asia to 95% bright sunshine in southern Sudan and Egypt (GRiSP 2013) and saturated soil with no standing water on mountain terraces of Asia to deep floodwaters fields of Thailand and Bangladesh. The worldwide diffusion has established rice as the most important human food on earth. Rice farmers feed more people than that of any other important cereal crop like wheat, barley, and maize.

Rice for Food Security

Food security emerges in a country when all citizens have physical and economic access to sufficient nutritious and safe food for dietary purposes. It has three important components, such as availability, access, and absorption of food. These components are linked to factors like socio-economic activities, environment, and human rights. The factors, when increasingly disturbed, shore food scarcity and price rise. The rise in food prices is bound to have escalatory consequences on poverty, malnutrition, health care expenses, and the economic burden of citizens.

At the present level rice is consumed by half of the world's population mostly living in Asia. Just after World War II, population growth surged in Asia. Two failed monsoons over the Indian sub-continent in 1965 and 1966 affecting rice production generated considerable pessimism for global food security among food policy analysts. The demand for consumption far outstripped the level of rice production; the widening gap between the levels of production and consumption-led experts to predict starvation. The only option available for food security was to revolutionize rice cultivation and production. Food and Agriculture Organization of the United Nation wisely declared 1966 the year of rice and many rice- growing nations adopted measures to improve rice production, grain quality, and marketing. Rice production increased because of the adoption of high-yielding cultivars, the extension of land area under irrigation cover, application of chemical fertilizers, and mechanization of the farm in the Green Revolution. The transitory success of the revolution in the alleviation of rural poverty through enhancement of rice production vanished in the subsequent 3–4 decades. Several constraints for rice production surfaced anew, such as high level of environmental pollution, increasing cost of the labour force, yield barrier of high yielding rice, shrinkage of *per capita* land area for rice cultivation, and diminishing resources of water. Added to the disadvantage was the unchecked and unabated rate of growth of population in Asian countries. The demand for rice consumption also increased because of changes in the lifestyle of modern society. Easy-going and wealthier urban people prefer rice for less cumbersome cooking and good taste. In consideration of the high importance of rice as human food, United Nation launched the International Year of Rice 2004 on 31 October 2003. Rice is the only food crop that has been honored by United Nation twice (Rice is Life, IRRI 2005). This unique tribute bestowed on rice by mankind is an elaborate expression of gratitude for survival.

Origin and Biodiversity

The wide biodiversity of rice and the successes of its production in contrasting habitats, encompass an enormous range of geographic-ecologic adaptation in the development of an efficient system that enhances the scope for biomass production as well as active physiological functions in most of the ecosystems on the earth. The absence of this essential attribute in other cereals makes them less suitable for cultivation as resilient crops are challenged by different physicochemical stresses. In contrast, rice holds promise to encounter the onslaught of climate change-induced environmental stresses, the tenacity of which may become more threatening shortly. Although antiquity of cultivation has not been firmly established hitherto, domestication of rice is one of the most important developments akin to human civilization (Sampath 1985). Rice farming is more than 10,000 years old (GRiSP 2013), there is a consensus that the plant has fed more people on earth over a longer time than any other crop despite the incidence of several natural calamities. The resilience of rice-based production systems in drought-, flood- and salinity-prone environments is unparalleled.

The line of descent for Asian rice *Oryza sativa* is most probably from a wild perennial stock to wild annual and onto annual cultigens (Fig. 1.1). The process of origin and domestication was initiated from the perennial wild rice *Oryza rufipogon*

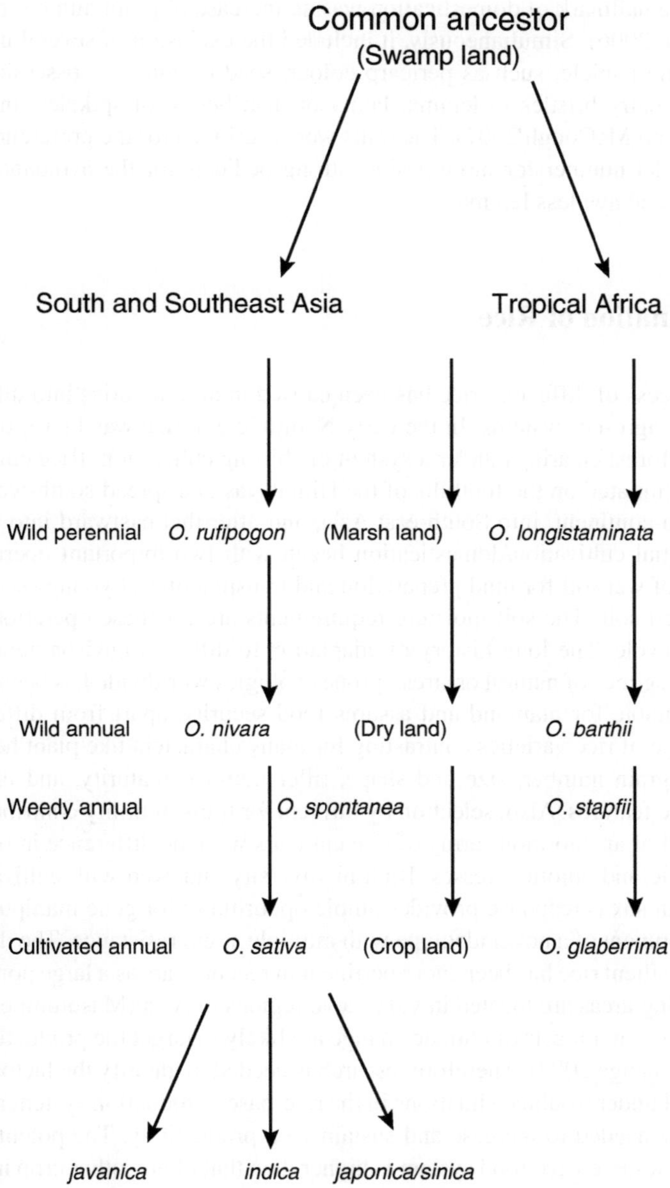

Fig. 1.1 Origin of rice from wild ancestors

(Cook and Evans 1983). In the long process of evolution, this putative progenitor advanced into the wild annual form. This development facilitated delineation between perennial and annuals forms, where the new form was named *Oryza nivara*. In comparison to perennial, the annual wild *Oryza nivara* propagated by seed that could have been the immediate progenitor of annual cultigens *Oryza sativa* (Chang

2003). The hallmark of domestication was an increase of grain number per panicle (Tian et al. 2006). Simultaneously, it included the exclusion of several undesirable traits of the panicle, such as pericarp colour, seed dormancy, possession of long awns and hairy bristles in lemma, laxity of distribution of spikelets in a panicle (Sweeny and McCough 2007). The traits were sacrificed for the preference given to high spikelet number for more grains, strong pedicels for the avoidance of grain shedding, and awnless lemma.

Dissemination of Rice

In the process of diffusion, rice has been carried in all directions into different climates and agro-ecosystems. In the early Neolithic era, rice was first grown direct-seeded in forest clearings under a system of shifting cultivation. Rice cultivation in Asia was initiated on the foothills of the Himalayas and spread south-west into the Indian sub-continent, into South-east Asia, and after that eastward into China and Japan. Initial cultivation/domestication began with two important operations like puddling of wet soil for mud preparation and transplanting of young seedlings into the puddled soil. The soil moisture requirements are for these operations and the plant life cycle. The long history of adaptation to different environments and soil moisture regimes of natural or stress-prone ecologies worldwide has been economically profitable for mankind and assures food security, apart from differentiating into a range of rice varieties contrasting for many characters like plant height, panicle size, grain number, size and shape, tiller number, maturity, and many other phenotypic features. Also, selection by farmers for traits of utility contributed to the production of an enormous array of rice cultivars with the difference in response to many biotic and abiotic stresses. Rich biodiversity endowed with cultivar-specific variation in stress response provides ample opportunity for gene manipulation and the development of a novel ideotype with multiple stress resistance. The demand for climate-resilient rice has been more pertinent in recent years as a large portion of the rice-growing areas are located in vulnerable regions of Asia (Masutomi et al. 2009) and the adverse impacts of climate change are likely to target the production system (Climate Change 2007). Therefore research is needed to identify the factors limiting grain yield under troubled situations in the rice-based production system and design technology needed to increase and sustain crop productivity. The potential of rice for a climate-defensive food system is higher than that of any other crop in the present context.

In a nutshell, the significance of the rice crop has been summarized into the following points.

(a) Rice cultivation and consumption are associated with human food security.
(b) Small and marginal farmers with farm sizes lesser than one hectare and the number of such farms worldwide are 144 million. Planting on minuscule plots is labor- intensive.

(c) Rice is the most important source of food, employment, and income for the rural poor. For these people living mostly in the tropics, there is no choice for a cereal crop other than rice. The annual rains limited for 2–3 months continuously are heavy and they flood agricultural farms making them unsuitable for wheat, barley, or maize cultivation.
(d) Rich biodiversity ensures diffusion into habitats contrasting for land topology, moisture regime, and other cultural conditions unparalleled for any crop of economic importance.
(e) Long history of cultivation is intricately related to food habits and human culture with important influence on social order, religion, and traditional customs.

Demand for More Production from Rice Lands

Making the rice-based production system sustainable is crucial for the future food security of mankind. Rice habitats worldwide are classified mostly on two factors, altitude (low land and upland) and water source (rainfed and irrigated). In the lowland, the rice field has saturated soil moisture with ponded water for 20% of growth duration. In comparison, ponded water remains in the area for 80% of growth duration in irrigated lowland. On a finer level, rice habitats worldwide are classified into four different categories (Bouman et al. 2007), such as irrigated lowland 79 million ha, rainfed lowland 54 million ha, flood-prone 11 million ha, and upland rice 14 million ha. The levels of production from the habitats however vary considerably. More than 75% of rice produced globally comes from 50% of the world's rice lands that are irrigated (Khush and Peng 1998). Irrigated rice lands are relatively more stable in comparison to the other habitats where the ecosystem is vastly fragile and grain production is subservient to environmental fluctuations. With the scope of yield enhancement being less likely in the other habitats, rice scientists look for opportunities to manipulate factors controlling grain yield in the irrigated conditions.

Increase in Rice Production and Consumption

The Green Revolution of the late 1960s represented a significant improvement in rice crop productivity through the introduction of new genotype miracle rice IR8 (Evenson and Gollin 2003). Indeed the level of production which was about 200 million tons in the early 1960s increased to more than 600 million tons in 2010. During this period, the success of new high- yielding cultivars also depended on in tandem improvement of the cultural conditions that assured stability of irrigation, soil fertilization, and eradication of pests. The world population was 3 billion in the 1960s and most of the rice consumers were underfed and under - nourished. The revolution brought about by the introduction of miracle rice effectively solved the interacting problems of food production, rapid population growth, and rural poverty, especially in the developing countries of Asia (Seck et al. 2012). Rice production in Asia increased at an annual rate of 3% during 1967–85, which far exceeded the population growth rate of 2.14% (IRRI 1995). In the 1960s per capita, rice consumption in Asia was 85 kg; it increased to 100 kg in 2010. The world population clock reports the presence of nearly 7.4 billion people on earth in 2015; compared to other countries the growth is very sharp in rice- growing nations of Asia. The level of growth is predicted to cross 9 billion by the year 2050, compeling the need for a multifaceted global strategy to ensure sustainable food security of rice consumers (Godfray et al. 2010).

This enormous growth of population will put a projected demand for the production of 900 million tons of rice. According to GeoHive (2015) world, rice production was 316 million tons in 1970 and it increased to 396 in 1980, 518 in 1990, 600 in 2000, 701 in 2010, and finally 745 million tons in 2013. Rice production in 2017 is estimated to be 758.9 million tons (FAO, April 2017). But the progression of growth to a level of 900 million tons by the year 2050 would be more daunting than before unless rice yields are augmented substantially through innovative research. This economic estimation has been corroborated by Alexandratos and Bruinsma (2012) forecasting a 50% rise in rice production for food security of the rising global population. Sheehy and Mitchell (2011a, b) have analyzed the global concern for food security and rice research necessary to prevent the catastrophic consequences of population growth and climate change. There are two subspecies of cultivated rice *Oryza sativa* L., namely *japonica* and *indica*; the latter is grown extensively under warm tropical climate in south China, South-east Asia, and South Asia occupying nearly 70% of the world's rice-producing area. *Japonica* rice grows in East Asia under a temperate climate that is relatively more stable compared to the unpredictable weather conditions of tropical areas. Considering the vast area of production and rapid growth of rice consumers dependent upon a fragile and unsustainable agro-ecosystem, the development of *indica* sub-species rice with high yield potential will contribute to global food security more than a similar improvement of *japonica* rice.

In recent years, global consumption of rice has remained strong driven by rapid population and economic growth in some Asian and African countries (Fig. 1.2). Although per capita consumption has declined marginally in some high-income Asian countries, it has increased substantially in low-income countries. Also, it is expected that rice consumption will increase significantly in some non-Asian countries for changes in dietary preferences. Despite significant development over the recent years, the poverty of rice consumers and farmers is still widespread in several south Asian nations. Overall poverty alleviation depends on significant improvement of rice production that will generate income for the poor people on earth. Statistical enumerations have projected a rise of 116 million tons of milled rice by the year 2035 from the level at 2010 to meet the demand (Fig. 1.3).

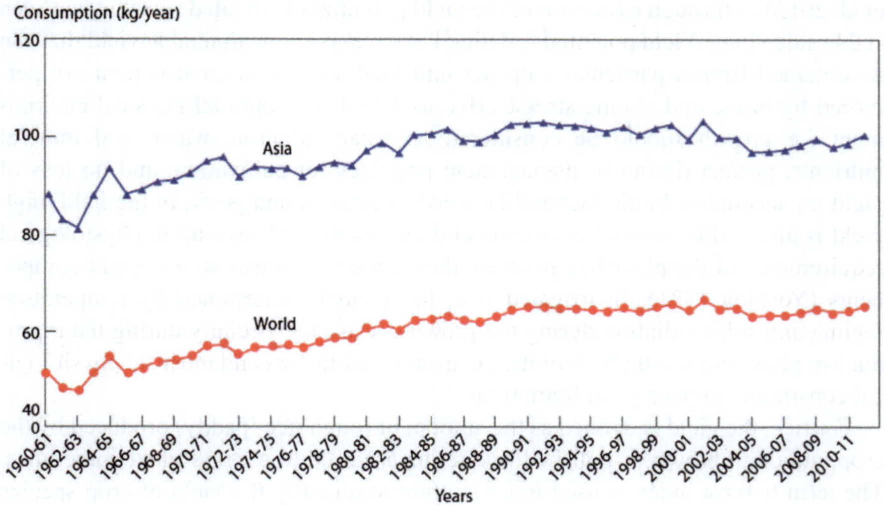

Fig. 1.2 Global and Asian per capita rice consumption. Sources: PSD online database USDA and FAOSTAT population database, FAO

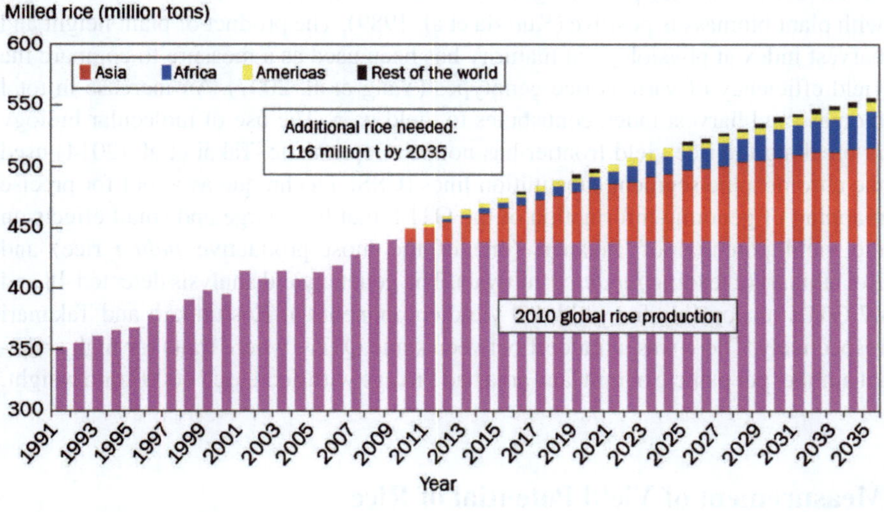

Fig. 1.3 Global rice production increases are needed to meet demand by 2035 (GRiSP 2013)

Breaking the Yield Frontier of Rice

The proposed enhancement of rice yield must be achieved by producing more grains with no increase in land area, irrigation water, fertilizer, and cost of labor. The key to solving the problem is to augment rice yield by closing the gap between the yield potential and average farm yield in rice-growing nations of the world (Van Wart et al. 2013) or through a revision of the yield potential of irrigated rice that has been stable since long. Yield potential is defined as the maximum attainable yield that can be obtained from a particular crop per unit land area in an environment not perturbed by biotic and abiotic stresses (Evans 1996). The optimal physical environment for growth should be considered on solar radiation, water, and mineral nutrients, perfect timing of management practices for cultivation, and no loss of yield on account of biotic factors like weeds, diseases, and pests. In the field, high yield is affected by several environmental factors that infringe upon physiological requirements of the plant-like photosynthesis, mineral nutrition, and yield components (Yoshida 1981). In irrigated rice, the yield is determined by temperature regime and solar radiation during the growing season, especially during the reproductive phase of growth; both of the environmental factors and intrinsic physiological constraints impose yield limitations.

For rice, the yield is defined as the amount of rough rice (paddy) produced by the crop, although brown or polished rice terms are used to express yield sometimes. The term harvest index is used in agriculture to quantify the yield of crop species including rice. It is the ratio of grain weight to total dry weight. In general, the above-ground dry matter of the plant is considered for the purpose. Dry grain yield equals harvest index multiplied by total dry weight. There is a negative relationship of harvest index with plant height (Yoshida 1981), but the relationship of height with plant biomass is positive (Kuroda et al. 1989). The product of plant height and harvest index at physiological maturity has been used as a measure to compare the yield efficiency of various rice genotypes (Yang et al. 2007). An increase in total biomass and harvest index contributes to yield gain. The use of molecular biology in breaking the rice yield frontier has not been optimistic. Takai et al. (2014) used the chromosome segment substitution lines (CSSL) technique as a tool for precise mapping of genomic information of the QTLS that have large and small effects on the yield potential of Takanari (one of the most productive *indica* rice) and Koshihikari (a leading *japonica* cultivar). The genetic yield analysis detected 48 and 47 QTLs responsible for yield and yield components of Koshihikari and Takanari respectively. There was a tradeoff between some QTLs, where traits for high spikelet number per panicle or unit land area had linkage with decreased 1000 grain weight.

Measurement of Yield Potential of Rice

The phenotypic performance of a crop for biomass production or grain yield is a function of both genetic and environmental factors and their interaction. According to Long et al. (2006), the yield potential of a grain crop is the seed weight obtained per unit area under optimum environmental conditions free from biological and physicochemical stresses. Yoshida (1981) proposed that the maximum yield of rice is predetermined by an interaction of yield potential of the variety and environment. The estimation can be obtained from the product of maximum photosynthetic net production per day or per unit of incident solar radiation and the number of days for grain production. There are two methods for the estimation of maximum photosynthetic output. In method 1, net production is obtained by subtracting respiration losses from gross primary production, whereas method 2 records the maximum photosynthetic efficiency of the crop. It is assumed in both the ways that all photosynthates produced are carbohydrates that are used in grain growth.

Method 1

Assumptions

(a) The period for yield production is 40 days after heading
(b) The average daily solar radiation is 400 cal/cm^2 per day of which is 45% is photosynthetically active.
(c) Of the photosynthetically active radiation, 5.5% is lost by reflection at the canopy surface and another 10% through absorption by inactive tissues.
(d) Eight photons are required to reduce 1 molecule of carbon dioxide, corresponding to an efficiency of 26% energy conservation.
(e) The loss due to light saturation in the upper leaves is 17% at 400 cal/cm^2 per day.
(f) The conversion factor for dry matter is 3900 cal/g
(g) The respiration loss is the sum of 1.5% of dry weight/day (1.5 kg/m^2 in this case) and 25% of the gross photosynthesis.
(h) The ratio of husk weight to the dry grain weight is 20%.

Calculation

$$\begin{aligned}\text{Maximum grain yield} &= \left(400\times10^4\times0.45\times0.945\times0.90\times0.26\times0.83\times0.75-1500\times0.015\times3900\right)\\ &\quad\times40/3900\\ &=1640\,\text{g}/\text{m}^2\left(\text{dry weight}\right)\\ &=19.1\,\text{t}/\text{ha}\left(\text{brown rice at 14\% moisture content}\right)\\ &=23.8\,\text{t}/\text{ha}\left(\text{rough rice at 14\% moisture content}\right)\end{aligned}$$

Method 2

In this equation net gain in biomass of crop has been related to photosynthetic efficiency in the equation.

$$W=\frac{E_u\times T\times\bar{S}}{K}\times10^4$$

Where W = dry matter (g), *Eu* = photosynthetic efficiency (net gains of chemical energy/total incident solar radiation in percentage), $\overline{S}$ = average daily incident solar radiation (cal/cm^2 per day), *K* = heat of combustion (cal/g). *T* = effective grain filling period in days.

The maximum yield potential calculated with assumptions.

(a) The effective grain filling period is 35 days for temperate and 25 days for tropic regions.
(b) Two *Eu* values were used, 3.5% for high efficiency and 2.5% for moderately high efficiency.
(c) The heat of combustion of grains is 4000 cal/g.
(d) The ratio of husk to the dry grain weight is 20%.
(e) The contribution of carbohydrates stored before flowering is negligible for the achievement of maximum yield potential.

When appropriate corrections are made for the weight (x 10^{-6}), area (x 10^4), husk (x1/0.8), and moisture content (x 1/0.86), grain yield will be computed as.

$$\text{Grain yield at 14\% moisture content}\left(\text{t / ha}\right) = \frac{E_\mu \times T \times \overline{S}}{K} \times 10^4 \times \frac{10^4}{10^6} \times \frac{1}{0.8} \times \frac{1}{0.86}$$

The results of this calculation have been shown in Fig. 1.4.

In method 1, E_μ can be computed as:

$$E_\mu = \frac{1{,}640 \times 3{,}900}{400 \times 40 \times 10^4} = 0.04 = 4.0\%.$$

Substituting 4.0% for *Eμ*, the value in the equation gives 20.4 t/ha when T is 35 days, 23.2 t/ha when T is 40 days.

Both the methods suggested above give identical values for grain yield at 14% moisture content for maximum potential yields. The equations suggest the benefits of a longer period of solar radiation on grain yield potential. In fact, on a per crop yield basis, estimated potential yield in the temperate region having a longer grain filling period is higher than that of the tropics (Fig. 1.4); in both conditions, there is a linear relationship between incident solar radiation, irrespective of variation in photosynthetic efficiency values (3.5% for high and 2.5% for moderately high) and potential yield (Yoshida 1981).

New Calculations for Yield Potential

Assuming all environmental factors optimal, Sheehy and Mitchell (2013) proposed an equation for calculating the maximum yield of rice, based on three principles, which follows the transformation of solar into chemical energy of carbohydrates in

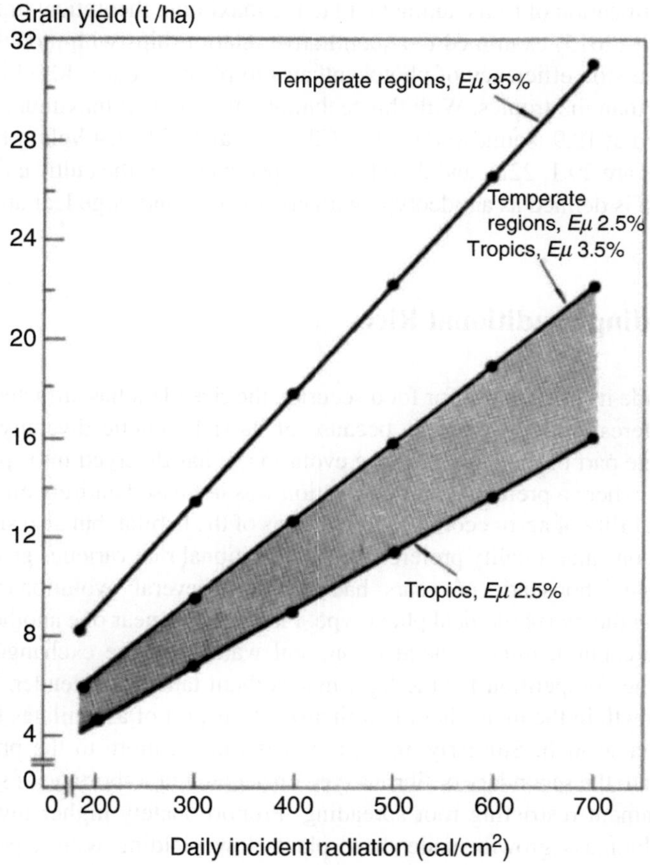

Fig. 1.4 The relation between potential yield and incident solar radiation during grain filling period of rice (Yoshida 1981 – with IRRI permission)

photosynthesis, conversion of carbohydrates into structural biomass, and partitioning of a part of the biomass to grains.

Y = Yield of rice (t ha^{-1}) at harvest with 14% moisture content on a fresh weight basis.
Y = 0.01 W gmax/ 0.86, (multiplication with 0.01 converts g m^{-2} to t/ha^{-1}).

Wgmax is the maximum grain as dry matter (g m^{-2}), and 0.86 is the fraction of dry matter in yield at 14% moisture content.

The maximum amount of grain is calculated from the equation Where W gmax = H Wsmax

H = Harvest Index (fraction of above-ground shoot biomass appearing as grain)
Wsmax = maximum shoot biomass in gDM m^{-2} at the end of growth duration, (DM = Dry Matter)

For identification of traits taking yield to the maximum theoretical limits Sheehy and Mitchell (2015) examined the quantitative relationships with models and suggested conversion efficiency of photosynthates to plant biomass 40% higher in the sub-tropics than the tropics. With this technique, the tropical maximum rice yields are predicted at 12.9 (semidwarf), 14.4 (C3 Vela), and 17.9 (C4 Vela) t ha^{-1}, where these values are 20.1, 22.4, and 25.0 t ha^{-1} respectively, for the cultivars in the sub-tropics. Vela is defined as an ideotype with erect leaves and large leaf areas.

Low Yielding Traditional Rice

Keeping aside its importance for food security, the rice plant has attracted unending research interest among scientists because of its rich genetic diversity spreading across a large part of the globe. During evolution, it has diverged into specific varietal groups, where a preference for cultivation was not based entirely on grain yield and compatibility of agro- ecological conditions of the habitat, but also significantly laced with consumer quality preference. The traditional rice varieties grown in high densities over thousands of years had acquired several evolutionary features encrypted on the morphological phenotype. Plants grown near one another compete for light perception, mineral acquisition, soil water, and the exchange of gases; especially the competition for the light makes them taller and slender. Preference for stem growth in the main shoot marginalizes transport of assimilates for branching and tiller growth. Similarly, root growth is limited more to the primary root system than to the secondary or fibrous type on account of a shortage of space in the soil environment restricting root spreading. Proportionately higher investment in vegetative biomass growth made these plants low yielding with a poor harvest index. Adaptations to the habitat over a long period, however, have made them relatively more climate-resilient compared to the new cultivars. Hence, the traditional rice varieties, although user friendly for cultivation on the part of the rice growers became gradually unwanted for poor economic return.

The First Approach to Improve Yield Potential

The yield potential of rice has improved enormously by genetic breeding twice in the recent past, the first time by the introduction of miracle rice IR8-parented cultivars and the second time through selection of hybrid vigor (Peng et al. 1999). These revolutions were adopted for irrigated rice only because irrigated rice contributes more than 75% of rice production and accounts for 55% of the land area under rice cultivation. Yoshida (1981) proposed that yield increase in rice can be achieved by either improving total biomass production or harvest index or both. The relationship of harvest index with grain yield is often closer than that of biomass (Laza et al. 2003). In pursuance, the first strategy for yield improvement embraced manipulation of both the factors through reduction of plant height for biomass partitioning more to grains than to vegetative structures and profusely tillering character for high biomass production. However, the manipulations became controversial because of some limitations. The increase of harvest index correlates inversely with biomass production affecting plant height and the overall yield of grains. Under the circumstances, the first strategy involved extensive inbreeding of elite high performing cultivars for refinement of traits ideal for higher grain yield. The approach was based on the concept that a self- pollinated crop like rice possesses a large part of its genetic variance as additive variance and the reduced margin of variance in elite cultivars provides an opportunity for the selection of superior genotypes. In this approach genotypic modification resulting in the release of semi- dwarf photo-insensitive high yielding IR8-parented rice was an instant success in the mid- 1960s.

Semidwarf high-yielding rice IR8 was first developed in IRRI in 1962 by Drs Peter Jennings and Henry Beachell crossing of dwarf variety Dee-geo-woo-gen of China with tall variety Peta of Indonesia. Infield trial conducted by agronomist Dr. S K DeDatta, IR8 produced 5 tons per hectare without nitrogen and 10 tons with nitrogen, i.e. almost 10 times higher than traditional rice. After testing in the Philippines, IR8 was released to different countries in 1966. The success of semidwarf rice depended on several key attributes. Semidwarf stature increased harvest index and lodging resistance, while photoperiod insensitivity permitted cultivation all-round the year. Under favorable irrigated conditions, yields of 9 to 10 t ha^{-1} were reported for IR8 (Chandler Jr 1969). This dwarfing breeding technology of the Green Revolution increased the level of production dramatically. More numbers of high-yielding *indica* inbred cultivars like IR36, IR 64, and IR72 were developed later (Peng and Khush 2003), but maximum yield never exceeded beyond 10 t ha^{-1} (Peng et al. 2000). The *indica* inbred approach to revise yield potential of irrigated rice was not successful after the 1970s because key features like semidwarf stature and photo-insensitivity could not be exploited further.

New Plant Type Rice

Grain yield of the high-yielding semidwarf *indica* inbred rice cultivars stagnated some years after the initial success of the Green revolution (Peng et al. 1999). It was construed that the stagnation could be the result of the plant type bearing some undesirable characters. Hence, a new breeding method was contemplated. This ideotype approach to plant breeding is in sharp contrast to the traditional empirical breeding approach of defect elimination and selection for higher yield benefit. Dr. S Yoshida, the most respected and eminent rice physiologist reported in 1972 that the intricate association between some morphological traits and yielding ability in response to nitrogen application led to the idealized plant type concept as a guide for breeding improved rice varieties.

In the semi-dwarf plant, active tillers and excess leaf area reduce canopy photosynthesis and panicle size because of mutual shading (Dingkuhn et al. 1991). Thus high tillering capacity, small panicle size, and lodging susceptibility were considered undesirable. Hence, rice breeders explored the possibilities of modifying the semi-dwarf high-yielding plant type and the physiological processes that limit potential yield (Khush and Peng 1996; Peng et al. 1999) to achieve another quantum jump in rice production. In this approach, ideotype selection was done on the capacity for high biomass production, which improved the final yield without any upgradation of harvest index (Furbank et al. 2015). Dingkuhn et al. (1991) proposed an idealized new plant type through modification of several traits of semidwarf cultivar, which included characters, as reviewed by Peng et al. (2008). The traits modified are both morphological and physiological.

(a) enhanced leaf growth and low tillering at early vegetative growth stage (3–4 tillers when direct-seeded and few unproductive tillers)
(b) reduced leaf growth and high nitrogen content of upper leaves for an increased ability for photosynthesis, especially at the reproductive stage of growth.
(c) high carbohydrate storage capacity of stem for remobilization at grain filling stage.
(d) large panicle with higher sink capacity.
(e) an extended period of grain filling.

The objective of new plant-type rice was to out yield IR8-parented semidwarf rice through the improvement of biological yield and partitioning of assimilates to sink organs. The breeding work started in 1989 with 2000 germplasm from IRRI gene bank to screen and identify donors for desired traits like low tillering, large panicles, thick stems, vigorous root system, and short stature. These traits selected from donors in tropical *japonica* germplasm named "bulu" or *javanica* germplasm increased panicle size with larger numbers of spikelets, but not yield. Crop sink size (spikelets per m^2) increased with the increase of panicle size (spikelets per panicle) (Peng et al. 1994). The first generation NPT was developed in a record period of fewer than five years and tested for a field trial in late 1993 (Peng and Khush 2003). These new elite cultivars did not express yield advantage better than *indica* inbred lines because of poor grain filling (Khush and Peng 1996; Zhu et al. 1997).

The second-generation NPT breeding was initiated in 1995 by crossing the first generation NPT lines with elite *indica* parents (Yang et al. 2007). Yang et al. (2007) have rigorously compared the second generation new plant type rice with the highest yielding inbred *indica* and hybrid cultivars and identified the reasons for the failure of expression of the superiority of new plant type cultivars over the others. Primarily the reasons for constraints in yield improvement in NPT rice were abysmal improvement in harvest index or biomass production in the absence of sufficient tillering and poorer grain filling percentage compared to *indica* cultivars. Hence, the intent for second-generation NPT was to design for traits with sufficient tillering capacity favoring high biomass and leaf area production and smaller panicle size without change of panicle length encouraging lax orientation of spikelets for high grain filling. In practice, genes pyramided from *indica* background, not only reduced panicle size and grain density and increased tillering capacity but also improved grain quality and disease resistance in the second generation new plant type. Some of the second - generation NPT lines were tested for observational trials first in 1998 (Peng et al. 2004). The agronomic trials were initiated later in 2002. The trials revealed the utility of the second- generation NPT better than the first generation NPT. The yields were higher, there were more spikelets per m^2 and grain filling percentage was higher thanks to genealogy from elite *indica* parents. However, the yield advantage gained by the second generation NPT could not match the yield progress achieved in the *indica* inbred breeding program. Besides, the targets to make a 10% rise in yield potential of irrigated lowland rice in the tropical dry season are 330 panicles per m^2, 150 spikelets per panicle, 25 mg oven-dried grain weight, 22 t/ha above- ground biomass (at 14% moisture content) and 0.5 harvest index (Peng and Khush 2003), but the second-generation NPT lines failed in the expression of yield potential by scoring less than 115 spikelets per panicle (Peng et al. 2004).

The Second Approach to Improve Yield Potential

The second approach for revising grain yield potential through exploitation of hybrid vigor was initiated in the late 1970s, while the development of new plant-type rice began nearly ten years after. In this approach, the strategy was to emphasize diversity in the genealogy of parental materials, because the crossing of parents with small genetic variation precludes the possibility of obtaining desirable genetic combinations. Since heterosis relies on genetic differences, the wider the gap between parental lines better is the performance of the progenies. Because genotypic compatibility is crucial for breeding, the selection of parental lines was confined to the sub-species of the *Oryza sativa* complex with AA genome only. Hybrid rice production based on genetic cytoplasmic male sterility requires three types of breeding lines, namely (A) a cytoplasmic male sterile line, (B) a maintainer line, and (C) a restorer line. While the male sterility precludes self-pollination, the genetic distance between A and C lines is directly correlated with the production of high-performing hybrids. The efforts made to exploit heterosis through hybrid rice breeding in either inter-specific type of *indica*: *japonica* or inter-varietal *indica*: *indica* crosses for an increase of yield potential have not been entirely successful always. Hybrid rice is grown in China since 1976 and as of today, more than 50% of rice-growing areas account for hybrid rice there (Yuan 2003). Hybrid rice holds a yield advantage of 15% more than inbred rice (Yuan 1994). The yield advantage subsists on early growth vigor for the rapid extension of leaf area, efficient sink formation and greater mobilization of stem and leaf sheath assimilates to grains. Once more, Yuan (2003) used a new inventory to combine an ideotype approach with inter-sub-specific heterosis in the new generation super rice in China. The super rice varieties were developed through the breeding of inbred and/or hybrid varieties. The breeding program was initiated by Professor Longping Yuan in 1998 with the strategy to assemble an ideotype approach of NPT with inter-subspecific heterosis. The F1 hybrid combination was developed by crossing intermediate type between *indica* and *japonica* with an *indica* parent.

Many of the super rice cultivars became famous because of high grain yield and good grain quality (Cheng et al. 2007). The hybrids expressed better grain yield compared to *indica* inbreds and NPTs because of the superiority of sink size (Yang et al. 2007). Hybrid rice development in China has been summarized and reassessed by Wang and Deng (2018), wherein the three-line hybrid rice system based on cytoplasmic male sterility of the 1970s was attributed as first-generation and the double line hybrid rice system based on photoperiodic and thermogenic sensitive male sterile lines of 1980s was considered the second generation. The researchers proposed and implemented a third-generation hybrid rice breeding system that is based on a transgenic method to propagate and utilize stable recessive nuclear male sterile lines. This hybrid was assumed to overcome undesirable traits of the first and second-generation hybrid rice and advance grain yield.

Demerits of Hybrid Rice

In a review, Fischer (2015) opined that the progress in yield gain worldwide is proximate to concurrent growth of yield potential and gradual closure of the gap between yield potential and farm yield through adoption by farmers of known innovations faster . Although both super and NPT breeding using the ideotype approach have been successful in raising the yield barrier, seldom an improvement of yield potential is noticed else were outside China. Poor grain filling has consistently undermined the inherent high yield potential of this large-panicled rice. In China itself, the reputation of hybrid rice is miserable. The crop is susceptible to diseases and pests and is mostly used for feeding livestock and poor people.

Nevertheless, the approach to raise yield potential by the expansion of sink size through the increase of spikelet number per panicle has been a genuine effort. Meng et al. (2016) compared four large panicle rice cultivars two each belonging to high and low grain filling traits and identified some morpho-physiological factors beneficial for yield advantage, such as greater leaf photosynthetic rate and root function during grain filling stage, higher biomass accumulation, and assimilate partitioning in the post-anthesis period, and larger, thicker and more erect upper three leaves for the efficient perception of irradiance.

The strategies for rice breeding with innovating techniques as surmised above may be successful in the years to come, but there has been a genuine apprehension for a new calamity. It is noticed that both dwarfing and hybrid breeding, have promoted genetic uniformity at the cost of divergence and the newly released cultivars become susceptible to pests and diseases. The cultivars are also wasteful of increasingly scarce commodities like fertilizers, water, and fossil fuels. Rampant misuse of the limited resources has a significant environmental footprint in the form of pollution of soil and water and global warming threatening not only food security but also the survival of mankind on earth. At present, the problem of food security in Asia is serious, although not precarious; half a billion inhabitants of the region go hungry each passing day. Thus, genetic modification of rice as envisaged in a super green revolution has to take care of both yields as well as resistance to the existing and predicted physicochemical stresses.

Phenotypic Modification of Rice with the Progress of Time

In tandem with the demand for more food production rice agro-ecosystem has undergone considerable modification over time, where there has been continued pressure for new phenotypic features amenable for higher grain yield. In pursuance old traditional rice biotypes have given way for newer cultivars like high yielding semi-dwarf and new plant type rice (Fig. 1.5).

Fig. 1.5 Diagrammatic features of traditional, semi-dwarf, NPT rice

Salient Features of Different Rice Cultivars

Traditional Rice

A monsoonal crop grown from June to December.
Maturity duration: Short (60 days), medium (110–120 days), long (130–150 days)
Cultural diversity: Upland, low land, irrigated low land, and deep water.
Photosensitivity: Short photoperiod.

Plant type: Tall plant nearing six feet height, susceptible to lodging, low tillering, high leaf number, leaf upright, but droopy some time, low leaf area index, large panicle, moderate grain number, long history of cultivation confers resistance to mineral deficiency and other biotic and abiotic stresses, species biodiversity extremely high on account of adaptation to a wide spectrum of ecological niches around the globe, strong apical dominance. Using the Shannon-Weaver diversity index Rabara et al. (2014) examined 307 traditional rice cultivars and reported a phenotypic diversity index averaging at 0.73 and 0.45 for quantitative and qualitative traits respectively. The traditional cultivars being a repository for such wide diversity indices for various traits augur well for rice breeding and the development of new plant types.

Grain yield: Panicle size is large. The size increases with an increase in plant height (Yoshida 1981). Grain yield is low and low harvest index <0.3.

Semidwarf High Yielding Rice

Both wet and dry season crop, released in 1966.
Cultural habitat: Irrigated low land.
Photosensitivity: Insensitive

Plant type: Plant height 1 m- semidwarf achieved through the introduction of a major recessive gene *sd1*, short sturdy stem, resistance to lodging, profuse tillering, low apical dominance, erect leaves for the efficient perception of light energy, high leaf area index, highly responsive to fertilizers, species diversity low, names of semidwarf cultivars are Taichung Native 1, Guang-chang-al, Al-jio-nan-f'e, Guang-hwal-al and IR8 (De Datta 1981). IR8 is the miracle rice of the Green Revolution. It has been derived from a cross with tall *indica* rice Peta of Indonesia and semidwarf Dee-geo-woo-gen of China and has been introduced widely into various rice-growing nations. The plants are sensitive to mineral deficiency and environmental stresses.

Grain yield: High, Harvest index ≥ 0.5. It out-yields traditional rice when fertilizers and pesticides are used under irrigated conditions. The reduction of plant height depresses biomass production and increases carbon partitioning to grains.

New Plant Type Rice

Both wet and dryland crops.
Cultural habitat: Irrigated low land, developed at IRRI, the Philippines in the late 1980s.
Photosensitivity: Insensitive

Plant type: Based on the concept of ideotype breeding approach (Donald 1968), a new plant type was developed at IRRI, Philippines. The cultivars were designed to yield an advantage 20–25% over the semidwarf rice. The first generation NPT lines were generated from tropical *japonica* background in 1993. They had large panicles with numerous spikelets and no unproductive tillers. But the yield was poor because of low biomass and poor grain filling.

Morphological features- Plant height 90–110 cm, low tillering with 4–5 effective tillers, thick and sturdy stem, resistant to lodging, heavy panicle with 200–250 grains, thick erect and dark green leaves, vigorous root system, 100–130 days growth duration; low biomass production; susceptible to diseases and pests; compact panicle.

Grain yield: High; harvest index 0.6; poor grain filling

The first-generation NPT lines were found not suitable for tropical conditions because of the limitation of tillering and biomass production. The spikelet number per panicle was reduced and plant height and tiller numbers were increased in the second-generation new plant type. But they were no better than the inbred indica rice (Peng et al. 2008).

Hybrid Rice

Rice flower is bisexual and self-pollinating. One flower does not depend on another for pollination and fertilization. Therefore self-pollination in rice is a natural and normal process. For cross-pollination, the female partner is either male sterile or emasculated and super rice varieties were developed through the breeding of inbred and/or hybrid male varieties. A breeding program was initiated by Professor Longping Yuan in 1998 with the strategy to assemble an ideotype approach of NPT with inter-subspecific heterosis. The F1 hybrid combination was developed by crossing intermediate type between *indica* and *japonica* with an *indica* parent. The inbreeding process resulted in the production of progenies bearing characters similar to the parents. To develop new inbred rice, breeders first cross two or more cultivars and select traits suitable for ideal characters through subsequent inbreeding.

Both wet and dry season crop; released in 1976 in China.
Cultural habitat: Irrigated low land.
Photosensitivity: Insensitive

Plant type: Plant height 100–120 cm. High yield potential beyond traditional rice cultivars was achieved first through the exploitation of hybrid vigor by breeding for a three-line *indica* F1 hybrid in 1976 (Yuan 1994); A line for male sterility, B line for maintaining male sterility of A line (maintainer) and, R line for restoring the fertility of A line and produce hybrid seeds (restorer). Erect to semi-erect plants; long flag leaf; early maturity with long and dense panicle; sturdy stem resistant to lodging, high nitrogen use efficiency, long and bold grains.

Grain yield: Grain sink size larger than NPT, High yield, Harvest index 0.6

Super Rice

Both wet and dry season crop.
Cultural habitat: Irrigated low land.
Photosensitivity: Not sensitive

Plant habit: Semi-dwarf cultivars developed through the combination of improvement in plant type and utilization of growth vigour. The objective was to improve yield potential 100 Kg grain ha^{-1} day^{-1} (Yuan 2001) or 12% more in super hybrid as compared to ordinary hybrid and inbred cultivars (Huang et al. 2017). The Super hybrid breeding program was initiated by Professor L Yuan in China using the strategy to combine the ideotype approach with the utilization of inter-subspecific heterosis between *japonica* and *indica* subspecies in 1998 (Yuan 2001). Plant type had a close relation with single panicle weight, grain yield, and choice for heavy panicle were accompanied by changes in some morphological features. The morphological traits were moderate tillering capacity (270–300 panicles/m^2), dense (5 g/panicle) and droopy panicle at maturity, plant height 100 cm, flag leaf 50 cm, second and third leaves 55 cm, all three leaves above panicle height, all leaves remain erect until maturity, narrow and V-shaped leaves, thick leaves with a specific leaf weight of 55 g/m^2 for all three top leaves, leaf area index of top three leaves = 6.0, harvest index = 0.55. Compared to medium or low panicle weight types, the heavy panicle rice bears several other important characteristics. The LAI (leaf area index) in them was lower, but the length, width, thickness, and area of the top three leaves were significantly higher than the smaller weight panicle plant types, and these features correlated positively with spikelet and grain number and grain weight per panicle and overall grain yield (Jun et al. 2006).

References

Alexandratos N, Bruinsma J (2012) World agriculture towards 2030/2050: the 2012 revision. ESA Working Paper No. 12–03, FAO, Rome

Bouman BAM, Lampayan RM, Toung TP (2007) Water management in irrigated rice: coping with water scarcity. International Rice Research Institute, Phillines. 54 p

Chandler Jr RF (1969). Plant morphology and stand geometry in relation to nitrogen. Agronomy & Horticulture – Faculty Publications. 196. Chapter 12, University of Nebraska, Lincoln, USA. pp. 265–289

Chang TT (2003) Origin, domestication and diversification. Chapter 1.1. In: Smith CW, Dilday RH (eds) Rice: origin, history, technology, and production. John Wiley & Sons, Incl, pp 3–25. ISBN 0-471-34516-4 © 2003

Cheng SH, Cao LY, Zhuang JY, Che S, Zhan X, Fan Y, Zhu D, Min S (2007) Super hybrid rice breeding in China: achievements and prospects. J Integr Plant Biol 49:805–810

Climate change (2007) the physical science basis: summary for policymakers. Contribution of Working Group I to the Fourth Assessment Report of the Intergovernmental Panel on Climate Change, IPCC , Paris

Cook MG, Evans LT (1983) Some physiological aspects of domestication and improvement of rice (Oryza spp.). Field Crop Res 6:219–238

De Datta SK (1981) Principles and practices of rice production. Wiley, New York

Dingkuhn M, De Vries FP, De Datta S, Van Laar H (1991) Concepts for a new plant type for direct seeded flooded tropical rice. In: Direct seeded flooded rice in the tropics. International Rice Research Institute, Los Banos, Philippines. pp. 17–38

Donald CM (1968) The breeding of crop ideotypes. Euphytica 17:385–403

Evans LT (1996) Crop evolution, adaptation and yield. Cambridge University Press, London

Evenson RE, Gollin D (2003) Assessing the impact of the green revolution, 1960 to 2000. Science 300:758–762

Fischer R (2015) Definitions and determination of crop yield, yield gaps, and of rates of change. Field Crop Res 182:9–18

Fitzgerald MA, Mccouch SR, Hall RD (2009) Not just a grain of rice: the quest for quality. Trends Plant Sci 14:133–139

Furbank RT, Quick WP, Sirault XR (2015) Improving photosynthesis and yield potential in cereal crops by targeted genetic manipulation: prospects, progress and challenges. Field Crop Res 182:19–29

Geohive (2015) World: rice production in metric tonnes. In: http://www.geohive.com

Godfray HCJ, Beddington JR, Crute IR, Haddad L, Lawrence D, Muir JF, Pretty J, Robinson S, Thomas SM, Toulmin C (2010) Food security: the challenge of feeding 9 billion people. Science 327:812–818

GRiSP (Global Rice Science Partnership) (2013) Rice almanac 4th edition. International Rice Research Institute, Los Baños, Philippines

Huang M, Tang Q, Ao H, Zou Y (2017) Yield potential and stability in super hybrid rice and its production strategies. J Integr Agric 16:1009–1017

Huke RE, Huke EH (1990) Rice: then and now. International Rice Research Institute, Philiines, 44p

IRRI (1995) World Rice Statistics 1993–95. Chapter 1.Rice in the world. In http://www.fao.org/3/w8439e/w8439e08.htm

Jackson MT (1995) Protecting the heritage of rice biodiversity. GeoJournal 35:267–274

Jun M, Ming D-F, Yang S-M, Zhu Q-S (2006) Characteristics of rice plant with heavy panicle. Agric Sci China 5:911–918

Khush G, Peng S (1996) Breaking the yield frontier of rice. In: Increasing yield potential in wheat: breaking the barriers. Editors-Reynolds MP, Rajaram S, McNab A. Proceedings of a workshop held on 26–28 Narch 1996 in Ciudad Obregon, Sonora, International Maize and Wheat Improvement Center, Mexico. pp. 36–51

Khush GS, Peng S (1998) Improving yield potential by modifying plant type. IRRI Discussion Papers 287587, International Rice Research Institute, Philippines

Kuroda E, Ookawa T, Ishihara K (1989) Analysis on difference of dry matter production between rice cultivars with different plant height in relation to gas diffusion inside stands. Japanese J Crop Sci 58:374–382

Laza MR, Peng S, Akita S, Saka H (2003) Contribution of biomass partitioning and translocation to grain yield under sub-optimum growing conditions in irrigated rice. Plant Production Sci 6:28–35

Long SP, Zhu XG, Naidu SL, Ort DR (2006) Can improvement in photosynthesis increase crop yields? Plant Cell Environ 29:315–330

Mahapatra I, Singh K, Pillai K, Bapat S (1985) Rice soils and their management. Indian J Agronomy 30:1–41

Masutomi Y, Takahashi K, Harasawa H, Matsuoka Y (2009) Impact assessment of climate change on rice production in Asia in comprehensive consideration of process/parameter uncertainty in general circulation models. Agric Ecosyst Environ 131:281–291

Meng T-Y, Wei H-H, Chao L, Dai QXK, Huo Z, Wei H, Guo B, Zhang H (2016) Morphological and physiological traits of large-panicle rice varieties with high filled-grain percentage. J Integr Agric 15:1751–1762

Mohanty S (2013) Trends in global rice consumption. Rice Today 12:44–45

Peng S, Khush GS (2003) Four decades of breeding for varietal improvement of irrigated lowland rice in the international Rice research institute. Plant Production Sci 6:157–164

Peng S, Khush G, Cassman K (1994) Evolution of the new plant ideotype for increased yield potential. Proceedings of the Breaking the Yield Barrier: Proceedings of a Workshop on Rice Yield Potential in Favorable Environments. International Rice Research Institute, Los Banos, Philippines, 1994, pp. 5–20

Peng S, Cassman KG, Virmani S, Sheehy J, Khush G (1999) Yield potential trends of tropical rice since the release of IR8 and the challenge of increasing rice yield potential. Crop Sci 39:1552–1559

Peng S, Laza R, Visperas R, Sanico A, Cassman KG, Khush G (2000) Grain yield of rice cultivars and lines developed in the Philippines since 1966. Crop Sci 40:307–314

Peng S, Laza RC, Visperas RM, Khush GS, Virk P, Zhu D (2004) Rice: progress in breaking the yield ceiling. Proceedings of the 4th International Crop Science Congress, 26 Sept-1 Oct. 2004, Brisbane, Australia. pp. 1–11

Peng S, Khush GS, Virk P, Tang Q, Zou Y (2008) Progress in ideotype breeding to increase rice yield potential. Field Crop Res 108:32–38

Rabara RC, Ferrer MC, Diaz CL, Newingham MCV, Romero GO (2014) Phenotypic diversity of farmers' traditional rice varieties in the Philippines. Agronomy 4:217–241

Sampath S (1985) History, origin and antiquity of rice. In: Jaiswal PL (ed) Rice research in India. Indian Council for Agricultrual Research, New Delhi, pp 1–4

Seck PA, Diagne A, Mohanty S, Wopereis MC (2012) Crops that feed the world 7: Rice. Food Security 4:7–24

Sheehy J, Mitchell P (2011a) Developments in rice research: visions and pragmatism. World Agric 2:1–19

Sheehy JE, Mitchell P (2011b) Rice and global food security: the race between scientific discovery and catastrophe. In: ACCESS NOT EXCESS—the search for better nutrition. Ed. Pasternak C. Smith- Gordon, Cambridgeshire, pp. 81–90

Sheehy JE, Mitchell P (2013) Designing rice for the 21st century: the three laws of maximum yield. Discuss Paper Series, International Rice Research Institute, Philippines, 48, 19p

Sheehy JE, Mitchell P (2015) Calculating maximum theoretical yield in rice. Field Crop Res 182:68–75

Sweeny M, McCough S (2007) The complex history of the domestication of rice. Ann Bot 100:951–957

Takai T, Ikka T, Kondo K, Nonoue Y, Ono N, Araia-Sanoh Y, Yoshinaga S, Nakano H, Yano M, Kondo M (2014) Genetic mechanisms underlying yield potential in the rice high-yielding cultivar Takanari, based on reciprocal chromosome segment substitution lines. BMC Plant Biol 14:295. https://doi.org/10.1186/s.12870-014-0295-2

Tian F, Zhu Z, Zhang B, Tan LB, Fu YC, Wang X, Sun CQ (2006) Fine mapping of a quantitative trait locus for grain number per panicle from wild rice (Oryza rufipogon Griff.). Theoretical Appl Genetics 113:619–629
Van Wart J, Kersebaum KC, Peng S, Milner M, Cassman KG (2013) Estimating crop yield potential at regional to national scales. Field Crop Res 143:34–43
Wang H, Deng X (2018) Development of the third generation hybrid rice in China. Geneomic, Proteomics Bioinformatics 16:393–396
Yang W, Peng S, Laza RC, Visperas RM, Dionisio-Sese ML (2007) Grain yield and yield attributes of new plant type and hybrid rice. Crop Sci 47:1393–1400
Yoshida S (1981) Fundamentals of rice crop science. International Rice Research Institute, Philippines
Yuan L (1994) Increasing yield potential in rice by exploitation of heterosis. Hybrid rice technology: New developments and future prospects. In: Hybrid rice technology: new developments and future prospects. Selected papers from the International Rice Research Conference, Virmani SS (Ed). International Rice Research Institute, Philippines. pp. 1–6
Yuan L (2001) Breeding of super hybrid rice. In: Peng S, Hardy B (eds) Rice research for food security and poverty alleviation. International Rice Research Institute, Philippines, pp 143–149
Yuan L (2003) Recent progress in breeding super hybrid rice in China. In. Science Progress in China. Elsevier, pp. 231–236. https://doi.org/10.1016/B978-008044970-8/50048-8
Zhu Q, Zhang Z, Yang J, Cao X, Lang Y, Wang Z (1997) Source-sink characteristics related to the yield in intersubspecific hybrid rice. Sci Agric Sin 30:52–59

Chapter 2
Botany of Rice Plant

Abstract Cultivated rice *Oryza sativa* L. is a semi-aquatic annual grass belonging to the family *Poaceae* of class *Monocotyledonae* of angiosperm plants. Genus *Oryza* has another cultivated species, namely, *Oryza glaberrima* Steud. and about 20 wild species. *Oryza* species have been delimited into three sections, i.e. *Oryza, Ridleyanae,* and *Granulata*. *Oryza* group has two complexes, *Oryza sativa* and *Oryza officinalis*. *Oryza sativa* complex has seven species including the two cultivated rice and they belong to the AA diploid genome with 24 chromosomes. In Asia, *O. sativa* is differentiated into three sub-species based on geographical distribution, namely *indica, japonica,* and *javanica*. The plant is native to semi-aquatic conditions, where the presence of inter-organ aerenchyma facilitates the exchange of respiratory gases in submerged plant parts. Rice is a bisexual, self-pollinated crop having a growth period of 3–6 months. Plant height varies from 0.4 to more than 5 meters in the case of floating rice. There are three distinct growth phases; they are vegetative, reproductive, and ripening. The vegetative phase covers a period from germination to panicle initiation. Tillers are produced from stem nodes in acropetal sequence during this period. The reproductive phase includes the period from panicle initiation to anthesis/heading and finally, the ripening phase extends from heading to grain maturity. Each productive tiller bears a terminal panicle type inflorescence. Spikelets borne on pedicels on the primary and secondary branches of panicle are individual flowering units that produce one seed each at maturity. Atop pedicel, a spikelet bears two rudimentary glumes, two sterile lemmas, and one fertile floret borne on a short stem axis named rachilla in acropetal succession. The fertile floret has six stamens and one uni-ovulate pistil with a bifid and feathery stigma enclosed between lemma and palea. The seed consists of a caryopsis (brown rice) and a hull. Caryopsis has two parts, embryo, and endosperm. Hull comprises lemma, palea, and rachilla.

Keywords *Oryza sativa* · *Indica* · *Japonica* · Panicle · Spikelet · Caryopsis · Hull

P. K. Mohapatra, B. B. Sahu, *Panicle Architecture of Rice and its Relationship with Grain Filling*, https://doi.org/10.1007/978-3-030-67897-5_2

Taxonomy of Rice Plant

Rice plant belongs to the class *Monocotyledonae* of angiosperms, the order *Glumiflorae*, the family *Poaceae/Gramineae,* and the genus *Oryza*. In the grass family *Poaceae*, the genus *Oryza* is grouped in the tribe *Oryzineae* of the sub-family *Bambusoideae* or *Ehrhartoideae*. *Oryza* originated about 14 million years ago in Malesia (GRiSP 2013). In course of evolution, speciation led to diversification and dispersion of *Oryza* into various agroecosystems of the tropics. *Oryza* has mainly two cultivated species, namely *Oryza sativa* L. and *Oryza glaberrima* Steud, and about 20 related wild species (Vaughan 1994). *Oryza* species have been delimited and divided into three sections, such as *Oryza, Ridleyanae,* and *Granulata*. *Oryza* group consists of two complexes, i.e. *Oryza sativa*, and *Oryza officinalis*. The former has eight species including both the cultivated species and they belong to the AA diploid genome group with 24 chromosomes on the chromosome pairing in the meiosis. In the case of the latter, there are nine wild rice species, where the chromosome number is 24 or 48 (Table 2.1). Out of the ten, four species are diploid, four species tetraploid and one both diploid and tetraploid. The genome of diploid species is CC, BB, DD, or EE and tetraploid species is BBCC or CCDD. *Ridleyanae* and *Granulata* sections have 4 and 2 wild species respectively. The cultivated and wild rice species are closely related to bamboos than to other major cereals like wheat, maize, and sorghum belonging to the grass family *Poaceae*. Lack of genetic relation makes rice incompatible for hybridization with other cereals, although opportunities are fairer for crossings between wild and cultivated species of the genus.

Of the *Oryza* species, the *Oryza sativa* complex is pan-tropical and has been studied in great detail for genetic lineage. This complex includes cultivated species *Oryza sativa* L. and *Oryza glaberrima* Steud; the former is commonly known as Asian rice while the latter is named African rice. *O. sativa* species is grown throughout the world in many agro- ecosystems, while *O. glaberrima* is mostly endemic to South Africa.

Karyotype of the Rice Genome

Oryza sativa L. is a grass with a genome size of 430 Mb spread across 12 chromosomes (2n = 24). The nuclear content of DNA has more than 50% repetitive sequences. Usually, chromosomes of rice species are small bearing few morphological markers, which renders them difficult to discern and identify. The smallness of genome size makes it amenable for genetic modification compared to other major cereals. In Asia, *O. sativa* is differentiated into three sub-species based on geographical distribution, namely *indica, japonica,* and *javanica*. Sub-species *indica* includes tropical and sub-tropical varieties grown in South and South East Asia and Southern China. In contrast, *japonica* sub-species refer to varieties adapted to the

Table 2.1 Classification and distribution of species in the genus *Oryza* according to GRiSP 2013; (Vaughan 1994)

Taxa	Genome	Number of chromosome (n)	Distribution	Comments/ Alternative classification
Section *Oryza*				
Series *Oryza*: *sativa* species complex				
O. sativa	**AA**	**12**	World wide	
O. glaberrima	**AA**	**12**	West Africa	
O. nivara	**AA**	**12**	Tropical Asia	Annual ecotype of *O.rufipogan*
O. rufipogan	**AA**	**12**	Tropical Asia to northern Australia	
O. meridionalis	**AA**	**12**	Northern Australia	
O. barthii	**AA**	**12**	Africa	
O. longistaminata	**AA**	**12**	Africa	
O. glumaepatula	**AA**	**12**	South America	South America *O. rufipogan* *O. glumaepatula*
Series *Latifoliae: officinalis* species complex				
O. minuta	**BBCC**	**24**	Philippines, Papua New Guinea	
O. officinalis	**CC**	**12**	Tropical Asia to Papua New Guinea	
O. rhizomatis	**CC**	**12**	Sri Lanka	
O. malampuzhaensis	**CCDD**	**24**	India	Tetraploid race of *O. officinalis*
O. schweinfurthiana	**BBCC**	**24**	Africa	Tetraploid race of *O. punctata*
O. eichingeri	**CC**	**12**	West, central and East Africa, Sri Lanka	The only species found in both Africa and Asia
O. alta	**CCDD**	**24**	Central and South America	
O. grandiglumis	**CCDD**	**24**	South America	
O. latifolia	**CCDD**	**24**	Central and South America	
O. punctata	**BB**	**12, 24**	Africa	
Section *Australiensis*				
O. australiensis	**EE**	**12**	Australia	Member of *O. officinalis* complex
Section *Brachyantha*				
O. brachyantha	**KK**	**12**	Africa	
Section *Padia* Basal or primitive section of *Oryza*				
O. *schlechteri*	**HHKK**	**24**	Indonesia and Papua New Guinea	
O. coarctata	**KKLL**	**24**	South Asia to Myanmar	

(continued)

Table 2.1 (continued)

Taxa	Genome	Number of chromosome (n)	Distribution	Comments/ Alternative classification
Series *Ridleyanae: ridleyi* species complex				
O. longiglumis	**HHJJ**	**24**	Indonesia and Papua New Guinea	
O. ridleyi	**HHJJ**	**24**	Southeast Asia to Papua New Guinea	
Series *Meyerianae: meyeriana* species complex *granulata* species complex Variety of *O. meyeriana*				
O. granulata	**GG**	**12**	South and Southeast Asia	
O. meyeriana	**GG**	**12**	South and Southeast Asia	
O. neocaledonica	**GG**	**12**	New Caledonia	

temperate climate of Japan, China, and Korea. Some of the *javanica* varieties are grown in Indonesia along with the *indica* cultivars.

In a recent classification, the *Oryza* genome has been classified into 11 groups from AA to LL (GRiSP 2013). Most of the *Oryza species* are delimited into four genome complexes in two sections namely *Oryza* and *Padia* (Table 2.1). Two diploid species with no major relatives are individually placed in their section of the genus, such as *O.australiensis* and *O. brachyantha*. The wild rice species hold important traits of economic importance and can contribute to enhancing the ability of cultivated rice not only for higher grain yield but also climate resilience and disease resistance (Thomas et al. 2017). Because wild relatives have furnished genes for hybrid rice revolution; careful checking of species identity, maintenance of herbarium specimens, and conservation in gene bank with accession number has been advised (Vaughan et al. 2003).

Characteristics of the Rice Plant

Cultivated rice *Oryza sativa* is a semi-aquatic annual grass. The crop has dispersed into different geographical locations around the world and subsequent adoption in localized agriculture has led to diversification and speciation. The *indica, japonica,* and *javanica* sub- species are native to widely contrasting climates of the world. The geographic isolation that followed the introduction and regional domestication of the sub-species led to the development of genetic segregation and incompatibility. Genetic variation apart, the ecogenetic differences in morphological and

Table 2.2 Morphological and physiological characteristics of *indica, japonica*, and *javanica* subspecies of *Oryza sativa*

Fast growth rate in *indica*	Slow growth rate in *japonica* and *javanica*
Broad to narrow, light green leaves	Broad, stiff, light green leaves
Long to short, slender, somewhat flat grains	Long, broad, thick grains
Profuse tillering	Low tillering
Tall to intermediate plat stature	Tall plant stature
Mostly awnless	Long awned and awnless
Thin, short hairs on lemma and Palea	Long hairs on lemma and Palea
Easy shattering	Low shattering
Soft plant tissue	Hard plant tissue
Varying sensitivity to photoperiod	Low sensitivity to photoperiod
23–31% amylose	20–25% amylose
Variable gelanization temperatures (low and Intermediate)	Low gelanization temperature

physiological attributes are enormous between them (Table 2.2). The tropical rice, spread over the warm climate of South-East Asia has been named *indica* race. In comparison, the *japonica* race is native to cool temperate climates of China and Japan where vegetative growth duration and yield potential are much greater than the *indica* race. The third ecogenetic race *javanica* has a large plant size, slow growth rate, and large and bold grains.

In the tropics, ecogenetic diversification of *indica* rice reached its peak level and this race served as a precursor for the other two races i.e. *japonica* and *javanica*. All rice cultivars are annual and they primarily propagate by seed production. Nevertheless, some of the cultivars survive as perennial by producing new tillers from residual nodes of stem after harvest by ratooning. The habitat of the rice plant is usually semi-aquatic condition. However, the plant is highly adaptable to variations of habitat and climatic conditions. Upland rice is grown in a low moisture regime, whereas low land varieties are cultivated in submerged conditions with water levels varying between 1.5 and 5 m. The presence of aerenchyma tissue in leaf, culm, and root facilitates diffusion of oxygen from aerial parts to roots under submergence and helps adaptation to aquatic conditions (Fig. 2.1). Plant height in rice varies by cultivar and environmental variations. Upland rice is small-statured with a height of less than 0.5 m. In contrast, some of the deepwater rice varieties can elongate up to five meters to match the rising level of pool water.

Morphology of Rice Plant

Rice is a bisexual, self-pollinated crop, growth period of 3–6 months depending on the variety and environment in which it is grown. There are three distinct growth phases; they are vegetative, reproductive, and ripening. The vegetative phase covers

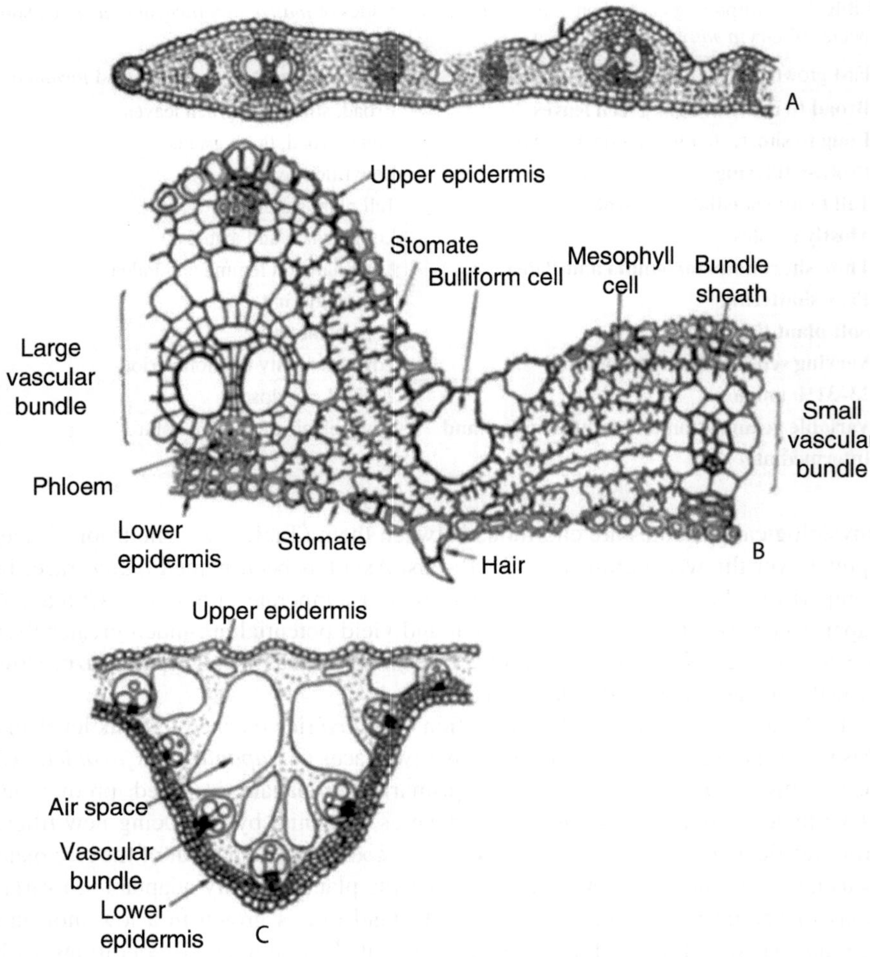

Fig. 2.1 Cross-section of rice leaf and its parts, stem, and root showing the presence of aerenchyma (**a–d**) (Yoshida 1981 – with IRRI permission)

a period from germination to panicle initiation. Tillers are produced from stem nodes in acropetal sequence during this period. The reproductive phase includes the period from panicle initiation to anthesis/heading and finally, the ripening phase extends from heading to grain maturity. The potential yield capacity of the plant is determined in the reproductive period, whereas actual yield is realized by grain filling in the ripening stage. Normally, a rice plant with a growth duration of 120 days spends 60 days in vegetative and 30 days each in reproductive and ripening stages under a tropical environment (Fig. 2.2; Yoshida 1981).

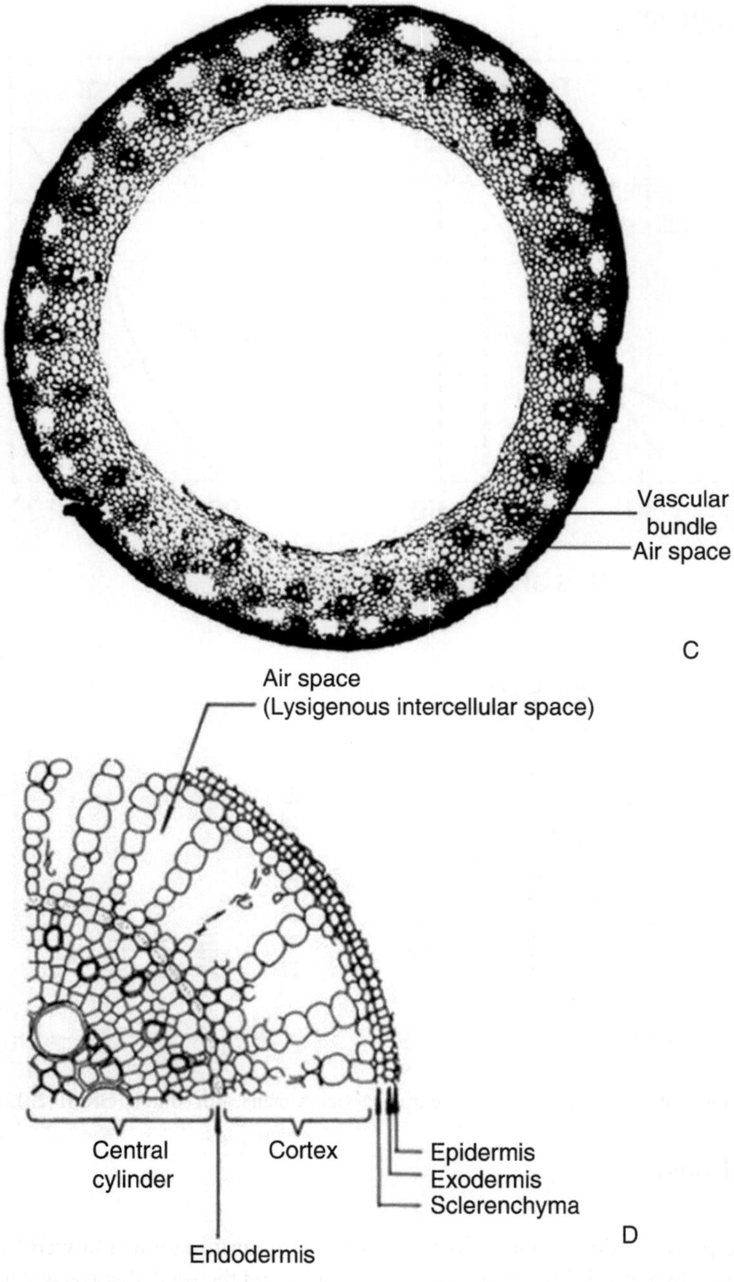

Fig. 2.1 (continued)

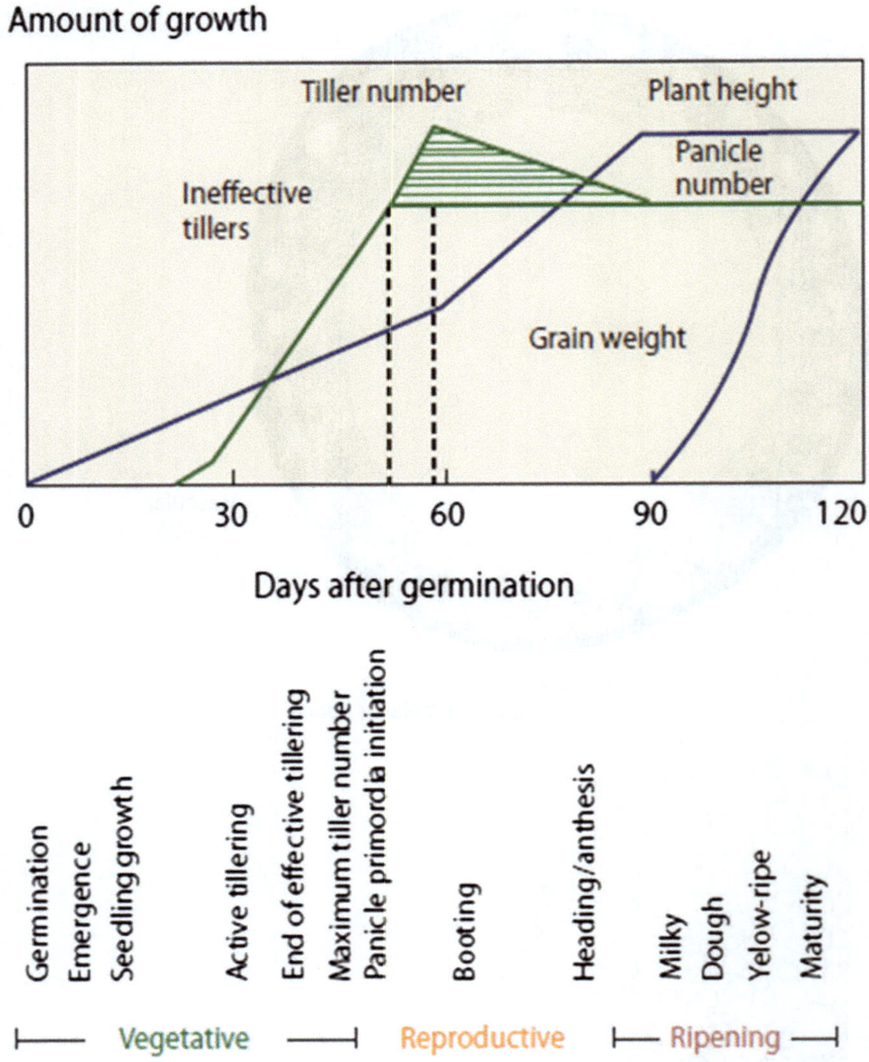

Fig. 2.2 Schematic growth of a rice variety in the tropics Rice Almanac 4th edition (GRiSP 2013)

Vegetative Phase

The vegetative phase of development in rice begins with germination followed by seedling growth when physiological dormancy is broken and the seed absorbs water. In germination, the radicle ruptures the seed coat first. The first seminal root arises through the coleorhiza and it is followed by the emergence of coleoptile (Fig. 2.3). Further, the lateral roots arise from the developed seminal roots. Organic and phosphatic manures promote root development. Nodal roots absorb necessary nutrients

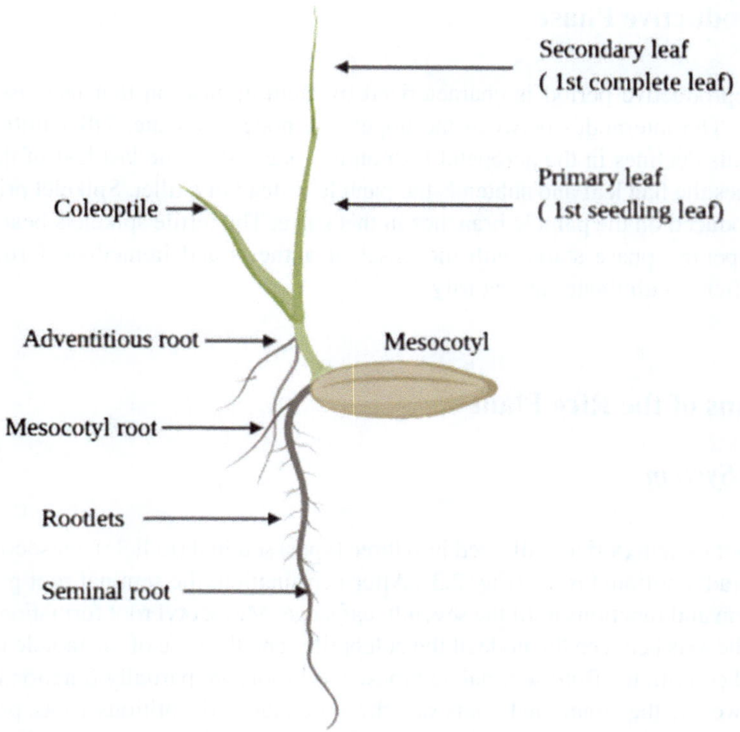

Fig. 2.3 Parts of a young rice seedling germinated in dark (GRiSP 2013)

from the surrounding water. In underground germination, a short stem, known as mesocotyl, develops first and it lifts the coleoptile upwards for emergence. Once above ground, coleoptile splits and forms the first primary leaf of the seedling.

In the vegetative stage, the plant produces a root system and many tillers (arise from buds in an alternate pattern), extends plant height by the production of nodes (solid part), and increases leaf number at regular intervals. Tillering is initiated when the main stem has 5–6 leaves. The early formed tillers (may be considered as independent plantlets) on lower stem nodes are productive. Each productive tiller bears a terminal flowering head or panicle. The capacity for panicle production and grain yield declines progressively with time in tillers on successive nodes (up to twenty). The vegetative stage culminates with the panicle primordia initiation and onset of the reproductive period.

Reproductive Phase

The reproductive period is characterized by stem elongation that increases plant height. The internodes between the upper five nodes elongate. Tiller initiation in leaf axils declines in the acropetal fashion of node order. The last leaf of the culm becomes the flag leaf and subtends the panicle instead of a tiller. Spikelet primordia are produced on the panicle branches in this stage. The fertile spikelets bear grains. The ripening phase starts with the onset of anthesis and immediate fertilization thereafter. It culminates at maturity.

Organs of the Rice Plant

Root System

The root system of rice is divided into three types, seminal (radicle), mesocotyl, and nodal (adventitious) roots (Fig. 2.3). After germination, the seminal root grows up to 15 cm and functions until the seventh leaf stage. Mesocotyl root formation occurs from the axis between the node of the coleoptiles and the base of the radicle in deep-seeded conditions. Both seminal and mesocotyl roots are partially functional.

However, the functional roots are the secondary adventitious roots produced from the lower part of the internode of the basal nodes of the stem generally present below the surface of the soil, which also gives rise to a tiller. At each node, 5–25 roots are developed. They are also called crown roots. Roots directly developing out of the stem nodes are primary roots. In course of growth, they produce secondary roots and secondary roots produce tertiary roots to generate an extensive fibrous system of root structure (Fig. 2.4). The root diameter decreases with the increase in the branching order (Yoshida 1981). The network of fibrous roots better developed horizontally rather than vertically helps exploration of minerals from the surface layer of soil under submergence more efficiently than a tap root system. A taproot system is used for penetration deeper underground in quest of water and other resources. The upper part of the internode of stem nodes above the soil surface also produces adventitious roots, but they are thinner than those produced from the lower part. Soil environment affects root hair formation on the roots; in aerobic soil, a large number of hairs are produced, but soil submergence discourages its production. Additionally, the root length of many rice cultivars exceeds 1 m in soft upland soil, but it seldom goes beyond 40 cm in flooded soil. The matured roots possess large air spaces allowing air passage from root to shoot in aquatic conditions.

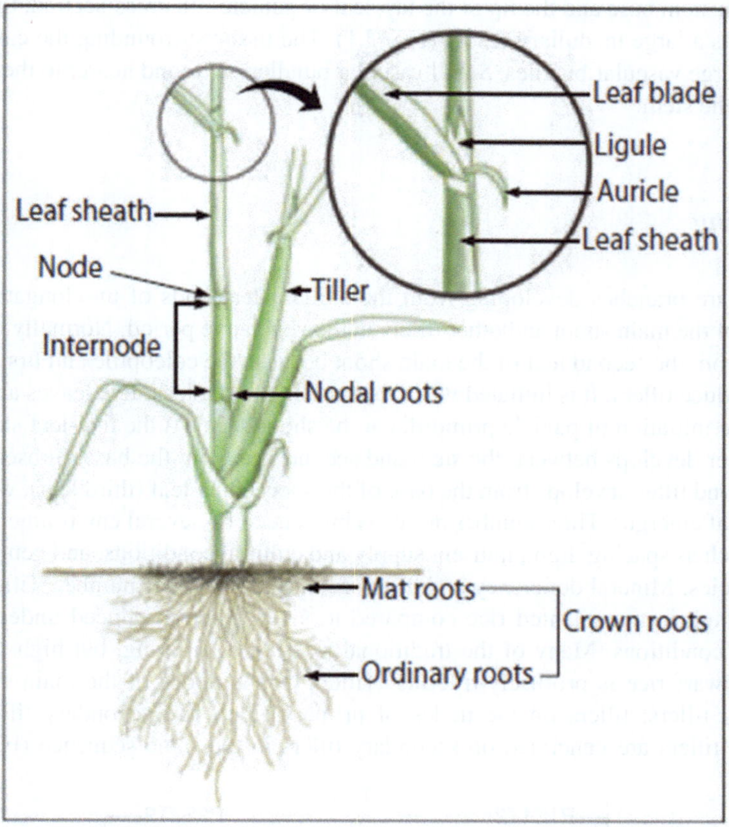

Fig. 2.4 The parts of rice stem and tillers (GRiSP 2013)

Shoot System

Rice stem is erect, cylindrical, and hollow with the absence of pith except in nodes (Fig. 2.1). It is composed of several nodes and internodes. Early duration cultivars have smaller internodes compared to late duration cultivars. The number of internodes may vary from 10 to 20 depending on genotype and environmental conditions. Each node bears one leaf in alternate phyllotaxy and a bud in its axil. The bud may become a tiller. The total number of nodes on the main shoot is the addition of plus two to leaf number; plus two that accounts for the nodes for coleoptiles and panicle. Internode extension occurs in a hierarchical pattern in the acropetal direction with each succeeding internode becoming longer than the preceding one (Pattanaik and Mohapatra 1988). Usually the upper five internodes of the culm elongate; the uppermost one is the longest of all. Some deepwater rice cultivars increase lower internode lengths by 30 cm each to match with the rising level of water under flooding. Internode elongation accounts for plant height (distance between stem base and the tip of the last leaf or panicle). In cross-section, the internode has a large medullary cavity (Fig. 2.1). The tissue surrounding the cavity has many large vascular bundles. Small vascular bundles are found nearer to the epidermis of the stem.

Tillering

Tillers are branches developing from the axillary leaf buds of un-elongated stem nodes of the main shoot and other tillers in the vegetative period. Normally tillering starts from the second leaf of the main shoot because the coleoptile and first leaf do not produce tillers. It is initiated when the seedling has at least five leaves and stops with the initiation of panicle primordia at the shoot apex. At the five-leaf stage, the first tiller develops between the stem and second leaf from the base. Subsequently, the second tiller develops from the base of the succeeding leaf (third leaf), when the sixth leaf emerges. Tiller number in rice is influenced by several environmental factors, such as spacing, light, nutrient supply and cultural conditions, and genotype of the species. Mineral deficiency and deep seeding reduce tiller number. Tiller number is high in transplanted rice compared to a few tillers produced under direct-seeded conditions. Many of the traditional rice is low tillering, but high-yielding semi- dwarf rice is profusely-tillering. Tillers on the nodes of the main stem are primary tillers; tillers on the nodes of primary tillers are secondary tillers and tertiary tillers are generated on secondary tillers in the same sequence (Fig. 2.5).

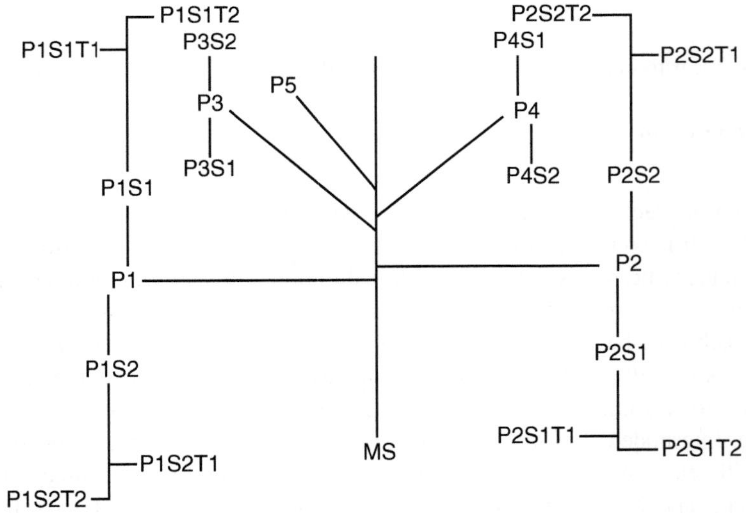

Fig. 2.5 Tillering pattern of *Oryza* species. MS, P, S, and T represent main shoot, primary, secondary and tertiary tillers respectively. P1: first primary tiller, P5: fifth primary tiller, P1S1: first secondary tiller emerged on P1, P2S1: first secondary tiller of P2, P1S1T1: first tertiary tiller of P1S1, and so on. (Mohapatra et al. 2011)

Tiller development in rice is asynchronous, but there is synchrony in tiller senescence and maturity. Hence tiller growth duration shortens in late-tillers that limit the size of panicle and yield owing to a reduced capacity of assimilate production from a limited number of leaves on the stem (Mohapatra and Kariali 2008). Fertile tillers bear panicle. Many of the late-tillers are sterile and die in the reproductive phase. Although tillers are branches of the main shoot, at later stages they develop own roots and become independent.

Leaf

Each node of stem bears one leaf at an angle. The first leaf of the plant is the coleoptile leaf. The second leaf emerges through the coleoptile; it is smaller in size and has no lamina. The other leaves are normal except the flag leaf, which is slightly modified. Leaf size increases with node order on the culm up to the third leaf below the flag leaf. Most of the early to medium-duration rice cultivars produce 10–18 leaves on the main shoot. All leaves of the plant do not have the same life span. The upper leaves are longer spanned compared to the lower leaves; the flag leaf (topmost) has the longest span of all. At any given stage between panicle primordial initiations to heading, only the upper three leaves are physiologically active, while the other leaves below function poorly and senesce early. The contribution of flag leaf in providing photosynthetic assimilates for grain filling is maximal.

Parts of the rice leaf: A matured leaf has different parts, such as a sheath, lamina, ligule, and two auricles (Figs. 2.6 and 2.7; Yoshida 1981).

Leaf-Sheath

Leaf-sheath originates from the stem node. Its growth occurs from the base (swelling; pulvinus) may continue after the lamina has attained the maximum length. The sheath is a cylinder-shaped ribbon-like structure and surrounds the younger parts of the shoot to protect them from physical damage. As the base of the sheath exceeds the length of the internode, it enwraps the succeeding stem node and a part of the upper leaf sheath on the basal part of the plant. However, internodes become progressively longer with an increase of node order on the stem, and leaf sheaths reduce elongation and become increasingly shorter. Leaf-sheath connects the lamina to the stem node and helps in the passage of photosynthetic assimilates. By itself, the leaf sheath contributes very little to photosynthesis. It provides mechanical support to breaking strength before initiation of stem elongation. Besides, it works as temporary storage for photosynthetic assimilates before heading along with the culm. The starch and sugar (non-structural carbohydrates) accumulated in the vegetative

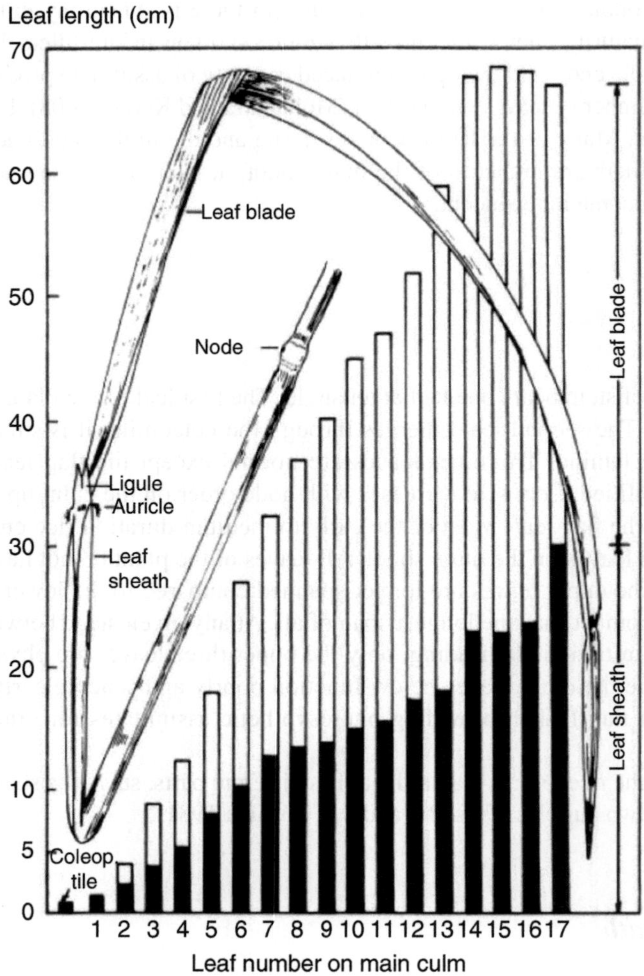

Fig. 2.6 Form and dimensions of rice leaf components according to Yoshida (1981 – with permission from IRRI)

structures are remobilized into the grains ripening particularly under exigency of environmental stresses.

Lamina

The lamina is the distal part of the leaf longer than the leaf sheath. It is flat, narrow, and expanded the upper part of the leaf joined with leaf sheath at the thick collar position. The lamina is narrow, hispid, or pubescent with a distinct midrib and

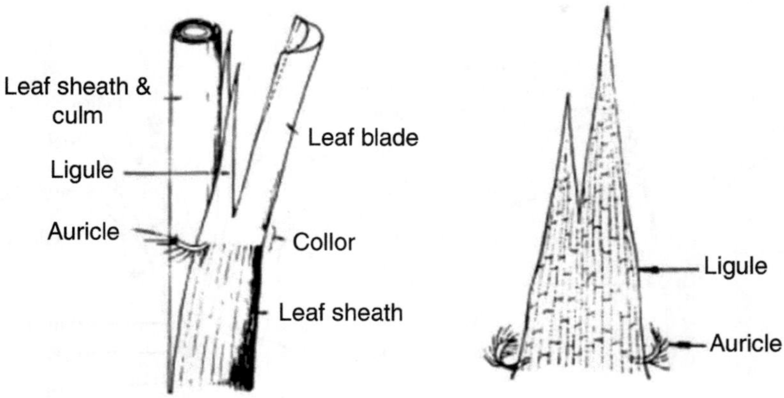

Fig. 2.7 Morphology of the area between blade and sheath of rice leaf according to Yoshida (1981 – with permission from IRRI)

parallel venation. The last leaf, i.e. the flag leaf, has a lamina shorter and broader than others. At the booting stage, the lamina of the flag leaf is parallel to the panicle axis, but after panicle emergence, it falls down forming an acute angle with panicle axis. Lamina is green and photosynthetic and responsible for the primary production of assimilates.

Auricles

At the boundary between the leaf sheath and lamina, there is a pair of claw-like appendages named auricle, which encircles the stem. It helps to bend the lamina towards the adaxial side. The surface of the auricles is covered with coarse hairs. Some rice cultivars have no auricles. The ligule is a thin acuminate membranous (papery) structure found above the auricles on the opposite side of the lamina and represents a continuation of the sheath. The ligule tends to split as it develops. The size, color, and structure of the ligule differ between rice cultivars (Fig. 2.7).

Panicle

Structural Organization

In the grass family *Poaceae*, the flowers are borne in a seed head (inflorescence) found on the uppermost internode of the culm. The inflorescence structure is defined by the spatial arrangement of flowers (florets). The florets are borne in small branches named spikelets. Spikelets are a unique characteristic of the grass family. Rice has one floret in each spikelet, while other members of the family like wheat

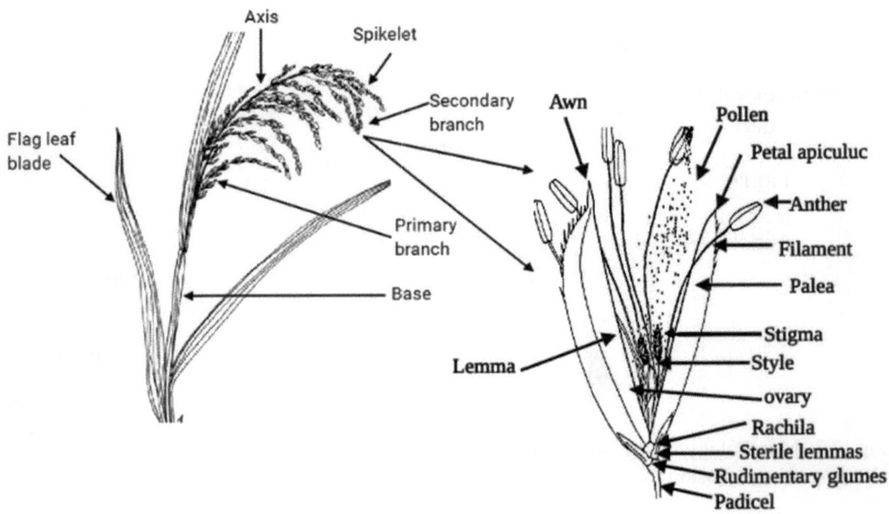

Fig. 2.8 Morphology of rice panicle and spikelets (GRiSP 2013)

and maize have more than one floret per spikelet. Xu and Vergara (1986) considered it a modified racemose branching inflorescence, scientifically named panicle, consisting of a central rachis with nodes bearing one primary branch per node. Panicle type of inflorescence is also common to some other members of the grass family *Poaceae*, like *Panicum*, *Avena,* and *Sorghum*. In comparison, rice panicle has a unique characteristic. It is a complex determinate type of inflorescence bearing pyramidal clusters of small wind-pollinated flowers (spikelets) in a racemose orientation where the individual spikelets are not directly attached to the main axis (rachis). The spikelets are borne on pedicels of lateral branches, both primary and secondary, and this type of branching is categorized as a raceme (Fig. 2.8). In contrast, the inflorescence of wheat and barley are spikes, where spikelets are directly attached to the main axis in the absence of pedicels. Panicle morphology is determined by the length of the rachis (neck node to the degenerated growing point at the base of the terminal primary branch), length and number of primary and secondary branches, and spikelet number (Ikeda et al. 2005). The panicle stands erect at blooming but becomes droopy at maturity. It has 8–10 loosely arranged lateral primary branches at 2–4 cm intervals, one branch borne on each node of the rachis. The length of the branches decreases in an acropetalous sequence that contributes to the pyramidal shape of the panicle (Fig. 2.9).

Some of the primary branches twist around the panicle axis to provide flexural rigidity while the others hang around it as concentrated loads to increase the local curvature (Silk et al. 1982). Lateral branches attached to a primary branch are the secondary branches. The first 2–3 nodes of the primary branch bear secondary branches. Spikelets are borne on the nodes of the secondary and residual part of primary branches, one at each node (Mohapatra 2013) in a fashion similar to the determinate type of distribution of primary branches on the rachis. The branch tip,

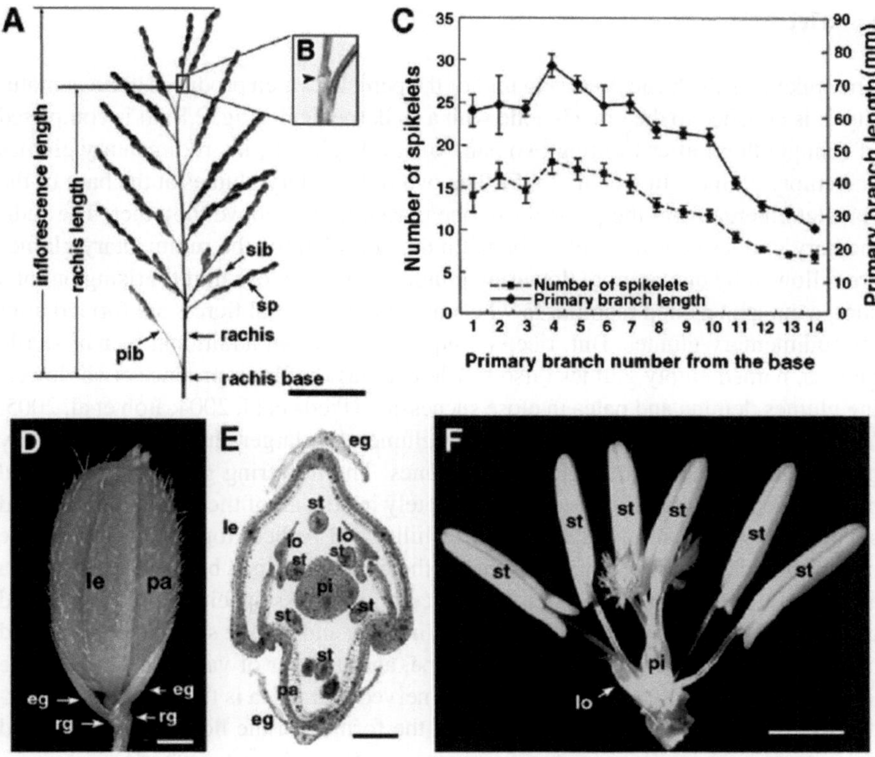

Fig. 2.9 (**A**) Mature inflorescence just after heading 13 primary inflorescence branches. pib: primary inflorescence branch, sib: secondary inflorescence branch, and sp.: spikelet. (**B**) Close up of the region surrounded by a square in (**A**), showing vestige of rachis meristem (arrowhead). (**C**) Positional change of primary branch length and the number of spikelets on each primary branch. The vertical bar represents standard error. (**D**) The external appearance of spikelet at maturity. (**E**) Cross-section of mature spikelet. (**F**) Mature flower (Glumes are removed). Spikelet comprises a pair of rudimentary glumes (rg), a pair of empty glumes (eg), two lodicules (lo), six stamens (st), and one pistil (pi). Bar = 5 cm for A, 1 mm for D, E, and F (Ikeda et al. 2004 with permission of the author)

secondary or primary, always terminates with a spikelet. Subject to filling, spikelet bears a grain at maturity. More recently Zhao et al. (2015) used a high- throughput quantitative image analysis method to ascertain the diversity of panicle architecture in rice and established a strong positive linear relationship between total lengths of the primary branch (summation of the length of all primary branches) with spikelet number per panicle. It indicated a strong linkage between the two traits of rice panicle.

Spikelet

The spikelet is the basic flowering unit of the panicle, which produces grain at maturity. It is attached to the branch node with a stalk (pedicel) (Fig. 2.8). It is comprised of a single floret after forming two pairs of sterile glumes, i.e. rudimentary glumes and empty glumes (Itoh et al. 2005). The two rudimentary glumes at the base of the spikelet emerge from the pedicel tip, one placed a little above the other. The rudimentary glumes bear no axillary buds. In the grass family, the rudimentary glumes are followed by one or more flowering glumes which enclose florets arising out of a short stem axis named rachilla. In wheat and barley, several florets are formed after the rudimentary glumes. But, rice is unique in having an additional pair of sterile glumes, named empty glumes (also sterile lemmas) before it produces two flowering glumes, lemma and palea in close succession (Ikeda et al. 2004; Itoh et al. 2005; Yoshida and Nagato 2011). The empty glumes are larger than the rudimentary glumes but shorter than the flowering glumes. The flowering glumes, lemma, and palea (bracts), encase the caryopsis completely inside; all of them held together and supported by the rachilla. The bracts, rachilla, and pedicel together comprise the hull, although major structural features of the hull are defined by the bracts (lemma and palea). The lemma, and palea, in particular, are grass-specific floret organs and they protect the caryopsis from pathogens, insects, and abiotic stresses. Lemma and palea differ significantly in their size, shape, and number of vascular bundles. The lemma is hard, rough, and awned and five nerved, but palea is three nerved; the latter is closely compatible, but shorter than the former. In the floret, the lemma and palea protect the tender reproductive organs present in the inner whorls.

There are six stamens (Fig. 2.10), each bearing a two-celled anther box on a slender filament. The stamens surround one uni-ovulate pistil having a style and a hairy and bifid (laterally extended) plumose stigma to facilitate aerial pollination. At

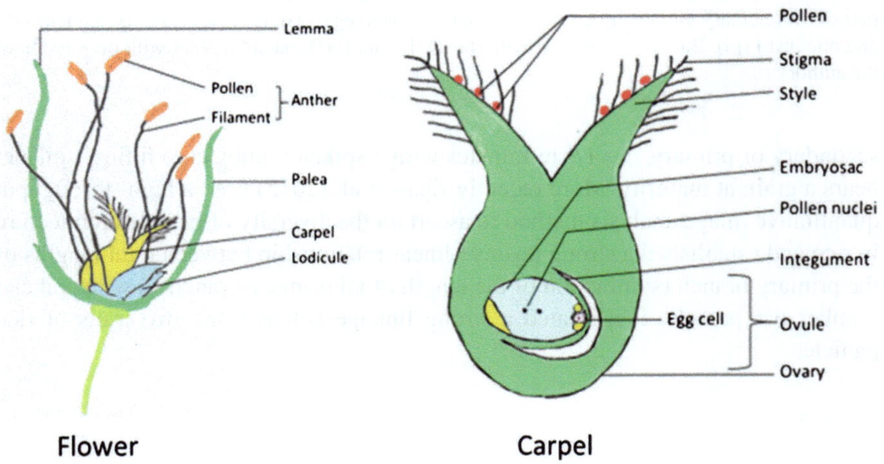

Fig. 2.10 Parts of flower and carpel of rice

the base of the ovary, two small, oval, and fleshy bodies named lodicules are present. They are scale-like organs placed asymmetrically in the second whorl primordium, just outside the stamen whorl, which is considered the equivalent of petals of eudicots. At the time of flowering, they become distended with water; the rapid swelling forces the opening of the flower through the separation of the lemma and palea.

Rice Grain

A spikelet bears a grain at maturity. It is a true fruit caryopsis enclosed in a hull. The caryopsis (brown rice) mainly comprises the embryo and endosperm surrounded by several thin layers of pericarp tissues. For *indica* rice, the hull consists of lemma, palea, and rachilla. In *japonica* the hull includes rudimentary glumes and a part of pedicel also. Grain diversity in structure, size, shape, color, food value and aroma in rice is extremely high and unparalleled in cereals. The grain weight varies from 10 to 45 mg at 0 moisture level. The fruit coat comprises of the quadrangular epicarp, two to three layers of compressed mesocarp, and the central single layered tube cells as endocarp. Testa and tegmen are not quite visible due to pressure generated from the developing seed. Aleurone and embryo cells are rich in protein bodies. The endosperm contains the starch grains that lie below the aleurone layer.

The embryo consists of scutellum, epiblast, plumule, and radicle (Fig. 2.11).

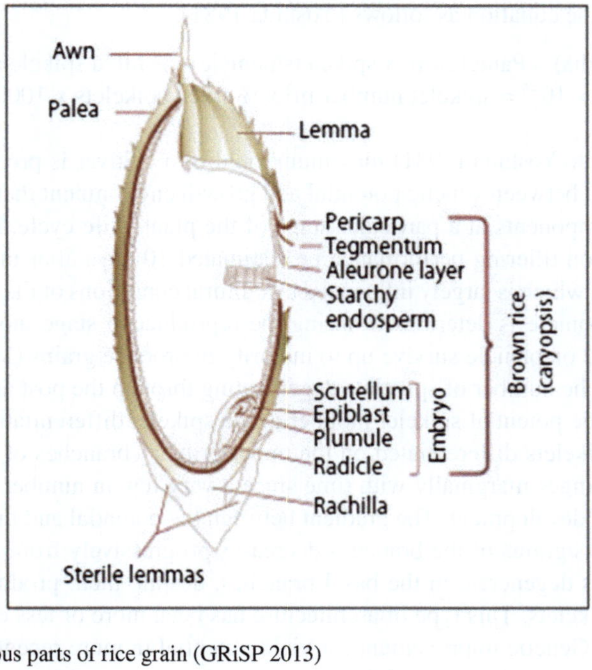

Fig. 2.11 Various parts of rice grain (GRiSP 2013)

Grain Yield

The grain yield of rice is usually reported as rough rice (paddy) at 14% moisture content. The efficiency of crop yield is measured in terms of biomass partitioning in favor of the reproductive parts at maturity, as indicated in the harvest index. The harvest index is the ratio of dry grain yield to total plant dry weight. It is an indicator for an economically useful fraction of the total biological yield (Yoshida 1981). For convenience, sometimes total plant dry weight is replaced by above ground biomass weight for avoiding shard efforts in extracting the roots from the soil for inclusion in total plant dry weight. The traditional tall varieties have a low harvest index of about 0.3 because of extensive biomass investment in vegetative organs. In contrast, semidwarf rice has a harvest index of 0.5 by promoting biomass partitioning into reproductive parts.

$$\text{Harvest Index} = \frac{\text{Dry grain yield}}{\text{Total plant dry weight}}$$

Grain Yield Components

The grain yield of the rice plant has four components, i.e. number of panicles/m^2, the number of grains/panicle, percentage of fertile grains, and grain weight. It is computed in the equation as follows (Yoshida 1981).

Grain yield (t/ha) = Panicles/m^2 x spikelets/panicle x % filled spikelets x 1000 grain weight (g) x 10^{-5} = spikelet number m^2 x % filled spikelets x 1000 grain weight (g) x 10^{-5}

According to Yoshida (1981) maximum yield of a cultivar is predetermined by the interaction between genetic potential and growth environment that influence the four yield components at a particular stage of the plant's life cycle. Panicles/m^2 is dependent upon tillering performance i.e. estimated 10 days after maximum tiller number stage, which is largely influenced by cultural conditions of the field. Spikelet number per panicle is determined during the reproductive stage, not all spikelets that originated on panicle survive up to maturity to produce grains (Mohapatra and Sahu 1991). The number of spikelets degenerating through the post anthesis period underscores the potential spikelet number at the spikelet differentiation stage. The number of spikelets differentiated on the upper primary branches of the panicle is low and it changes marginally with time since a very few in number degenerate at later stages of development. The gradient between the potential and the actual number of spikelets/grains of the branches decreases progressively from base to top as more spikelets degenerate in the basal branches, despite them producing a higher number of spikelets. This type of architecture has been more or less common to all rice panicles. Genetic improvement for yield potential in more recently introduced high-yielding rice has not altered the pattern as of today, as evident in the comparison of cultivars IR36, Jaya, and Akul (Fig. 2.12), where Akul is old traditional rice

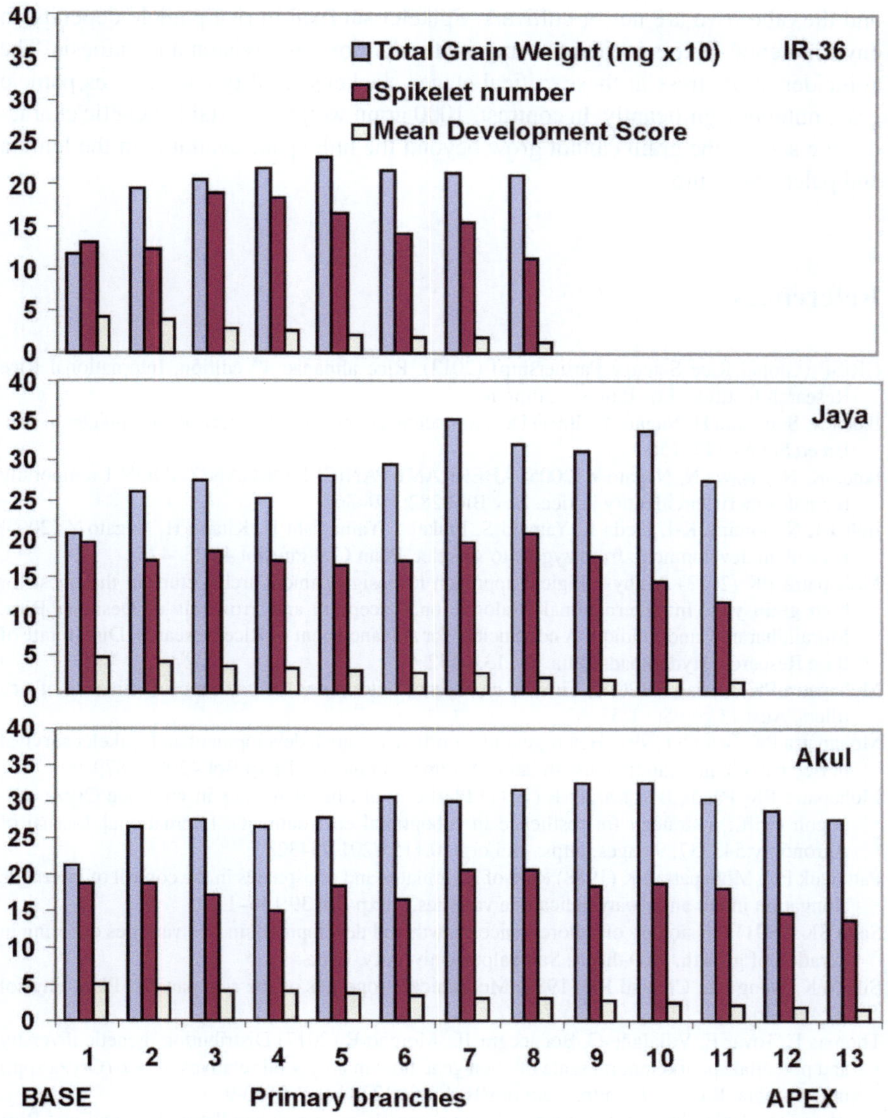

Fig. 2.12 Grain weight, spikelet number and mean developmental score of primary branches of two semidwarf (IR36 and Jaya) and one traditional tall *indica* (Akul) rice cultivars differing in panicle size grown during the dry and wet seasons of 1987 respectively (Sahu 1991). A high developmental score indicates less development and vice versa, cited from Mohapatra (2013)

and the other two are newer cultivars. Spikelet survival in rice panicle depends on environmental factors incident at stages of reduction cell division and anthesis. The coincidence of stress at these critical stages slackens fertility and reduces panicle grain number significantly. In contrast, 1000 grain weight is a stable genetic character; the size of the grain cannot grow beyond the hull space available in the lemma and palea enclosure.

References

GRiSP (Global Rice Science Partnership) (2013). Rice almanac 4th edition. International Rice Research Institute, Los Baños, Philippines

Ikeda K, Sunohara H, Nagato Y (2004) Developmental course of inflorescence and spikelet in rice. Breed Sci 54:147–156

Ikeda K, Nagasawa N, Nagato Y (2005) ABERRANT PANICLE ORGANIZATION 1 temporally regulates meristem identity in rice. Dev Biol 282:349–360

Itoh J-I, Nonomura K-I, Ikeda K, Yamaki S, Inukai Y, Yamagishi H, Kitano H, Nagato Y (2005) Rice plant development: from zygote to spikelet. Plant Cell Physiol 46:23–47

Mohapatra PK (2013) A physiological approach to design panicle architecture in the quest for high grain yield.In: International Dialogue on Perception and Prospects of Designer Rice. Muralidharan K and Siddiq E A eds.Society for Advancement of Rice Research, Directorate of Rice Research, Hyderabad, India. pp. 131–142

Mohapatra PK, Kariali E (2008) Time of emergence determines the pattern of dominance of rice tillers. Aust J Crop Sci 1:53–62

Mohapatra PK, Sahu S (1991) Heterogeneity of primary branch development and spikelet survival in rice panicle in relation to assimilates of primary branches. J Exp Bot 42:871–879

Mohapatra PK, Panda BB, Kariali E (2011) Plasticity of tiller dynamics in wild rice Oryza rufipogon Griff.: a strategy for resilience in suboptimal environments. International Journal of Agronomy:543237, 9 pages. https://doi.org/10.1155/2011/543237

Pattanaik PK, Mohapatra PK (1988) Role of assimilates and phosphates in the control of internode elongation in tall and dwarf indica rice varieties. J Exp Bot 39:136–1380

Sahu SK (1991) Physiology of inflorescence growth and development in rice varieties differing in duration of growth. PhD thesis, Sambalpur University, India

Silk WK, Wang LL, Cleland RE (1982) Mechanical properties of the rice panicle. Plant Physiol 70:460–464

Thomas E, Tovar E, Villafañe C, Bocanegra JL, Moreno R (2017) Distribution, genetic diversity and potential spatiotemporal scale of alien gene flow in crop wild relatives of rice (Oryza spp.) in Colombia. Rice 10:13. https://doi.org/10.1186/s12284-017-0150-9

Vaughan DA (1994) The wild relatives of rice: a genetic resources handbook. International Rice Research Institute, Philippines

Vaughan DA, Morishima H, Kadowaki K (2003) Diversity in the Oryza genus. Curr Opin Plant Biol 6:139–146

Xu X, Vergara BS (1986) Morphological changes in rice panicle development: a review of literature. IRRI Research Paper Series No 117. International Rice Research Institute, Philipines

Yoshida S (1981) Fundamentals of rice crop science. International Rice Research Institute, Phippines

Yoshida H, Nagato Y (2011) Flower development in rice. J Exp Bot 62:4719–4730

Zhao S, Gu J, Zhao Y, Hassan M, Li Y, Ding W (2015) A method for estimating spikelet number per panicle: integrating image analysis and a 5-point calibration model. Sci Rep 5:16241. https://doi.org/10.1038/srep16241

Chapter 3
Ontogeny of Organ Development in Rice Plant

Abstract During the vegetative stage, the shoot apical meristem of the rice plant repetitively produces leaves as lateral outgrowths. Leaf size varies with its position on the stem axis. Initiation of panicle primordial marks the transition of vegetative to reproductive sage. The shoot apical meristem, instead of producing a leaf or tiller initial, is converted to an inflorescence (rachis) meristem and the floral fate of the shoot meristem is established. Soon after primary branch meristems generate on the elongated rachis and followed by secondary meristems on primary branches. Spikelets are produced on the nodes and tips of primary and secondary branches giving rise to a determinate panicle structure. The young panicle becomes visible to the naked eye at the secondary branch differentiation stage when panicle length is 0.5–0.9 mm. The stages of early to late spikelet differentiation occur when panicle elongation extends from 1.0 to 15.0 mm. Panicle length is 1.0–5.0 cm during pollen mother cell differentiation and 5.0–20.0 cm during reduction division of pollen mother cell stages. Nearly 16 days after visual panicle initiation, the flag leaf sheath swells enclosing the growing panicle inside the boot. During the reproductive period, the upper internodes of the stem, undergo a rapid growth sport to increase length in a hierarchical acropetal succession causing the culm to shoot up panicle emergence (heading). Anthesis (flowering) of upper spikelets coincides with the heading or the day following. The onset of anthesis is faster and earlier in older tillers. A cultivar may take 10–14 days for completion of heading and anthesis of all tillers. Within rice panicle order of anthesis of spikelets is not identical; the spikelets on upper primary branches take time less than those of the basal branches.

Keywords Rachis meristem · Branch meristem · Panicle · Spikelet · Tillers · Anthesis · Booting

Vegetative Stage and Leaf Development

Rice plant bears 10–18 leaves, one on each node of the main culm in an alternate phyllotaxy. During the vegetative stage, the shoot apical meristem of the rice plant repetitively produces leaves as lateral outgrowths. Leaf size varies with its position on the stem axis; the first two leaves, smaller in size are the juvenile leaves, whereas leaf numbers 3–5 are intermediate and the rest are adult leaves (Itoh et al. 2005).

P. K. Mohapatra, B. B. Sahu, *Panicle Architecture of Rice and its Relationship with Grain Filling*, https://doi.org/10.1007/978-3-030-67897-5_3

In the juvenile phase, a leaf primordium arises from cell divisions initiated at some distance from the summit of the apex of the plant. These cell divisions lead to the formation of a simple ridge, which subsequently develops laterally to encircle the stem axis in a crescent-shaped fashion.

Follow-up growth of the newly formed leaf primordium by rapid cell divisions in the apical and marginal regions makes it hood-shaped. The hood encloses the shoot apical meristem completely. With further growth of the plant, leaf primordia in various stages of development accumulate at the base of the growing apex. As vegetative growth proceeds, the shoot apical meristem enlarges and eventually emerges from the close confines of the coleoptiles leaf, but continues within the protection of the leaf bases. The shoot apical meristem behavior changes, although not discretely, with the attainment of adulthood. In rice plant bearing 10 leaves, the shoot apical meristem increases in size gradually with an increase of leaf number. For example, the shoot apical meristem at the tenth leaf stage is two times larger than what it is at the second leaf stage.

Other leaf traits that contribute to differences in individual leaf size and shape are the decline of cell division frequency and plastochron (the rate of leaf production) of the shoot apical meristem with the increase of stem node order of the vegetative culm. The cell division frequency of the shoot apical meristem is higher at the second leaf stage than at the fourth leaf stage and the plastochrons of the second and third leaves are 1.6 days compared to 4 days for the fifth leaf (Itoh et al. 2005). Counting of leaves of the main culm gives a physiologically meaningful age for the rice plant, but numbering them is often mired with confusion (Yoshida 1981). The coleoptile leaf appears at germination and finally enlarges to 1 cm long. The first leaf grows about 2 cm and lacks a leaf blade. It is often named a prophyll or incomplete leaf and not included in the leaf count. The second leaf possesses a well-developed lamina and grows larger than the first one in size.

Phaseic Development

Phasic development from vegetative to the reproductive stage (Fig. 3.3A) occurs with the initiation of panicle primordia when physiological and environmental conditions are favorable for floral induction. The shoot apical meristem, instead of producing a leaf or tiller initial, is converted to an inflorescence meristem. Here leaf primordium development is suppressed and the floral fate of the shoot meristem is established firmly. The last foliage leaf primordium produced by shoot apical meristem is for the flag leaf. After this shoot apical meristem is converted to inflorescence meristem or rachis meristem. In cereals elongation of the dome, the apical region of the shoot apical meristem bare of leaf initials is the first visible sign of transition of shoot apex from a vegetative to a reproductive state (Williams 1975). This elongating dome continues to produce single ridge primordia alternately on its flanks. These primordia are subsequently converted to double ridges by the growth of the tissue in a position axillary to the first ridge. The growth of the lower of the

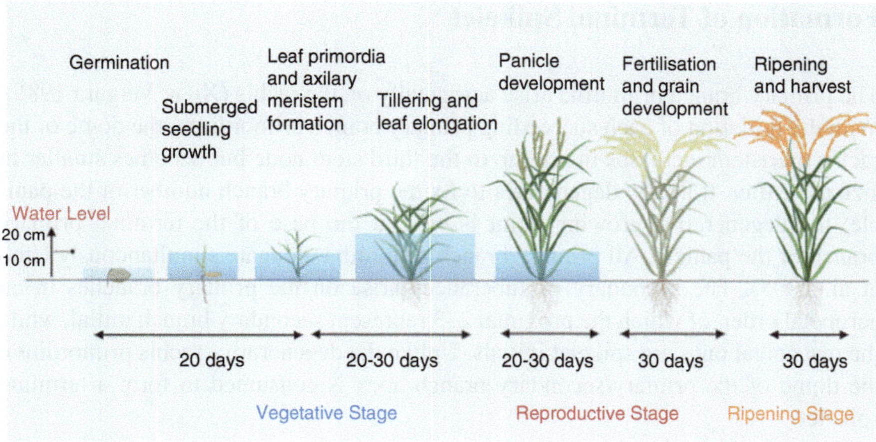

Fig. 3.3A Developmental stages of the rice plant

two ridges, the putative leaf initial (bract), is progressively suppressed, whereas, the upper ridge develops at an accelerating rate to form a spikelet, as seen in wheat (Fisher 1973). The extent of inhibition of growth of the lower ridge varies with the position of the primordium on the rachis, the bract primordium persisting longer in more basal positions.

Primordium Development

Transition to flowering in cereals accelerates the rate of primordium production on the apex and after several double ridges are established on the elongating apex, a gradient of spikelet development is established with the most advanced in the center of the lower middle part of the rachis (Bonnett 1966; Fisher 1973; Kirby 1974). The development of individual spikelets is graded both basipetally and acropetally to less differentiated spikelets at the tip and base. In rice, inflorescence development is more complex than that of wheat or barley, because initiation of primary and secondary branches precedes spikelet ontogeny. The order of development of primary and secondary branch primordia is not similar to that of spikelets. In rice panicle, there are two types of meristems i.e. rachis and branch meristems. During the inflorescence primordium elongating stage, which is similar to wheat, the rachis meristem in rice forms several double ridges on the reproductive apical cone; the upper ridge of which represents a primary branch meristem and the lower one corresponds to a bract. The bracts and their corresponding primary branch primordia appear simultaneously. The first bract primordium encircles the panicle primordium base, but subsequent primordia cover an arc less than 180°. With exception of the first two, growth is severely suppressed in the succeeding bracts, where numbers of whitish hairs are produced. This suggests that there is a graded transition from the vegetative to the reproductive state, as the shoot apical meristem develops.

Formation of Terminal Spikelet

The primary branch primordia arise acropetally on the rachis (Xu & Vergara 1986). With the initiation of each succeeding primary branch primordium, the dome of the rachis meristem increases in size up to the third stem node but becomes smaller in size thereafter. It finally degenerates to fix the primary branch number of the panicle; the degenerative growing point is seen at the base of the terminal primary branch of the panicle. All primary branch primordia elongate simultaneously (Itoh et al. 2005). The secondary protuberances arise on the primary branches in an acropetal order, of which the proximal 2–3 represent secondary branch initial, while the rest apical ones are spikelet initials. Unlike the degenerative rachis primordium, the dome of the primary/secondary branch apex is consumed to form a terminal spikelet.

Staging of Panicle Development

In rice, the staging of panicle development has been examined by various workers. Matsushima (1970), modified by Yoshida (1981), proposed seven developmental stages of rice panicle based on phenotypic estimates like leaf number index, the exertion of leaves, and length of panicle, but the relationship between panicle length and the developmental stage was considered more important. The young panicle becomes visible to the naked eye at the secondary branch differentiation stage when panicle length is 0.5–0.9 mm. The stages of early to late spikelet differentiation occur when panicle elongation extends from 1.0 to 15.0 mm. Panicle length is 1.0–5.0 cm during pollen mother cell differentiation and 5.0–20.0 cm during reduction division of pollen mother cell stages.

More precise staging of development has been proposed by Ikeda et al. (2004), where the developmental course of inflorescence and spikelet has been divided into nine and eight stages respectively. In this staging system, the recent results of genetics and molecular studies have been incorporated into anatomical and morphological investigations for the identity of landmark events of inflorescence and spikelet development. Stages of inflorescence development have been described in the table below (Table 3.1 and Fig. 3.3B). Late stages of panicle/spikelet development are described in Table 3.2 and Figs. 3.3C and 3.3D.

The expression of the *OSH1* gene marks indeterminate cells of the meristem; no expression in determinate cells or aborted meristem. *OsMADS* genes control organ development.

Booting Stage

Nearly 16 days after visual panicle initiation (6 days before heading) the flag leaf sheath swells enclosing the growing panicle inside it. The swelling of the flag leaf sheath is called booting and the 6 days following the event is the booting stage (Yoshida 1981).

Table 3.1 Staging of inflorescence development in rice

Stage Name	Meristem identity	Events	Inflorescence Length (mm)
Establishment of rachis meristem	Rachis meristem	Convertion of vegetative meristem to rachis meristem Enlargement of rachis meristem Formation of bract 1 primordium Onset of ~2stem internode elongation	0.05–0.1
Formation of primary branches I	Rachis meristem branch meristem	Rachis meristem reach maximum size Formation of first two primary branch primordial and bract 2 and 3 primordia	0.1–0.2
Formation of primary branches II	Rachis meristem branch meristem	Formation of primary branch primordial in spiral arrangement Abortion of rachis meristem at the end of this period Onset of ~1 stem internode elongation	0.2–0.4
Elongation of primary branches	Branch meristem	Simultaneous elongation of primary branches	0.4–0.6
Formation of higher0order branches	Branch meristem	Formation of secondary and tertiary branches Onset of uppermost stem elongation	0.6–0.9
Differentiation of glumes	Spikelet meristem Floret meristem	Differentiation of two rudimentary glumes, two empty glumes, lemma and Palea in ½ alternate arrangement	0.9–1.5
Differentiation of floral organs	Floral meristem	Differentiation of two lodicules, six stamens and a carpel in whorl	1.5–40
Rapid elongation of rachis and branches	Loss of meristem	Rachis and branches elongate exponentially Completion of anther and ovule development	40–220
Heading and flowering		Inflorescence protrudes from sheath of flag leaf	More than 220

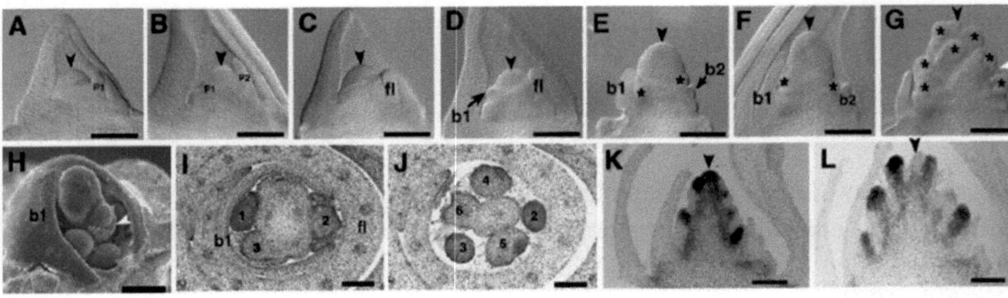

Fig. 3.3B Early stages of inflorescence development (Ikeda et al. 2004-with permission of the author). (**A**)–(**G**) Cleared shoot and inflorescence apices. (**A**) and (**B**): Mature vegetative shoot apices. In (**A**), the tips of P1 leaf primordium and shoot meristem are almost the same in height. When P1 leaf primordium has just protruded, the tip of the P2 leaf is higher than that of the shoot meristem (**B**). (**C**): Initial stage of rachis apex before first bract formation. Rachis meristem is larger than the vegetative shoot apex in (**A**). (**D**): First bract primordium formation. Tip of the flag leaf primordium is not higher than that of rachis meristem (**E**): Formation of two primary branch primordia. Rachis meristem is the largest at this stage (**F**): Rachis meristem forming several primary branch primordia. (**G**): Final stage of primary branch formation. In this rachis, about ten primary branches are present and bracts are differentiating hairs (white arrowhead). (H): SEM image of the young inflorescence. Tip of the bract become hairs (white arrowhead). (I) Rachis apex showing first three-branch primordia (1–3). Divergence angles of the first and second branch primordia are close to 180°. (**J**) Rachis apex showing 2nd to 6th branch primordia. Divergence angles of 4th to 6th primordia are approximately 144°. In (**A**) and (**B**), primary branch primordia are numbered from the oldest one. (**K**) and (**L**) Expression of *OSH1* in rachis apex. *OSH1* is expressed in the rachis meristem producing branch primordia (**K**), but not expressed in the rachis meristem after producing the final primary branch primordia (**L**). The expression of the *OSH1* gene marks indeterminate cells of the meristem and is not expressed in determinate or aborted cells. Meristems are indicated by black arrowheads. ★): primary branch meristem, fl: flag leaf, b1: first bract, and b2: second bract. Bar =150 μm for (**A**) to (**G**), 100 μm for (**H**) to (**L**)

Table 3.2 Staging of spikelet development in rice

Stage	
Name	Events
Formation of a pair of rudimentary glume primordial	Formation of two rudimentary glumes in ½ alternate phyllotaxy (Fig. 3.3D– A)
Formation of a pair of empty glume primordial	Formation of two empty glumes in ½ alternate phyllotaxy (Fig. 3.3D–B)
Formation of lemma primordium	Formation of the lemma in ½ alternate phyllotaxy Elongation of empty glumes (Fig. 3.3D–C)
Formation of Palea primordium	Formation of Palea in ½ alternate phyllotaxy elongation of empty glumes and lemma (Fig. 3.3D–D)
Formation of lodicule primordium	Formation of lodicules in whorl. Elongation of empty glumes lemma and palea (Fig. 3.3D–E and I)
Formation of stamen primordium	Formation of six stamen primordia in whorl (Fig. 3.3D–F and I)
Formation of carpel primordium	Formation of carpel. Differentiation of stamen into filament and anther (Fig. 3.3D–G and J)
Formation of ovule and pollens	Differentiation of integuments and embryo sac. Pollen differentiation (Fig. 3.3D–H)

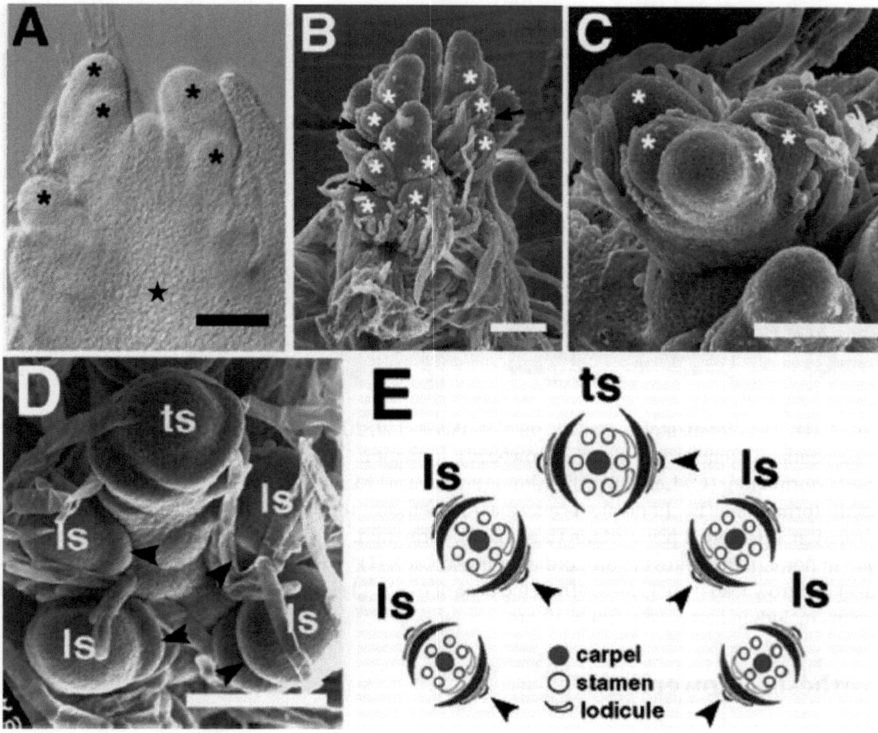

Fig. 3.3C Late stages of inflorescence development (Ikeda et al. 2004-with permission of the author). (**A**) Elongation of primary branches (★). (**B**) Differentiation of secondary branch primordia (☆). Distal cells of bract of secondary branches become hairs (arrows). (**C**) Top view of (**B**) showing biased distichous phyllotaxy of secondary branches (☆). (**D**) Spikelet formation in primary branch apex. Lateral spikelets (ls) are arranged with the divergence angle of about 110°. (**E**) Schematic presentation of terminal spikelet (ts) and lateral spikelet (ls) disposition in the primary branch apex. Terminal spikelet is distinguished from lateral spikelets by its direction of insertion and the position of glumes (arrowheads indicate lemma side). Bar = 100 μm

Heading Stage

The heading is marked by the exertion of panicle tip from the flag leaf sheath enclosure. The panicle continues its emergence until complete protrusion. The internodes elongate rapidly to expose the developing panicle outside from the flag leaf sheath enclosure. Elongation of the second internode from the top is completed a day or two before heading. Thereafter, the topmost internode extends rapidly to expose the panicle.

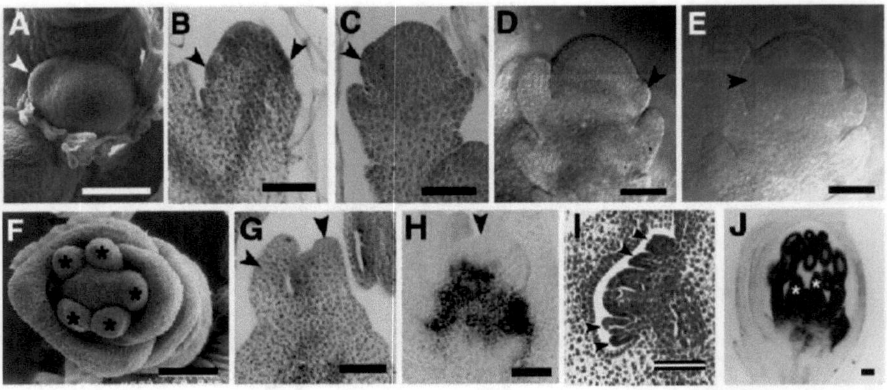

Fig. 3.3D Developmental course of spikelet (Ikeda et al. 2004-with permission of the author). (**A**) Rudimentary glume primordium (arrowhead) formation. (**B**) Empty glume primordia (arrowheads) formation. (**C**) Lemma primordium (arrowhead) formation. (**D**) Palea primordium (arrowhead) formation. (**E**) Lodicule primordia (arrowhead) formation. (**F**) Stamen primordia (☆) formation. (**G**) Carpel primordium (arrowhead) formation. (**H**) Down regulation of *OSH1* expression in floral meristem/ovule primordium (arrowhead). (**I**) Ovule formation. Inner and outer integuments (arrowheads) are growing from the base of ovule. (**J**) Expression of *OsMADS45* in all floral organs including carpel (★). Bar = 50 μm

Anthesis (Flowering)

Anthesis (flowering) of upper spikelets coincides with the heading or the day following. Compared to late tillers, the onset of anthesis is faster and earlier in older tillers. A cultivar may take 10–14 days for completion of heading and anthesis of all tillers. Therefore heading stage of the field-grown rice crop is defined as the time when 50% of panicles reach the stage. Under a tropical environment, anthesis occurs between 0800 and 1300 h (Yoshida 1981). Usually, spikelets open early in the day and close a few hours later to avoid heat stress-induced sterility. The time of the day of anthesis is not identical to all cultivars. The cultivars with early morning opening traits have been found beneficial for yield. However, the very early opening of spikelets may not be good for pollen development and pollen stability. The diurnal variation of anthesis time among cultivars is considered an escape route for thermal stress because heat reduces pollen fertility and pollen efficiency (Guo et al. 2015).

Events of Anthesis

Within rice panicle order of anthesis of spikelets is not identical; the spikelets on upper primary branches take time less than those of the basal branches (Figs. 3.3E, 3.3F). During anthesis, four events take place in the spikelet (Figs. 3.3F, 3.3G,

3.3H). In the first step tips of the lemma and palea begin to open. The separation of lemma and palea begins at the tip and extends downward. With the widening of the gap between the lemma and palea, anther filaments elongate and exert anther box outside the hull space. The bifid stigma opens to both sides before the splitting of the spikelet. The anthers remain above the stigmatic surface and scatter pollens on it. The pollination is cleistogamous in most cases. In step two, lemma and palea stretch to a maximum opening angle of 25–30°. The anther filaments elongate rapidly by longitudinal elongation of the filament cells leading to the exertion of anther box. The residual pollens are released into the air from the anthers. In the formation of the opening, the palea remains static, but the lemma moves to widen the opening. The lemma is pushed up because of the swelling of the two lodicules with the absorption of water from the surrounding tissue. The pressure generated causes the lemma to emerge outside unfastening the hook with its compatriot palea. The feathery tip of the stigma of the pistil becomes visible at this point. The spikelet remains open for a time period varying between half to a few hours during which self-fertilization occurs and then closes. The anther box dehisces just before or when spikelet opens allowing pollens grains fall on to the stigma surface. This stage is called the climax of anthesis.

In the third stage, the anther filaments start withering as the spikelet begins to close. The closure occurs because the lemma returns to its original position as lodicules lose water and become flaccid. The pistil is enclosed completely inside the spikelet in the fourth stage, while anthers are hanging outside (Fig. 3.3E). The anthers droop and die. In contrast to the fate of anthers, the spikelet tends to protect

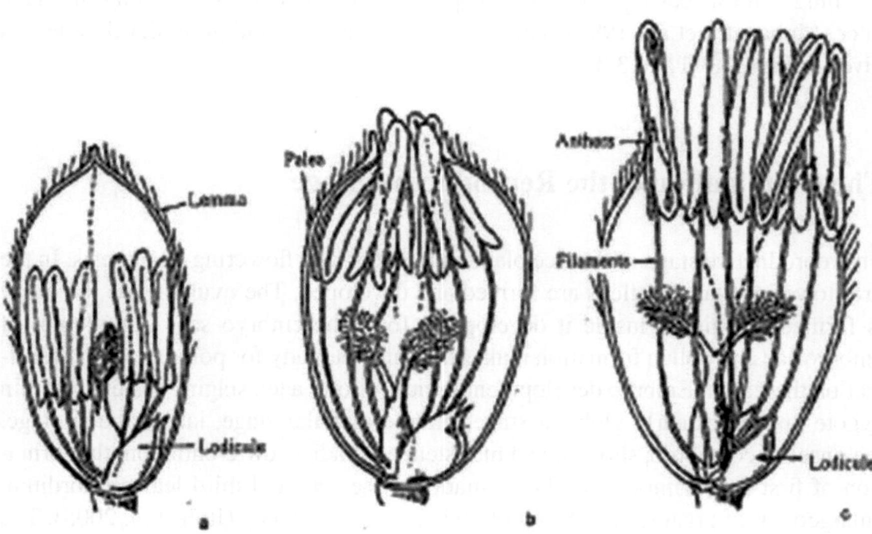

Fig. 3.3E Structure of rice flower through anthesis (Hoshikawa 1993 with permission)
(**a**) Immediately before anthesis. Lodicules begin to swell
(**b**) Anthesis begins. As soon as the filaments start elongation, pollination occurs
(**c**) Height of anthesis. Anthers emerge out of the spikelet (about 20 min after anthesis starts)

Fig. 3.3F Order of anthesis of spikelets within a panicle – Hoshikawa 1993 with permission. Numerical figure affixed to individual spikelet indicates anthesis day

the ovary. The whole event is closed in 20–30 min after the opening of the spikelet and anthesis comes to an end. All spikelets of the panicle do not reach anthesis at the same time. It takes a week time to complete the event (Fig. 3.3F). The spikelets reaching anthesis early produce better quality grains than those reaching the stage later (Mohapatra et al., 1993). The sequence of anthesis and grain development is given in Figs. 3.3G and 3.3H.

The Culmination of the Reproductive Stage

The reproductive stage of the rice plant culminates with flowering or anthesis. In the pre-flowering stage, pollens are formed and developed. The ovule inside the ovary is formed and tissue inside it develops to form the embryo sac. Completion of embryo sac and pollen formation makes the spikelet ready for pollination and fertilization thereafter. Embryo development in rice follows a ten staging system wherein zygote undergoes early globular stage, middle globular stage, late globular stage, the onset of coleoptile, shoot apical meristem and radicle differentiation, the formation of first leaf primordium, the formation of second and third leaf primordium, enlargement of organs, maturation of embryo and dormancy (Itoh et al. 2005). The expression of various unique genes at respective stages of the above development takes place.

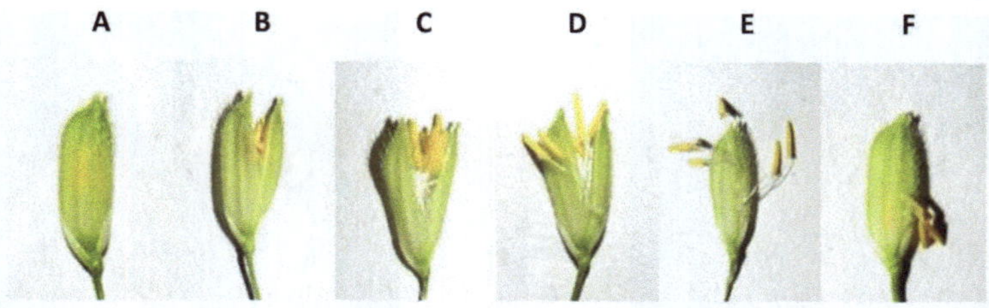

Anthesis steps in rice. A-Spikelet with closed lemma and palea, B- Opening of lemma and palea, C- Elongation of anther filaments, D- Anthers are exerted out, E- Closing of the Spikelet, F- End of Anthesis.

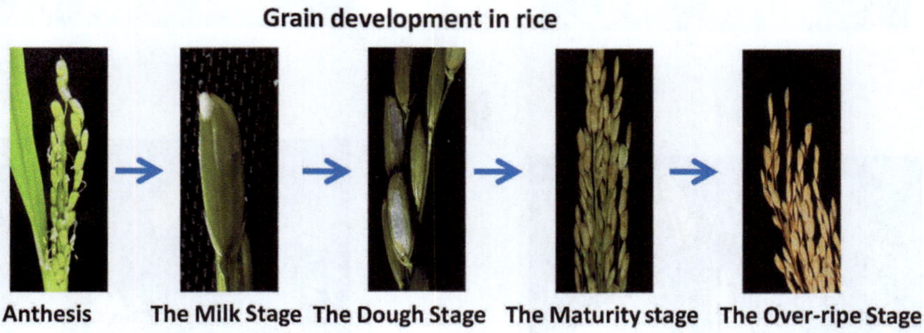

Fig. 3.3G Grain development in rice

Genes Involved in Panicle Organogenesis

The dynamic expression patterns of genes involved in the development of young rice panicles during the reproductive period have been elucidated by Wang et al. (2010). The authors have shown differential expression of genes represented by 2667 probe sets in the four consecutive stages of young panicle development from the differentiation of secondary branch primordium to pollen mother cell. At the early stage of panicle development, the positions of primary and secondary branches are determined, whereas the identity of flower organs and male and female gametes is established in late stages, coincidental to the temporal decline in activity of rachis meristem. The expression dynamics of genes controlling early- stage development are drastically opposite to that of the late group. Of the 2667 genes studied, expression levels of 215 gene probe sets belonging to the early development group decrease gradually from early to late stage, whereas in contrast, 794 gene probe sets of the late development group increase at the same time.

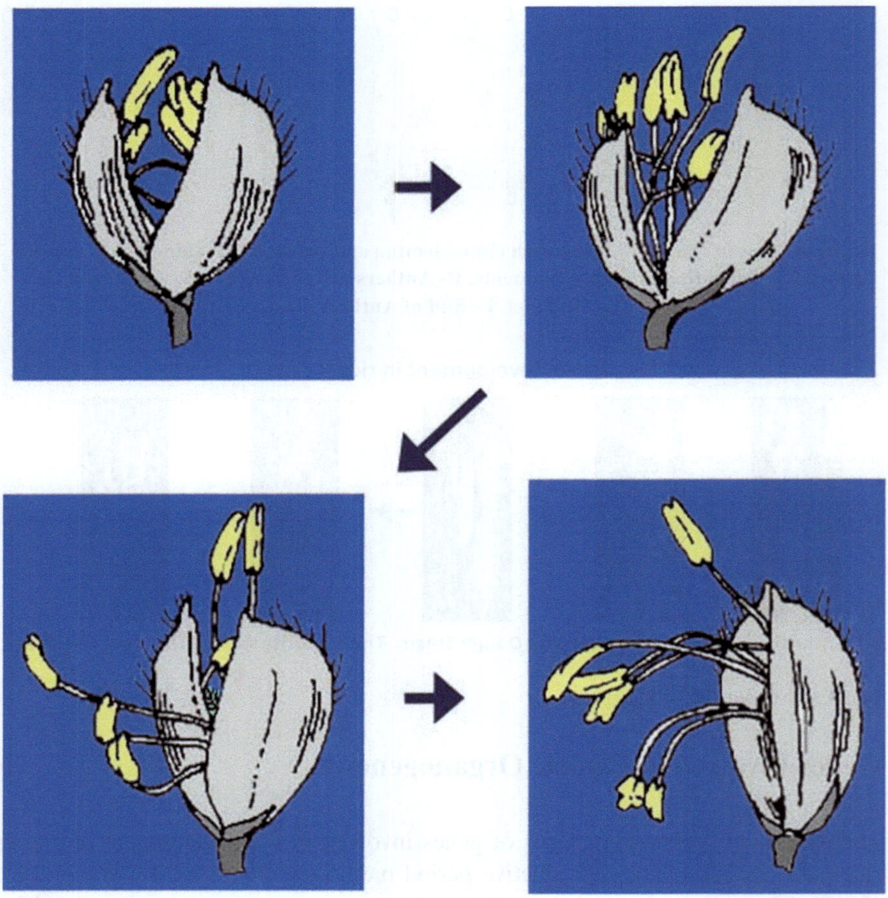

Fig. 3.3H Anthesis in rice

References

Bonnett OT (1966) Inflorescences of maize, wheat, rye, barley, and oats: their initiation and development. Bulletin no 721. University of Illinois (Urbana-Champaign campus), College of Agriculture, Agricultural Experiment Station

Fisher JE (1973) Developmental morphology of the inflorescence in hexaploid wheat cultivars with and without the cultivar Norin 10 in their ancestry. Can J Plant Sci 53:7–15

Guo W, Fukatsu T, Ninomiya S (2015) Automated characterization of flowering dynamics in rice using field- acquired time-series RGB images. Plant Methods 11:7. https://doi.org/10.1186/s13007-015-0047-9

Hoshikawa K (1993) Anthesis, fertilization and development of caryopsis, chapter 2. In: Matsuo T, Hoshikawa K (eds) Science of the rice plant, Morphology, vol 1. Food and Agriculture Policy Research Center, Tokyo, pp 339–376

Ikeda K, Sunohara H, Nagato Y (2004) Developmental course of inflorescence and spikelet in rice. Breed Sci 54:147–156

Itoh J-I, Nonomura K-I, Ikeda K, Yamaki S, Inukai Y, Yamagishi H, Kitano H, Nagato Y (2005) Rice plant development: from zygote to spikelet. Plant Cell Physiol 46:23–47

Kirby EJM (1974) Ear development in spring wheat. J Agric Sci 82:437–447

Matsushima S (1970) Crop science in rice: theory of yield determination and its application. Fuji Publishing, Tokyo, 379p

Mohapatra PK, Patel R, Sahu SK (1993) Time of flowering affects grain quality and spikelet partitioning within the rice panicle. Funct Plant Biol 20:231–241

Wang L, Xie W, Chen Y, Tang W, Yang J, Ye R, Liu L, Lin Y, Xu C, Xiao J (2010) A dynamic gene expression atlas covering the entire life cycle of rice. Plant J 61:752–766

Williams RF (1975) Shoot apex and leaf growth. Cambridge University Press, Cambridge. https://doi.org/10.1017/CBO9780511753404

Xu X, Vergara BS (1986) Morphological changes in rice panicle development: A research development. IRRI research paper series no 117. The Philipines

Yoshida S (1981) Fundamentals of rice crop science. Interational Rice Research Institute, Philippines

Chapter 4
Fertilization and Seed Development in Rice

Abstract At anthesis, stage spikelets remain open for 30–70 min. The anther box dehisces to release all pollens and become empty. The pollens released reach the feathery stigma. The synchrony between spikelet opening and anther dehiscence contributes to reaching self- pollination. The pollens germinate a few minutes after arrival and form pollen tubes. The contents of the pollen including a vegetative nucleus and two male gametes move into the pollen tube and penetrate the stigma. The elongating pollen tube enters into one of the two synergids through the micropyle end of the ovary to discharge its contents of two sperm nuclei and a vegetative nucleus. One sperm nucleus fuses with the egg nucleus and the other one moves out to combine with polar nuclei to effect double fertilization. The zygote of the egg nucleus undergoes multiple divisions to assume a globular shape by day two. Cell number increases rapidly in the embryo leading to the differentiation of initial shoot apex and root primordium by day five. Cell division continues to day 12. The embryo reaches the final size by day 25. Side by side, endosperm growth also occurs after the second nuclear fusion in the ovary. The two endosperm nuclei formed out of the first nuclear division endosperm mother nucleus, move away from each other and remain close to the antipodal side of the embryo sac. They do not form cell walls in between. The daughter nuclei do not rest and undergo further divisions to form a systematic multinucleate endosperm. In the second stage, endosperm enters into a period of cellular development, where cell divisions and multiplication occur in the peripheral cell layer that works like a cambium. Cell division is complete about 9–10 days after fertilization. At the end of cell division, the cells of the outermost endosperm cell layer develop to differentiate into aleurone layers. All cells derived from the periclinal division of the peripheral layer of endosperm are starch storing parenchymatous cells. They have multiple granular amyloplasts, each bearing several starch granules.

Keywords Anthesis · Pollen grains · Double fertilization · Gametes · Egg apparatus · Micropyle · Embryo · Endosperm · Rice seed

P. K. Mohapatra, B. B. Sahu, *Panicle Architecture of Rice and its Relationship with Grain Filling*, https://doi.org/10.1007/978-3-030-67897-5_4

Fig. 4.1 Structure of stamen

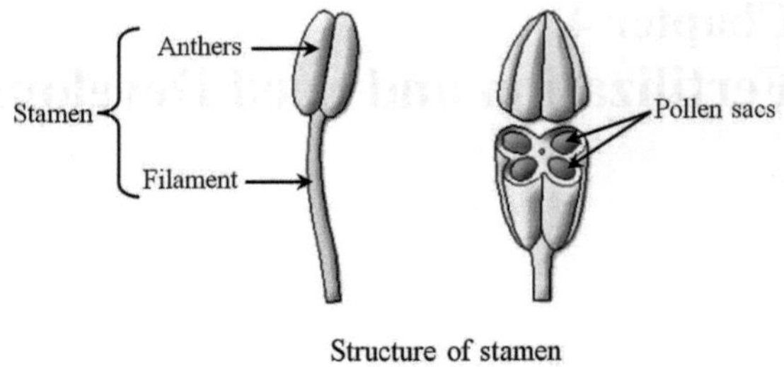

Dehiscence of Anther

Rice is a self-pollinating plant. The spikelet remains open for 30–70 min depending on weather conditions and closes after fertilization. Rice spikelet has six stamens; each stamen bearing one four locular anther on an elongated filament. The vascular bundle of anther leads to the filament. The wall of the anther is a four-layered tissue; the layers from outside to inside are epidermis, endothecium, middle layer, and tapetum. The tapetal cells break down at a later stage of pollen development and provide nutrition, whereas endothecial cells play a major role in the dehiscence of anther box. The four anther loculi are divided into two pairs, a small dehiscence cavity existing between each pair at the juncture. Anther filament begins to elongate just before the opening of the spikelet. At this stage, anther box dehisces from the upper and lower tips, and the dehiscence proceeds to the center (Figs. 4.1 and 4.2). The box releases all pollens and becomes empty. Anther withers and becomes dry thereafter. Endothecial cells regulate anther dehiscence. The cells have hairpin fibrous structures. The hairpin structures work as a spring in response to the change of turgor. The presence of the structures precipitates shrinkage of endothecium during drying that pulls the joint between anther loculi outwards. The fragile joint finally breaks and the splitting of anther wall outwards releases pollen grains. The pollens released in the air reach the feathery stigma just before the spikelet opening. The synchrony between spikelet opening and anther dehiscence contributes to self-pollination.

Pollination

In the megagametophyte, a style grows from the tip of the ovary, which branches into two plumose stigmata and they open up at the right angle to the axis of style at the time of anthesis (Fig. 4.3). The feathery structures are produced from the upper cells of style giving a brush-like appearance. The brush-like feature expands the area for catching pollen grains released into the air. The opening of stigmata lasts only for the anthesis stage and closes immediately after pollination and stigmata wither thereafter.

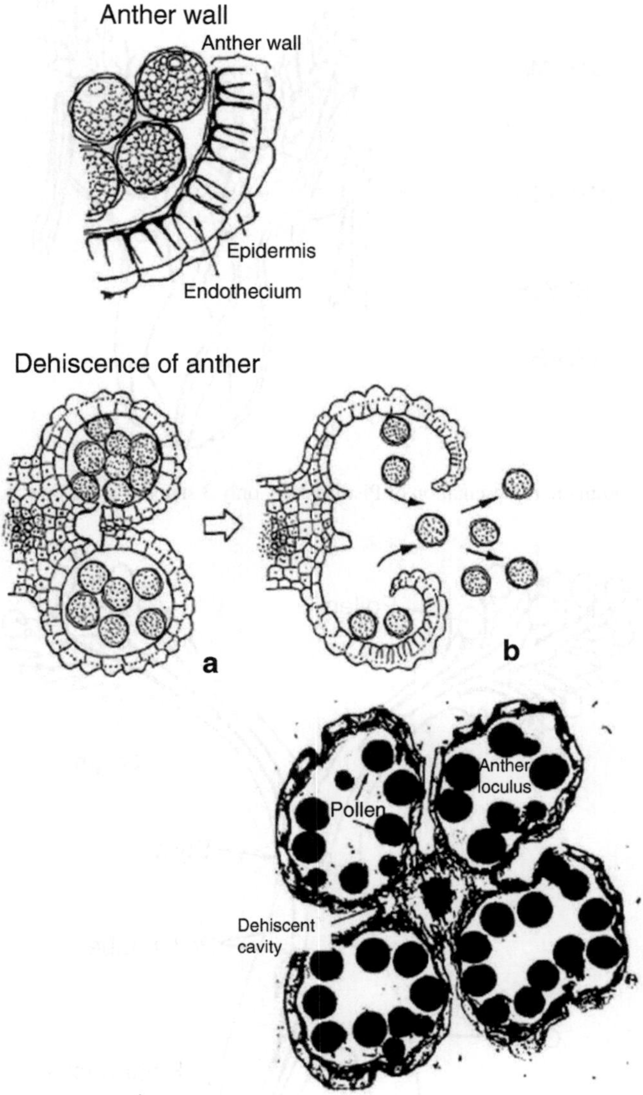

Fig. 4.2 Cross-section of the anther (Hoshikawa 1993, with permission)

Germination of Pollen

Pollen grains released from the anther box stick to the feathery stigmata. The surface of the stigmata receives many pollen grains before the opening of the spikelet. The number of pollens received by stigma ranges from several hundred to more than a thousand (Hoshikawa 1993). Many of the pollen grains on the sticky surface of stigma germinate a few minutes after arrival and form pollen tubes each (Fig. 4.4).

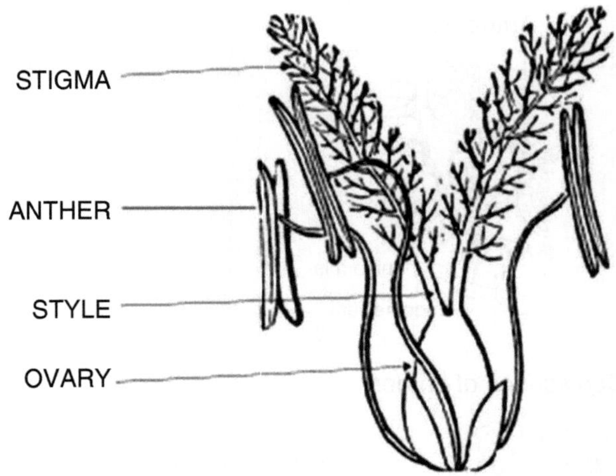

Fig. 4.3 Diagrammatic representation of Pistil of rice, only 3 stamens are shown in place of six

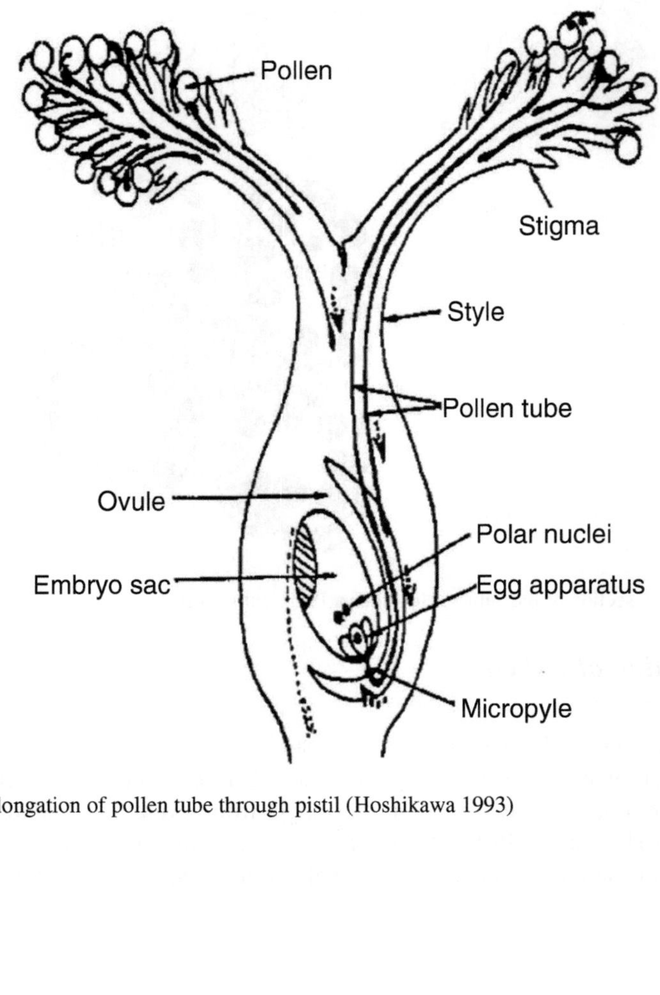

Fig. 4.4 Elongation of pollen tube through pistil (Hoshikawa 1993)

Some of the pollens do not germinate at all and pollen tubes formed in many others stop elongation in the style on way to reach the ovary. In pollen germination, the contents of the pollen including a vegetative nucleus and two male gametes move into the pollen tube through the germ pore; the pollen becomes empty gradually. The male nuclei remain at the end of the pollen tube followed by the vegetative nucleus just behind them.

The pollen tube penetrates the stigma through the gap between cells of its feathery structures and elongates downwards to the ovary (Fig. 4.4). As the tube extends and moves towards the ovary, cytoplasmic contents inside it recede from its base gradually and migrate towards the tip. The elongating pollen tube approaches the integuments of the ovary and finally explores the micropyle end on the lower part ovary. The integuments of the ovule are missing here and the embryo sac is exposed. Through the micropyle, the tip of the pollen tube penetrates the embryo sac. During pollen germination, many pollen tubes reach the micropyle, but only the one with the fastest growth penetrates the embryo sac. The other pollen tubes stop elongation as soon as the fastest tube enters the micropyle and degenerate later. The egg apparatus in rice megagametophyte is located in the constricted micropylar end of the embryo sac, in which a pyriform egg cell is superimposed by two overarching synergids. The pollen tube enters into one of the two synergids to discharge its contents of two sperm nuclei and a vegetative nucleus (Fig. 4.5). The synergids enlarge before the entry of the pollen tube and guide it for penetration into the micropyle. The elongated part of the synergid is called the filiform apparatus; the pollen tube enters into one of the two synergids through the filiform apparatus. The synergids begin to collapse and degenerate one after the other in quick succession at the time of fertilization. One of the sperm nuclei fuses with the egg nucleus, whereas the other one exits from the tip of the synergid to fuse with the polar nuclei and effects double fertilization (Figs. 4.6, 4.7 and 4.8). The fertilized pole nucleus becomes the endosperm mother nucleus.

Fertilization

Following entry of the pollen tube, the synergid swells immediately and its wall is torn. The male nuclei and vegetative nucleus of the pollen tube flow into the synergid. The synergid increases 2–3 times its original size and ruptures again on the side opposite to the micropylar end (antipodal side) bringing the male nuclei closer to the inner side of the egg apparatus. The two male nuclei get freed from the synergid and move in the opposite directions to each other. When the tip of the pollen tube is torn inside the synergid, the contents of the pollen tube mix with the cytoplasm of the synergid invoking a protein coagulation reaction. As a result many small protein lumps, known as X-bodies, appear. The X-bodies probably help in the free movement of the male nuclei. Soon after fertilization, the synergid shrinks in size and degenerates. The vegetative nucleus of the pollen tube, the nucleus of the synergid,

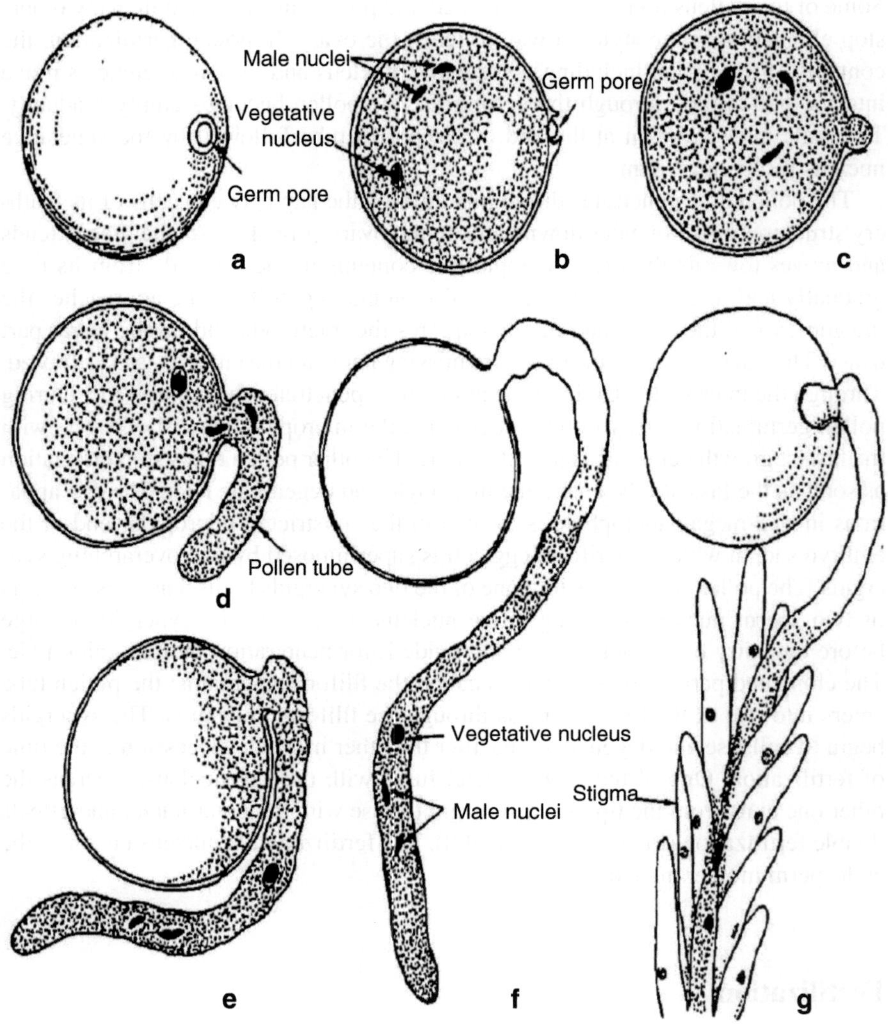

Fig. 4.5 The sequence of pollen germination on stigma (Hoshikawa 1993, with permission)

and X-bodies also degenerate and disappear after fertilization of the egg along with the synergid that does not participate in the delivery of pollen tube nuclei.

First Fertilization

One of the two male nuclei released from the ruptured synergid approaches the egg cell. The nucleus enters into the egg cell cytoplasm and fuses with its nucleus. The fertilized egg (2n) rests for about 10 h before initiating the first cell division. Cell

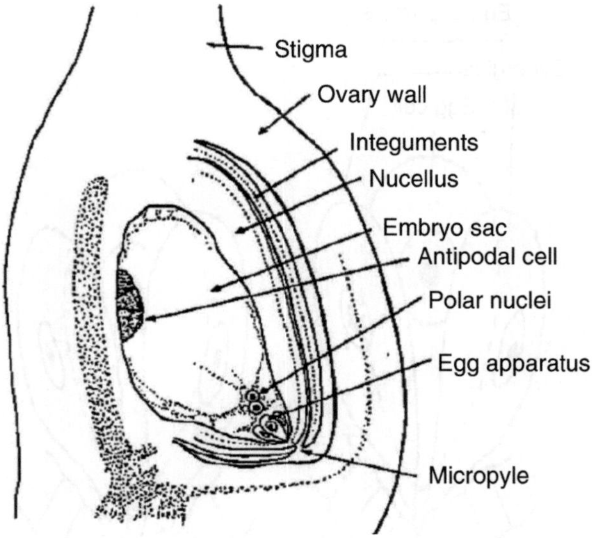

Fig. 4.6 A detailed account of fertilization and inner structures of rice ovary (Hoshikawa 1993, with permission)

division of daughter cells continues briskly for the formation of the embryo. The embryo proper is a diploid tissue.

Second Fertilization

The second male gamete released from synergid fuses with the binucleate polar cell of the embryo sac to form the primary triploid endosperm zygote in the second fertilization. The process of fusion begins with the pole nuclei approaching the egg apparatus after the synergid ruptures to release the male gametes. The pole nuclei attract the gamete and move away from the egg apparatus to the antipodal side of the embryo sac. In the first step, the male nucleus enters into one of the two polar nuclei and fuses with it. In the next step, the membrane separating the fertilized pole nucleus and the second pole nucleus disappear combining all three nuclei into one in triple fusion. After fertilization, the pole nucleus is 3n and called the endosperm mother nucleus. The zygote does not rest. The first mitotic division occurs 12 h after fertilization. Similar to other members of the grass family, the process of fusion of two sperm nuclei, one with an egg nucleus to create an embryo and the other with two polar nuclei resulting in endosperm formation is called double fertilization. Both of the fusions occur at the same time. The events of double fertilization are described below (Figs. 4.8 and 4.9) according to Hoshikawa (1993).

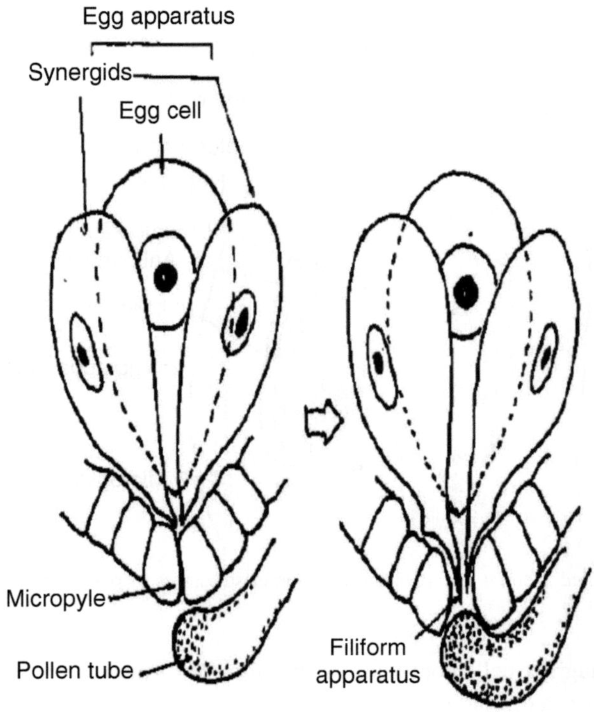

Fig. 4.7 Attraction of pollen tube to egg apparatus (Hoshikawa 1993, with permission)

Embryo Development

The zygote in the egg nucleus elongates soon after fertilization and divides transversely by an unequal division; the daughter cells are aligned at a right angle to the micropylar pole. In subsequent cell divisions, mainly on the antipodal side, there is no apparent pattern. By day 2 after fertilization, the embryo assumes a globular shape and exhibits distinct signs of differentiation with the appearance of a protoderm. The cell number of embryos increases rapidly to 100–200 by day three and they are polarized into basal and apical groups. Two-third of the cells of the embryo in the apical part are progenitors of the scutellum whereas the residual cells at the basal third form the embryo. By day four, a notch appears on the ventral side that differentiates the shoot apex initial (Hoshikawa 1993). The radicle meristem, the progenitor of the seminal root, also becomes visible in the embryo near the end of the tissue. A coleoptile primordium is formed on day five by swelling of tissue near the shoot apex. It is followed by the emergence of the first leaf initial below shoot apex on day sixth. The primordium of the second leaf appears at the upper position 180° away from the central axis of the first leaf. At this stage a bud scale developed on the lower part of the coleoptile and a ventral scale developed on the upper part of

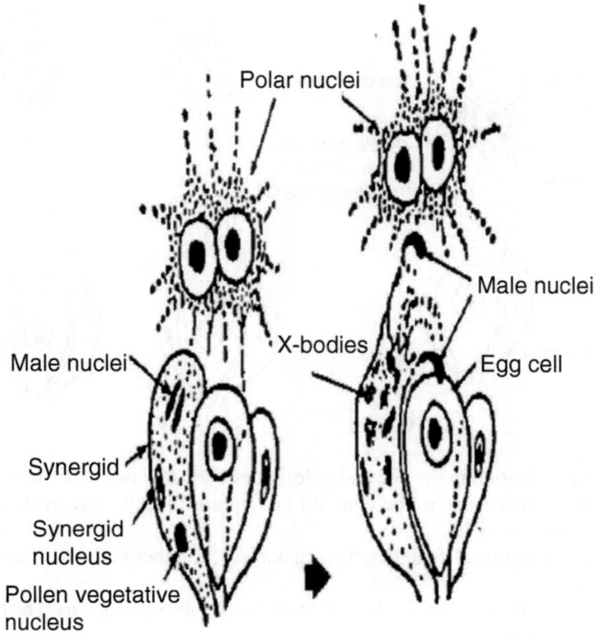

Fig. 4.8 Penetration of the pollen tube into synergid followed by the release of male nuclei (Hoshikawa 1993, with permission)

the coleoptile; they cover the coleoptile primordium. A root cap is differentiated at the tip of the seminal root also (Fig. 4.10).

The size of the embryo increases rapidly on day 10 when the primordium of the third leaf appears. Scutellum increases in size and vascular bundles in it are differentiated. By day 12, cell division as well as an increase in the size of the embryo stops. The cells at this stage continue to thicken only. The embryo gains the capacity for germination and growth at this stage. The primordium of the third leaf also stops growth and remains static. There is no further change in morphological differentiation and development and embryo tissue matures physiologically thereafter. In terms of cell division and differentiation, embryo development is synchronous with that of the endosperm. The overall growth and development of the embryo become very slow and by day 25 morphology of the embryo is completed. Between days 25 and 30 after fertilization, the tissue shrinks in size due to a decrease in water content of the whole grain and the grain is ready for harvest. At this stage, the embryo enters into a state of physiological dormancy.

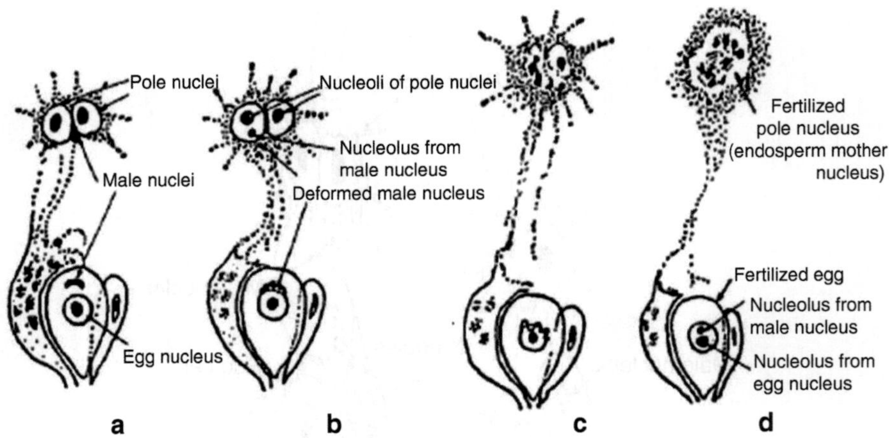

Fig. 4.9 Double fertilization in the egg and pole nucleus (Hoshikawa 1993, with permission)
(**a**) One of the male nuclei enters the egg cell and approaches the egg nucleus. Another male nucleus sticks to the pole nuclei.
(**b**) The male nucleus begins to penetrate the egg nucleus. Another male nucleus enters one of the pole nuclei.
(**c**) The male nucleus turns into string-like shapes and enters the egg nucleus. The membrane dividing the two pole nuclei starts disappearing.
(**d**) In the egg nucleus, two nucleoli coexist. The fertilized pole nucleus (endosperm mother nucleus) enters the prophase of nuclear division.

Endosperm Development

Nuclear Stage

The two endosperm nuclei formed out of the first nuclear division endosperm mother nucleus, move away from each other, and remain close to the antipodal side of the embryo sac. They do not form cell walls in between. The daughter nuclei do not rest and undergo further divisions to form a systematic multinucleate endosperm, where the cytoplasm surrounding the endosperm nuclei connects with the protoplasm covering the embryo sac nucleus. This stage is the nuclear stage of endosperm development. At the same time, the embryo sac grows longitudinally and expands in volume. The antipodal pole moves away from the nascent embryo. Endosperm nuclei division is very active in the area near the embryo, which recedes gradually towards the antipodal pole. The endosperm nuclei organize to form a single layer on the surface of the embryo sac differentiating into the peripheral layer of the endosperm. Over time the peripheral layer multiplies and extends itself towards the antipodal side. By day four after fertilization, endosperm nuclei near the embryo begin to form cell walls. The nuclei of the peripheral layer divide periclinally to form one more layer. The nuclei of the second layer also form cell walls converting the endosperm into a cellular tissue. The multiplications there-after occur by complete cell divisions. In the embryo sac, there are 3–5 antipodal cells at

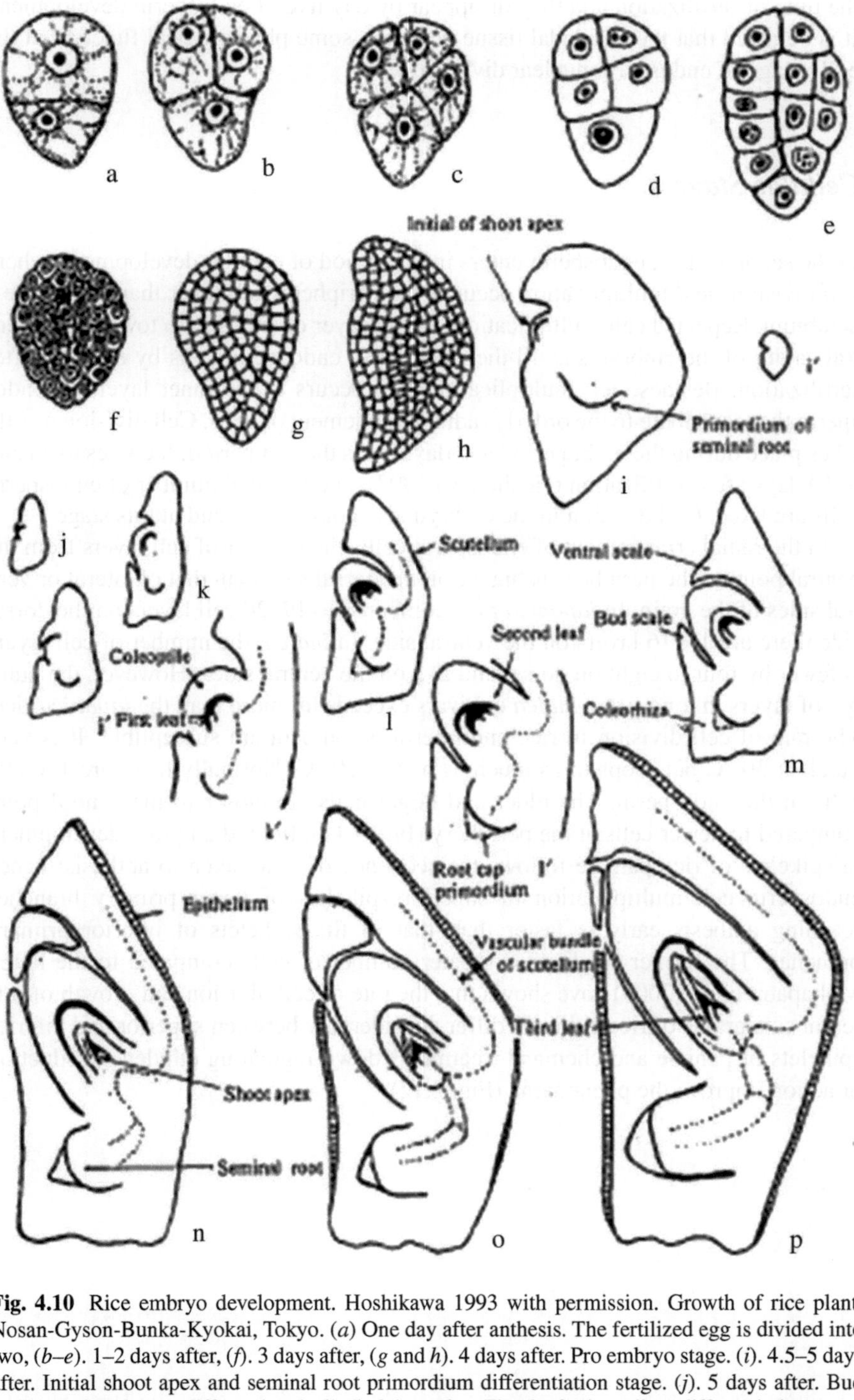

Fig. 4.10 Rice embryo development. Hoshikawa 1993 with permission. Growth of rice plant. Nosan-Gyson-Bunka-Kyokai, Tokyo. (*a*) One day after anthesis. The fertilized egg is divided into two, (*b–e*). 1–2 days after, (*f*). 3 days after, (*g* and *h*). 4 days after. Pro embryo stage. (*i*). 4.5–5 days after. Initial shoot apex and seminal root primordium differentiation stage. (*j*). 5 days after. Bud and ventral scales are differentiated. (*k*). 5–6 days after. First leaf primordium differentiation stage. (*l*). 6–7 days after. Second leaf primordium differentiation stage. (*m*). 8 days after. (*n*). 9–10 days after. Epithelia start differentiation. Third leaf primordium differentiation stage. (*o*). 11–12 days after. Cell division completion stage. (*p*). 25 days after. Embryo completion stage

the time of fertilization and they disappear by day five of endosperm development. It is believed that the antipodal tissue performs some physiological function at the early stage of endosperm nuclear division.

Cellular Stage

In the second stage, endosperm enters into a period of cellular development, where cell divisions and multiplication occur in the peripheral cell layer that works like a cambium. Repeated cell multiplications in this layer of endosperm towards the central cavity of the embryo sac fill the cavity with endosperm cells by day five after fertilization. Besides, cell multiplication also occurs in the inner layers of endosperm that contribute to the orderly radial arrangement of cells. Cell division mostly takes place during the night period and daytime is the rest period. It comes to an end 9–10 days after fertilization (Hoshikawa 1993), and the final number of endosperm cells are fixed. Cell division in the embryo also comes to an end at this stage.

In the radial arrangement of endosperm cells, the number of cell layers from the central point to the periphery is higher on the dorsal side than that of lateral or ventral sides of the grain. In *japonica* rice, compared to 19–20 cell layers on the dorsal side there are 14–16 layers on the ventral side. In *indica*, the number of cell layers is fewer by four to eight on dorsal and 2–3 on the ventral sides. However, the number of layers in long-grain *indica* cultivars exceeds far more than the *japonica* rice. The rate of cell division in rice endosperm is temperature susceptible. It is very quick at 30 °C but drops to as much as half at 20 °C. Normally there are 180,000 cells in the endosperm. The older and larger cells are closer to the central point compared to newer cells at the periphery (Fig. 4.11). Individual grain development in spikelets of rice panicle follows the sequence of time taken to anthesis; hence endosperm cell multiplication in superior spikelets of upper primary branches reaching anthesis early is faster than that of the spikelets of inferior primary branches. The former produces a greater number of cells compared to the latter. Mohapatra et al. (2009) have shown that the rate of cell division and growth of different size grains of rice cultivars differ considerably between superior and inferior spikelets of panicle and chemical treatments down-regulating ethylene production or action improve the phenomena (Fig. 4.12).

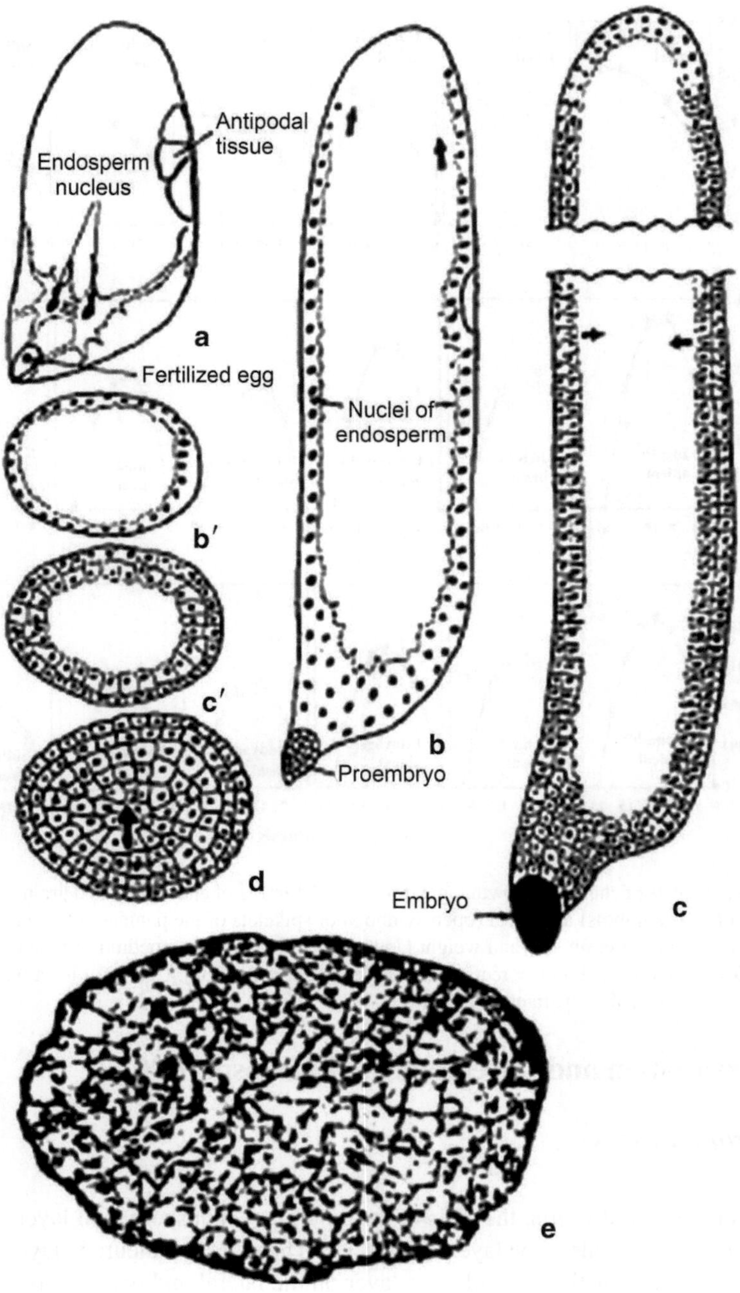

Fig. 4.11 Division and multiplication of endosperm tissue immediately after fertilization (*a*) Endosperm mother nucleus divides into two, and endosperm nuclei are formed. (*b*) Cross and vertical section of endosperm tissue 3 days after. One-layer peripheral layer of nuclei is formed. (*c*) 4 days after. Endosperm nuclei become two-layer tissues and are forming cell membranes. (*d*) 4–5 days after. Cavities in the endosperm cells, (*e*) Cross-section 5–6 days after. Cell tissues develop centrifugally with the central point (arrow, CP) at the center (X 98). (*f*) Cross-section 10 days after. The stage just after the multiplication of endosperm cells ends (X 55). (Hoshikawa 1993, with permission)

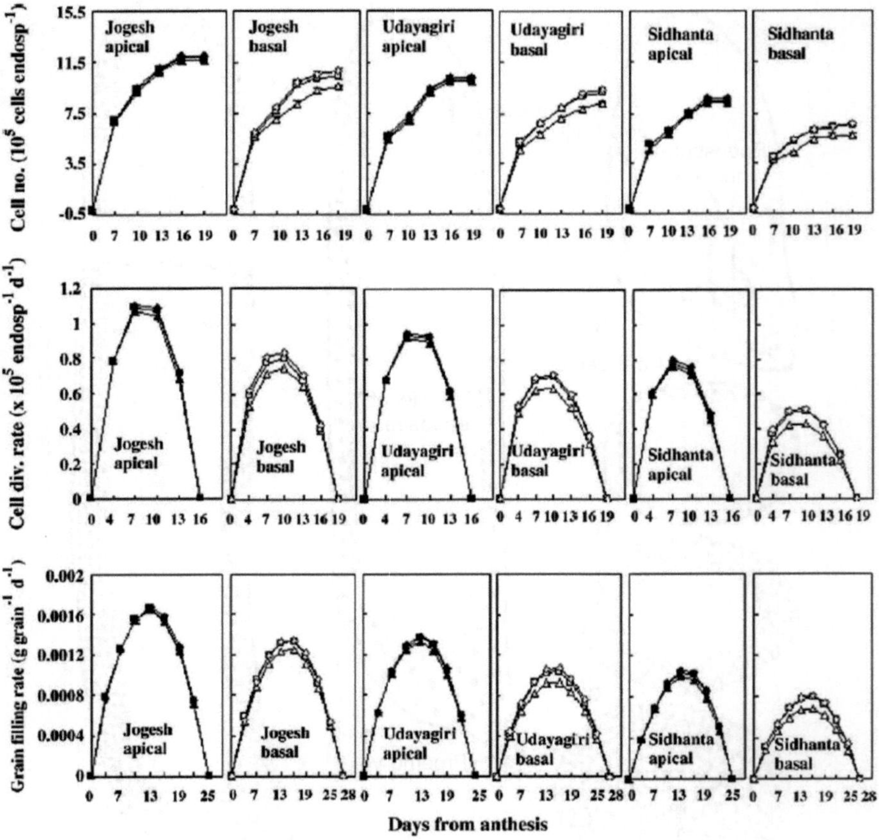

Fig. 4.12 Effects of chemical growth regulators on cell number of endosperm and grain filling rate of apical (close symbols) and basal (open symbols) of spikelets of the panicle in three *indica* rice cultivars differing in grain size and weight (Jogesh- large, Udayagiri- medium, Sidhanta- small). [$Co(NO_3)_2$ = rhombus, kinetin = rectangle, distilled water = triangle]. $Co(NO_3)_2$ is an inhibitor of ethylene synthesis (Mohapatra et al. 2009)

Differentiation and Development of Endosperm Cells

Aleurone Layers

At the end of cell division, the cells of the outermost endosperm cell layer develop to differentiate into aleurone layers (Fig. 4.13). There are 3–6 aleurone layers on the dorsal side, whereas there is only one layer on the lateral and one to two layers on the ventral side of the caryopsis. Aleurone tissue of the dorsal side becomes multi-layered because of the division and differentiation of peripheral cells. On day five after fertilization, one to two layers of cells in the periphery begin to differentiate into special shapes and divide thereafter once or twice to form more layers of cells.

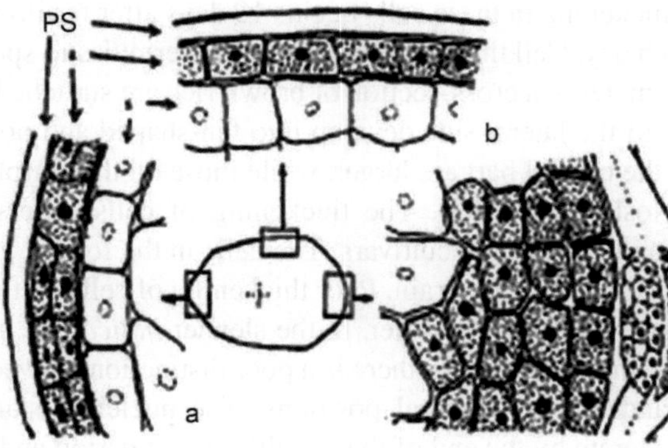

Fig. 4.13 The thickness of aleurone layers in different parts of the endosperm tissue. (*a*) ventral side, (*b*) lateral side, (*c*) dorsal side. PS- pericarp and seed coat, *a.* aleurone layer, *s.* starch storage tissue (Hoshikawa 1993, with permission)

On day six, these cells establish themselves as precursors of aleurone cells. Cell division and multiplication in the inner-most cell layer continuing up to day 10 after fertilization, form a multi-layered aleurone on this side unlike the others (Hoshikawa 1993). The aleurone layers of the dorsal side are near the conducting tissue of the seed coat. The aleurone cells are densely packed with lipid bodies and protein granules, but no starch at the ripening stage. The cells are very active during germination and synthesize enzymes for degradation of food reserve of endosperm during seed germination. The solubilized food reserves are delivered to the embryo through the aleurone. The role of aleurone cells is also crucial for assimilate transport into grain during the ripening period. The cell wall in them is very thick and has many plasmodesmata connections working as inlets for the transport of phloem assimilate from the conducting tissues. At later stages of ripening aleurone cells extend wall ingrowths and function as transfer cells for the inflow of assimilates into the endosperm cells. The cell layer inner to the peripheral aleurone layer finally becomes starch storage cells. The cells of this layer form the transport route for assimilates to inner starchy cells of endosperm tissue. The cells of aleurone become dead tissue by the end of the ripening stage.

Parenchymatous Tissue for Storage of Starch

All cells derived from the periclinal division of the peripheral layer of endosperm are starch storing parenchymatous cells (Fig. 4.13). The parenchymatous cells including those of the sub-aleurone layer do not undergo any differentiation and turn into starch storage tissue. The cells at the center are those formed earliest and most aged. Cell thickening starts from the central part and spreads gradually to the periphery. Thickening begins 5 days after fertilization and continues up to day 15. The cells of the sub-aleurone layer below the peripheral aleurone layer are the last to get thickened; thickening in these cells begins 10 days after fertilization and continues up to the 30th day. Cell thickening is also endosperm tissue-specific; cells on the dorsiventral diameter in a cross-section of brown rice are stretched out along the diameter, but cells of the lateral side develop into fan-shaped and polygonal structures. The cells of the central part are larger, while those on the periphery are much smaller in size (Hoshikawa 1993). The thickening of cells differs significantly between *japonica* and *indica* type cultivars. The cells in the former grow along the dorsiventral axis and form a thick grain. Poor thickening of cells on the dorsiventral axis makes the grain slender in the latter. In the slender *indica* rice grain, cells are arranged radially from the center and there is a poor distinction between cells on the dorsiventral axis and those on lateral positions. The nuclei of starchy cells are deformed and degenerate by the end of the cell thickening period and become dead at the time of grain ripening.

Storage of Food Reserve

The starch content of endosperm is more than 90%. It has multiple granular amyloplasts. Each amyloplast bears several starch granules. Storage of starch begins in the innermost cells of endosperm tissue from day four post-anthesis and spreads gradually to the outer cells centrifugally (Fig. 4.14). Yu et al. (2014) found no starch in the cells of endosperm on day 3 after fertilization. On day five, small spherical or elliptical amyloplasts appear and their number and size continued to increase temporally. The process of starch grain development is completed by day 15 in the inner cells, while it continues in the outer cells to day 30 post-anthesis. On day 15, the amyloplasts become polyhedric being fully packed with starch granules. The cells of the sub-aleurone layer are the last to complete starch grain development. The size of starch granules of rice averaging 3–8 μm, is the smallest of grain starch derived from plants. Usually, they are polygonal, but irregular in shape. Compound granules comprising 20–60 individual units in clusters possess a diameter up to 150 μm. The size of granules is about 40 μm in inner cells while it is small in the outer cells ranging between 10–20 μm. Irrespective of differences in the size of the starch granules between outer and inner cells, there is no difference in number. Inner cells are bigger in size and accommodate larger size granules, whereas it is just the opposite for the outer cells. The presence of air spaces between starch granules makes the *japonica*-type waxy rice opaque. The aleurone cells do not have starch granules at maturity.

The starch granules have numerous starch molecules, essentially polymer of glucose units in glucan structure of which linear type is amylose and highly branched type is amylopectin. Both of them are D-glucose homopolymers. The linear type amylose is composed of α-D-glucan units linked with one another by covalent type 1,4-glycosidic bonds. The branched chains of amylopectin structure also have many linear chains of α-D-glucan units, the highly branched glucans are linked with one another by α-1,6 glycosidic bonds.

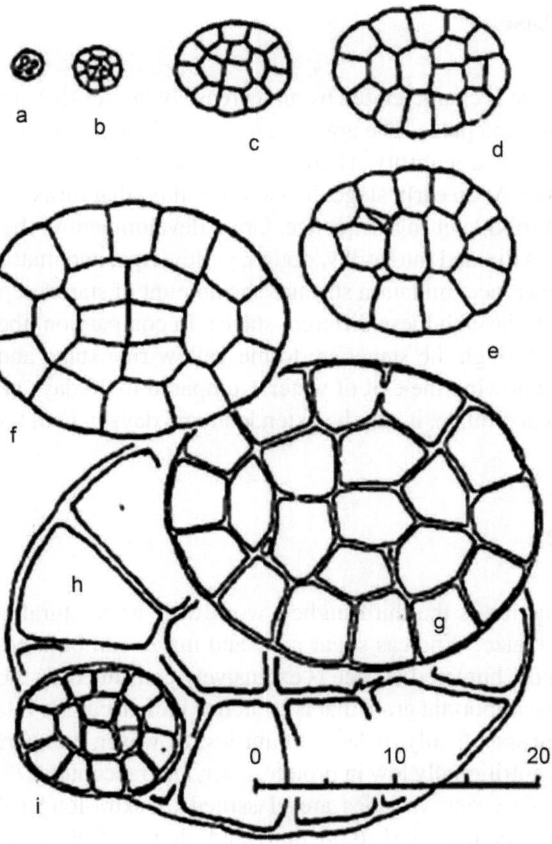

Fig. 4.14 Development of starch grains in the endosperm (Hoshikawa 1993, with permission) (**a**) Four days after fertilization (DAF), (**b**) 5 DAF, (**c**) 7DAF, (**d**) 10 DAF, (**e**) 12DAF, (**f**) 15 DAF, (**g**) 45 DAF, (**h**) Cells inside the endosperm tissue have large starch grains like this. But starch grains in the cells in the periphery of the tissue are far smaller

The protein content of rice endosperm varies between 5% and 8%. Protein grains are initiated in endosperm cells 6 days after anthesis. They occur as small spherical protein bodies inside amorphous proteoplasts of 1 μ diameter. Proteoplasts increase in size up to 3 μm in diameter with a protein body of 2 μm, until the stage of grain ripening. They are distributed more on the peripheral cells of endosperm being confined more to the sub-aleurone layers of tissue. In mature caryopsis, the protein bodies are distorted in spaces between amyloplasts because of their extensive growth in size (Yu et al. 2014). The number of these proteoplasts is larger on the dorsal than on the ventral side of the grain. The fat content of endosperm is very low. It mostly exists in the cells of the aleurone layer. The content is very poor in the starch storing tissue.

Ripening Phase

The ripening phase is characterized by an increase of kernel size and weight, change of color of lemma and palea. The grain is green in colour at the early stages of ripening but turns yellow at maturity. There are corresponding changes in the texture of the caryopsis also. At an early stage, it is a milky fluid that turns into the semi-solid dough and hard solid later in a sequence. Grain development in the spikelet following fertilization is divided into milky, dough, yellow ripe, and matured stages. Grain growth in dry mass accumulation sustains the amount of starch deposited primarily in the endosperm through these different stages. In comparison, the fresh weight of grain increases through the stages up to the yellow ripe stage and declines in the maturity stage following the exit of water. Compared to 30 days period of ripening in a warm tropical climate, it may be extended to 65 days in cool temperate regions.

Seed of Rice

Rice seed production is the third highest worldwide agricultural commodity, after sugar cane and maize. Whereas sugar cane and maize can be used for several purposes including the human diet, rice is exclusively consumed by the human population. It is the most important grain that is related to human calorie intake and nutrition worldwide, more specifically in Asian countries. However, compared to other cereals, rice grain is nutritionally low in protein, fiber, and fat contents (Table 4.1). Based on brown seed size, rice varieties are classified as extra-long (>7.50 mm), long (6.61–7.5 mm), medium (5.51–6.60 mm) and short (<5.50 mm) grained. Grain shape is characterized on length to breadth ratio; slender >3.0, medium 2.1 to 3.0, bold 1.1 to 2.0 and round <1.0. Seed quality differs considerably with the variation of seed size and accordingly the seeds are used for different types of food preparations in commensuration to grain quality. The long-grain seeds are high in amylose and remain stable after cooking. In comparison, medium-grain rice has high amylopectin and becomes sticky during cooking. Both long and medium grain rice is served plain as steamed rice i.e. the primary dish of rice consumers of south and

Table 4.1 Chemical composition of edible cereal grains

Property	Barley	Brown rice	Corn	Millet	Oat	Rye	Sorghum	Wheat
Carbohydrate (%)	55.8	64.3	63.6	63.4	62.9	71.8	58.0	69.7
Protein (N × 6.25) (%)	11.0	7.3	9.8	11.5	9.3	8.7	8.3	10.6
Fiber (%)	3.7	0.8	2.0	1.5	5.6	2.2	4.1	1.0
Fat (%)	3.4	2.2	4.9	4.7	5.9	1.5	3.9	1.9
Ash (%)	1.9	1.4	1.4	1.5	2.3	1.8	2.6	1.4
Net protein utilization (%)	62.0	73.8	58.0	56.0	59.1	59.0	50.0	53.0
Digestible energy (kJ/100 g)	1320	1550	1450	1440	1160	1330	1290	1360

Modified from Zhou et al. (2002)

south-east Asia. Short grain rice is used for pudding and some sweets dishes. They are commercially expensive because of low their grain yield characteristics.

Parts of Rice Grain and their Composition

In rough rice (paddy) the husk comprises 16–28% of dry weight. Removal of husk produces brown rice in milling. In brown rice final dry mass weight is distributed as pericarp 1–2%, aleurone, nucellus, and seed coat 4–6%, embryo 1%, scutellum 2%, and starchy endosperm 89–94%. Milling of rice for getting edible grain also yields two major byproducts, namely husk and rice bran. The husk has no food value but is used for several industrial purposes. Rice bran is the vitamin rice outer layers surrounding the endosperm that comprises 8% of the total dry seed weight, but it accounts for 60% of the nutrients found in the kernel. It includes pericarp, testa (seed coat), embryo, and aleurone layers of the seed (Fig. 4.15). The nutrient composition of different milling fractions of brown rice varies significantly.

Hull

Rice grain is an indehiscent single-seeded fruit in which the caryopsis coat composed of the seed coat, pericarp, and nucellar layers fuses tightly with the seed botanically named caryopsis (Bechtel and Pomeranz 1977). The outer enclosure of the caryopsis called hull (husk) consists of two interlocking bracts lemma and palea. The husk is inedible and provides insulation from temperature and moisture changes. The unilayer upper epidermis of the bracts has a cuticle on the outer surface. Silicic acid accumulates in the cells of the epidermis with time. The inner side of the epidermis has several parenchymatous cell layers, wherein chloroplasts develop in the inner 2–3 layers at anthesis. With the progress of grain development, chloroplasts degenerate and disappear completely at maturity. The parenchymatous tissue is confined to one or two cell layers. The cell walls of the layers are lignified and thickened. The cells of the unilayered lower epidermis also change at the same time as the other tissue. They collapse and lose their contents turning into a thin translucent membranous tissue. All cells of hull tissue die when it turns yellow at maturity.

Pericarp

The pericarp is the outermost layer (Fig. 4.16) developed from the ovary wall surrounding the caryopsis. The ovary becomes longer and thicker after fertilization and encloses the inner seed coat and nucellus, embryo, and endosperm. The cells of the upper epidermis of the pericarp (exocarp) divide first after fertilization and stretch

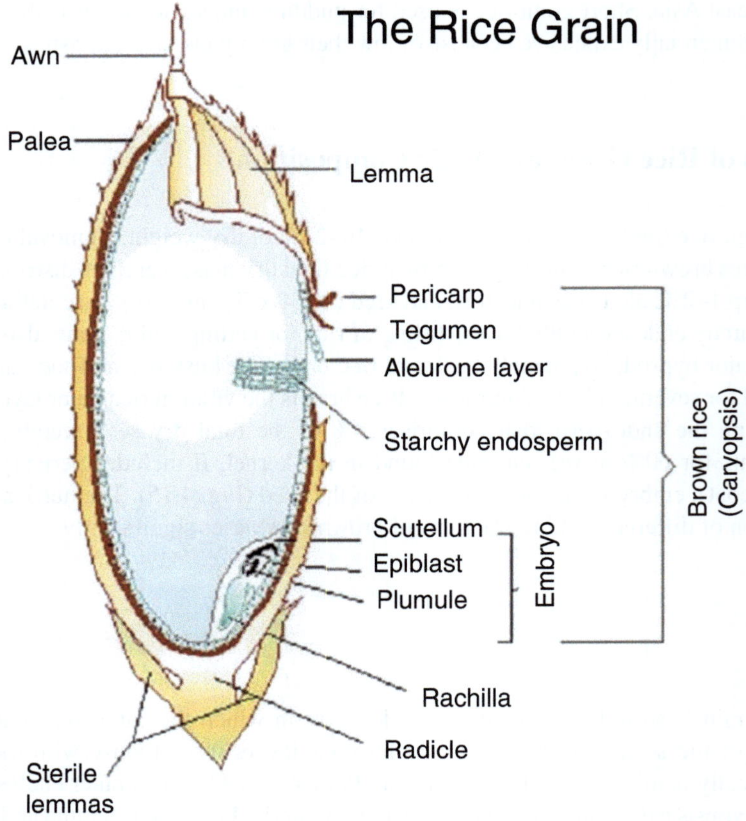

Fig. 4.15 Parts of rice grain at maturity (GRiSP 2013, with permission)

out in the longitudinal direction to become larger in size (Fig. 4.17). At the grain hardening stage, the cells are rectangular and thick-walled. Gradually they are lost with the progress of time and the surface of their walls finally turns into a cuticle. The cells of the middle layer (mesocarp) are the main part of the pericarp. This part has 5–6 layers of parenchyma cells. The cells accumulate starch by 3–5 days after flowering. The volume of the cells increases on day seven, and at the same time, the cells, as well as the starch grains, get lost due to programmed cell death (Yu et al. 2014). The mesocarp, hence, turns into a spongy tissue with broken cell walls. On day 15 after flowering starch grains completely disappear and the mesocarp has only one layer of cells. The cells of the lower epidermis of the pericarp (endocarp) have cross cells and tube cells. Some cells do not divide and elongate longitudinally to form tube cells, whereas some other cells multiply rapidly at the early stage of grain development and lengthen at a right angle to the longitudinal axis of the tube cells form cross cells (Yu et al. 2014). Chloroplasts are formed in these cells giving a green colour to the caryopsis at the early stage of grain development. Chloroplasts dissolve at a later stage when grain matures and the cell layers become hard and lignified. Cells of the lower epidermis do not divide; they lengthen in size and become lignified fibers at maturity.

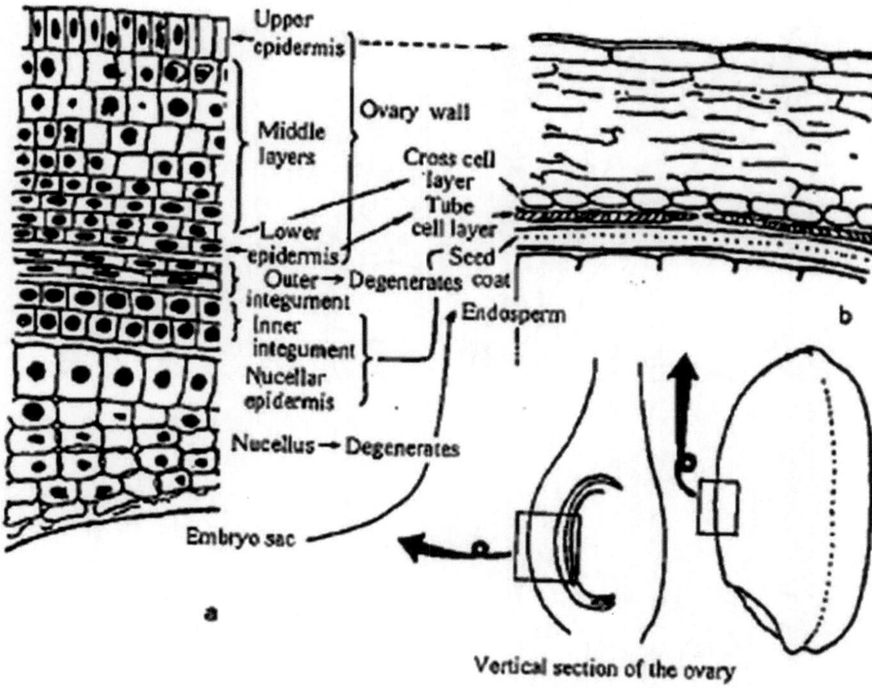

Fig. 4.16 Parts of rice pericarp and seed coat at (**a**) anthesis and (**b**) maturity (Hoshikawa 1993, with permission)

Seed Coat

Seed coat remains interior to the pericarp. Both of them are closely linked and difficult to separate from each other. While the ovary wall forms the pericarp, the seed coat is formed out of the integuments and nucellar epidermis of the ovule. At 3 days after fertilization, the integument has 3–4 cell layers. On day 5 after fertilization, some layers degenerate and get absorbed into the nucellus, but cells of the inner integument divide and lengthen to form unilayered cells. Interior to the inner integument, there is a multi-layered nucellus. On day 15 after flowering, the unilayered integument and nucellar epidermis are squeezed together due to the expansion of endosperm (Yu et al. 2014). Like cells of integuments, the nucellar cells develop walls in them that disappear later turning into a series of membranes called exosperm. The exosperm coheres with the integuments to form the seed coat in rice seed; cells of the seed coat are dead at the ripening stage.

Endosperm

The endosperm has external layers of aleurone and sub-aleurone tissues and it is followed by the inner starchy endosperm (Fig. 4.18). The aleurone is a specific tissue of rice endosperm and it is often referred to as peripheral endosperm. The

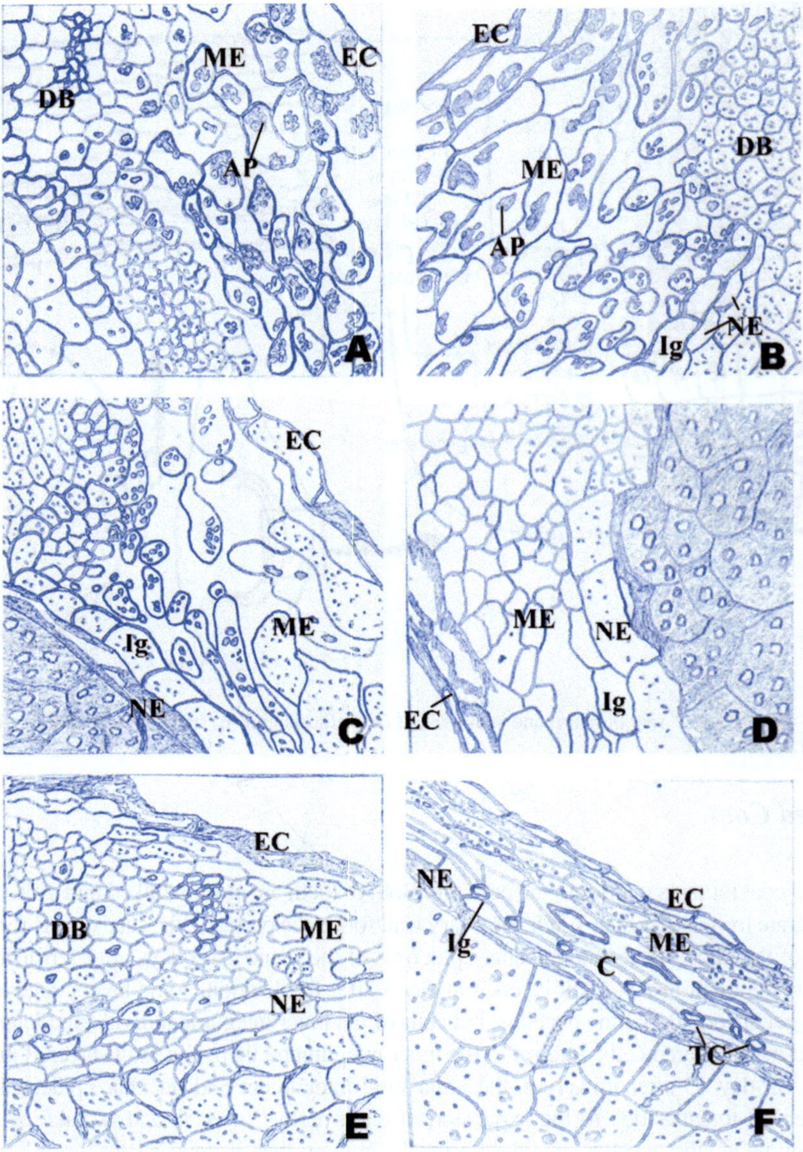

Fig. 4.17 Schematic drawing of rice seed coat and pericarp (**A**) diagram of 3 days after flowering (DAF); (**B**) diagram of 5DAF; (**C**) diagram of 7DAF; (**D**) diagram of 9DAF; (**E**) diagram of 15DAF; (**F**) Beside the lateral vascular bundle diagram at 15 DAF. DB, AP, Amyloplast; dorsal vascular bundle; EC, Exocarp; Me, Mesocarp; NE, nuclear epidermis; Ig, Integument; TC, Tube cell; C, Cross cells. (Adopted from Yu et al. 2014)

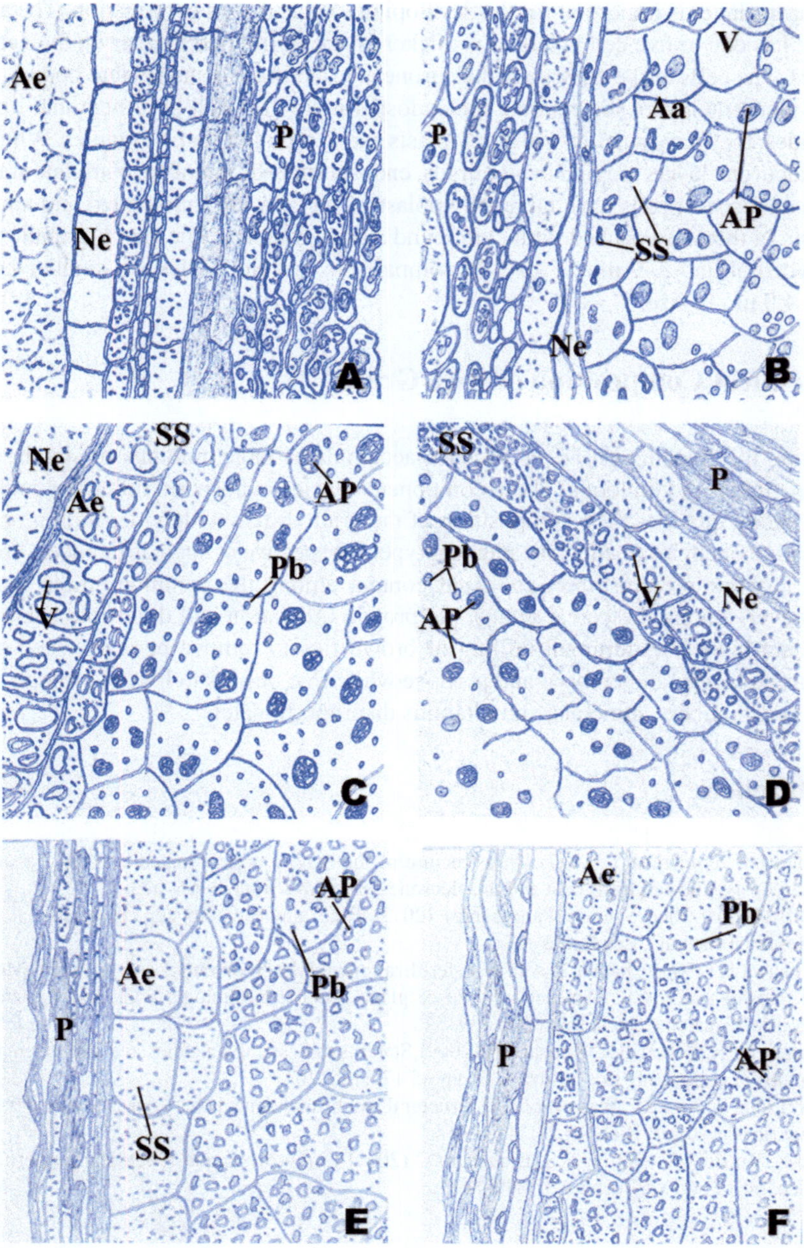

Fig. 4.18 Schematic drawing of rice aleurone. (**A**) diagram of 3DAF; (**B**) diagram of 5DAF; (**C**) diagram of 7 DAF; (**D**) diagram of 9 DAF; (**E**), diagram of 11 DAF; (**F**) diagram of 15 DAF. C, cross cells; AP, amyloplast; Ae, Aleurone; Ig, Integument; NE, nuclear epidermis; Me, mesocarp; TC, tube cell. (Adopted from Yu et al. 2014)

aleurone layer is important for the developing seed and seed germination. The aleurone has one to five cell layers; it is thicker on the dorsal and thinner on the ventral sides. The cells of the embryo and aleurone are rich in lipid and protein bodies. The sub-aleurone layers surrounding the endosperm are also rich in protein and lipids; besides, they have smaller size amyloplasts and compound starch granules. Whereas the aleurone layers are rich in oily lipids, endosperm cells mainly accumulate starch. The endosperm cells have large amyloplasts. The cells are thin-walled and amyloplasts in them are packed with compound starch granules. The starch granules are polyhedral and 3–9 μm in size. In comparison, protein bodies are smaller being 0.5–4.0 μm in size.

Chemical Composition of Rice Grain

During the ripening phase, endosperm accumulates large amounts of starch, proteins, lipids, and minerals. These components of rice endosperm influence cooking and eating qualities. The composition of the grain differs widely depending on the variation of environment, soil, and genotype. Irrespective of such differences, brown rice has low protein, fiber, and lipid content among the common edible grains. However, among the cereal grains, net protein utilization and digestible energy in rice are highest. Additional milling of brown rice to remove pericarp, seed coat, aleurone layers, and embryo and produce white rice, results in loss of proteins, lipids, sugars, fibers, minerals, and vitamins disproportionately.

References

Bechtel DB, Pomeranz Y (1977) Ultrastructure of the mature ungerminated rice (Oryza sativa) caryopsis. The caryopsis coat and the aleurone cells. Am J Bot 64:966–973

GRiSP (Global Rice Science Partnership) (2013) Rice almanac, 4th edn. International Rice Research Institute, Los Baños

Hoshikawa K (1993) Chapter 2: Anthesis, fertilization and development of caryopsis. In: Matsuo T, Hoshikawa K (eds) Science of the rice plant, vol 1. Morphology. Food and Agriculture Policy Research Center, Tokyo, pp 339–376

Mohapatra PK, Sarkar RK, Kuanar SR (2009) Starch synthesizing enzymes and sink strength of grains of contrasting rice cultivars. Plant Sci 176:256–263

Yu X, Liang Z, Xiong F, Wang Z (2014) Structural and histochemical characterization of developing rice caryopsis. Rice Sci 21:142–149

Zhou Z, Robards K, Helliwell S, Blanchard C (2002) Composition and functional properties of rice. Int J Food Sci Technol 37:849–868

Chapter 5
The Cellular Basis of Rice Seed Growth

Abstract Mature rice seed has three major components, the embryo, and nutrient storage tissue endosperm, both belonging to filial generation and seed coat (fused to pericarp) i.e. derived from ovule integuments and belongs to the maternal generation. An increase in cell number in them primarily contributes more to growth than to the rise in cell volume. The tissue development in endosperm has an additional advantage of rapid cell division and expansion of cell size through a massive accumulation of storage compounds. Cell cycle type changes with endosperm growth. Firstly, the primary triploid endosperm nucleus undergoes brisk acytokinesis divisions resulting in a syncytium, which cellularises later and cell number increases further with mitotic divisions. Cell volume increases and starch granules accumulate in cells after cessation of cell division. The transition to the storage phase occurs because of a gradual and asynchronous switch of mitotic to the endreduplication cell cycle. This cycle is characterized by one or more rounds of DNA replication without cytokinesis increasing the ploidy level of the nucleus. The progression of cells through their division cycle is operated by the activity of a serine/threonine cyclin (CDK) enzyme-dependent protein kinase. In sequence to the endoreduplication phase, endosperm enters into the terminal phase of programmed cell death. This phase is genetically programmed in which the assembly of cells is disorganized in course of cell maturation.

Keywords Endosperm · Cell cycle · Syncytium · Endo-reduplication · Starch · Programmed cell death

Phases of Seed Growth

The growth of rice seed, similar to multicellular plant organs or organisms' depends on the number and size of constituent cells. The growth through the aggregation of the number and size of different types of tissues is straightforward, but the dynamics of growth are unresolved in the absence of a causal relationship with cellular mechanisms controlling division and expansion of cells. In the mature rice seed there are three major components, the sporophyte (embryo) and nutrient storage tissue (endosperm), both belonging to filial generation and seed coat (fused to pericarp) i.e. derived from ovule integuments and belongs to the maternal generation. In each of

P. K. Mohapatra, B. B. Sahu, *Panicle Architecture of Rice and its Relationship with Grain Filling*, https://doi.org/10.1007/978-3-030-67897-5_5

the seed components, an increase in cell number primarily contributes more to growth than to the rise in cell volume. The tissue development in endosperm has an additional advantage of rapid expansion of cell size through a massive accumulation of storage compounds. Moreover, compared to the fertilized zygote, mitotic division of the primary endosperm nucleus is faster being potentiated by rapid cell cycle activity of the two polar nuclei. Hence rice seed growth in dry mass has been calibrated mostly by the growth of endosperm tissue.

Rice endosperm has been a key source for dietary calories in many food products and assumes significance greater than any other plant tissue. The commercial seed value of rice is determined by the endosperm both in qualitative and quantitative terms. On the whole, the accumulation of dry mass with time in seed growth comprises three distinct phases showing an S-shaped curve. Phase 1 initiates immediately after fertilization with a minimal rise in dry mass. This lag phase of growth is characterized by cell proliferation through rapid cell division. Phase I is critical for grain development and provides the license for final seed yield by fixing the potential cell number. Although cell number rises briskly in this period, the cells contribute very little to dry mass because of the smallness of size. This phase is completed within the first 12 days after anthesis (Table 5.1 – Patel and Mohapatra 1996). In

Table 5.1 The variations in endosperm cell number, dry weight, starch content, sucrose synthase and invertase activity of developing rice kernels on top and basal part of panicle of rice cultivar OR-776-8p-32 (Patel and Mohapatra 1996)

	Days after anthesis						
	8	12	16	20	24	28	32
Endosperm cell number per grain							
Top	16,450	20,067	20,190				
Basal	14,608*	16,005*	16,564*				
Dry weight (μg cell^{-1})							
Top	0.208	0.552	0.677				
Basal	0.169*	0.368*	0.526*				
Dry weight (mg grain^{-1})							
Top	3.427	11.070	13.670	14.950	16.019	17.157	17.380
Basal	2.481*	5.891*	8.650*	9.597*	9.683*	10.148*	10.230*
Starch content (mg grain^{-1})							
Top	0.975	4.553	6.497	7.403	8.338	9.212	9.311
Basal	0.437*	1.746*	2.719*	3.420*	3.537*	3.628*	3.710*
Sucrose synthase (nmol min^{-1} grain^{-1})							
Top	104.54	198.85	124.53	54.27	46.45	36.35	23.91
Basal	81.52*	135.90*	89.31*	23.77*	23.14*	18.29*	10.61*
Invertase (nmol min^{-1} grain^{-1})							
Top	11.48	14.92	7.36	4.16	3.87	2.91	2.42
Basal	12.69	13.06	8.94*	5.26*	2.95	2.23	1.89

Values are significantly different from the top spikelet at * $P < 0.05$ level, using two-tailed t-test. Cell dry weight = weight of the endosperm/cell number. The embryo and pericarp were removed from the grains for observations taken before day 20 of anthesis

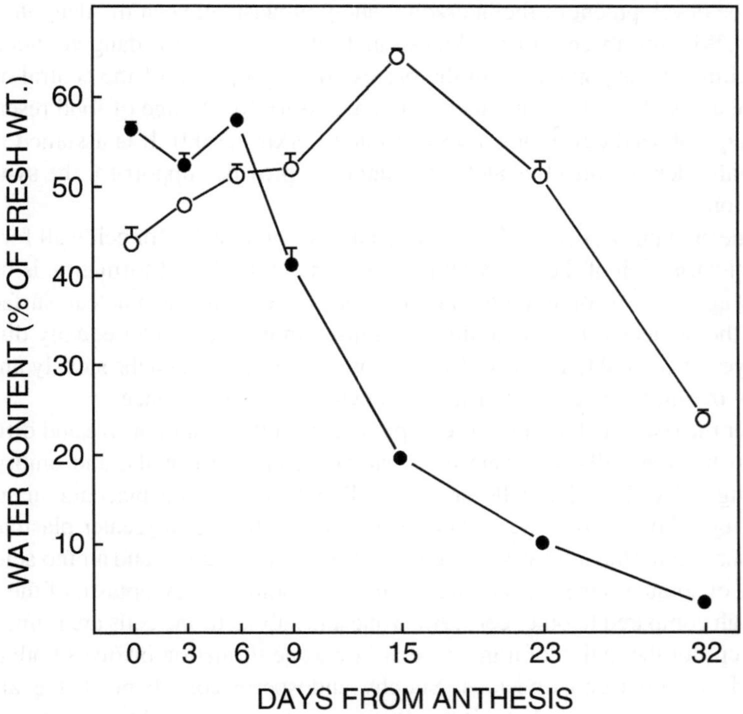

Fig. 5.1 Temporal changes in water concentration of superior (good grain- closed symbols) and inferior (poor grain- open symbols) spikelets of rice (cultivar Kajalghara) panicle in the post-anthesis period (Mohapatra et al. 1993)

phase II, the seed tissues are specified and differentiated for transfer of assimilates from the vascular maternal seed coat for the nourishment of the growing seed. This phase is responsible for linear weight gain, where existing cells enlarge and accumulate storage compounds. The cells expand with the entry of water and provide space for the transfer of assimilates in an aqueous medium. The phase is characterized by endo-reduplication that leads to polyploidy of endosperm cells, which makes cells metabolically stronger and garner storage compounds as food reserves. In general, cells that have undergone endo-reduplication are larger in size than comparable cells with no endo-reduplication. The final phase has a marginal gain of dry matter. Water concentration decreases significantly and the seed attains physiological maturity (Fig. 5.1). Maturation is associated with reduced metabolism and dormancy.

Cell Cycle Types During Endosperm Development

Cell cycle types have been studied in endosperm tissue because it is not only the most prominent seed component but also the most valuable product of modern agriculture. After triple fusion of two polar nuclei in the central cell of embryo sac by one sperm cell nucleus, the primary endosperm triploid nucleus undergoes acytokinesis divisions resulting in a syncytium. The cell cycle of the coenocytic endosperm lacks the development of the interzonal phragmoplast between the daughter nuclei formed. With the absence of cytokinesis and cell formation, the daughter nuclei tend to maximize their population in the pre-existing cytoplasm of the central cell and provide a base line of cell number that is necessary for storage of food reserve at a later stage of seed development (Sabelli and Larkins 2009). It is assumed that the maternally derived nucellus and antipodal cells provide support for the syncytium formation.

In the next phase, cell cycle activity changes significantly for cell wall formation and cellularisation of the syncytial nuclear domain. Cell wall formation is initiated by the aggregation of microtubules that originate from the nuclear surface and define the nuclear-cytoplasmic domain putting individual nuclei equally distanced in a layer of cells. After cell wall formation, nuclei divide synchronously and periclinally in mitotic type cell division with cytokinesis in sequence.

After the rise in cell number is completed, the cells expand in size and differentiate into transfer cells, aleurone cells, starchy endosperm cells, and embryo surrounding cells. Transfer cells are the cell layers near the placenta in nucellar projections. They have extensive cell wall invaginations and greater plasma membrane surface for the inflow of phloem assimilates like sucrose and amino acids. The density of solute transporters in the plasma membrane and cytoplasm of the cells is very high compared to other cells. Aleurone and sub-aleurone cells are found in layers except for the region of transfer cells. Aleurone formation involves both periclinal and anticlinal cell divisions. Starchy endosperm cells beneath the aleurone layers are smaller in size, but mitotically more active compared to inner large endosperm cells. Starch granule accumulation in the inner cells of endosperm begins after the end of cellularization. The cells increase in size and rapidly accumulate starch and storage proteins and the process is associated with high levels of ATP and high energy states. This high metabolic demand infers that the activities are energy-limited.

Endoreduplication

The transition of endosperm cells from the cell division phase to the storage phase occurs because of a gradual and asynchronous switch of mitotic to the endreduplication cell cycle. In eukaryotic organisms cell cycle has four sequential phases, S-phase involves DNA duplication and M-phase is mitotic division and in between, there are two lagging phases G1 and G2. M-phase is divided into prophase, metaphase, anaphase, and telophase followed by cytokinesis (Fig. 5.2).

In general, the successful completion of the M-phase is necessary before DNA duplication in S-phase. However, some specialized cells, such as endosperm cells, skip M-phase for endomitosis. This is a specialized cell cycle event where DNA synthesis is repeated without chromatin condensation, sister chromatid segregation, and cytokinesis. There is a spatial and temporal pattern in the switching of mitotic/endoreduplication transition; the small nuclei are found in the periphery and their size increases progressively inwards. DNA content, nuclear size, and cell size are positively correlated. Synthesis and storage of starch and proteins are rapid during this phase, which directly contributes to the linear growth of caryopsis. This cycle is characterized by one or more rounds of DNA replication without cytokinesis increasing the ploidy level of the nucleus.

Functions of Endoreduplication

According to Sabelli and Larkins (2009) there are three important functions of endoreduplication, such as (a) provision of more gene templates for high transcription to cope with high metabolic demand, (b) facilitating cell expansion and tissue growth with an increase of cell number, (c) increase of nucleotide pool to be utilized by the embryo during germination. For plant cells, endoreduplication is normally correlated with volume expansion, but its importance in the endosperm tissue development of rice has not been explored. After completion of endoreduplication and accumulation of storage compounds, the endosperm cells undergo programmed cell death involving extensive DNA degradation. The dynamics of endosperm cell transition from acytokinesis to cytokinesis and then to endoreduplication and its

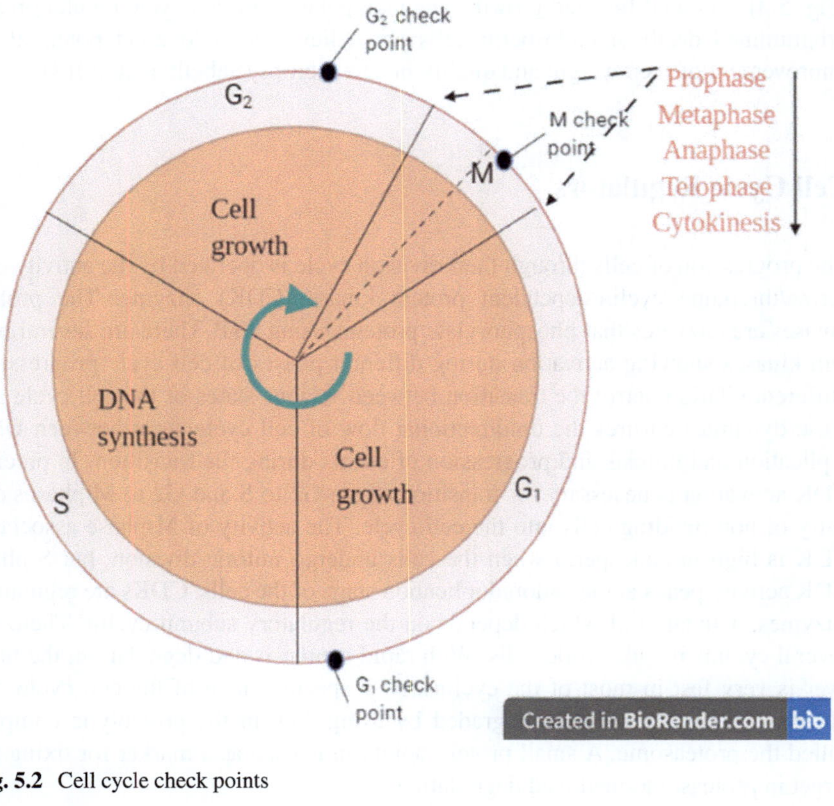

Fig. 5.2 Cell cycle check points

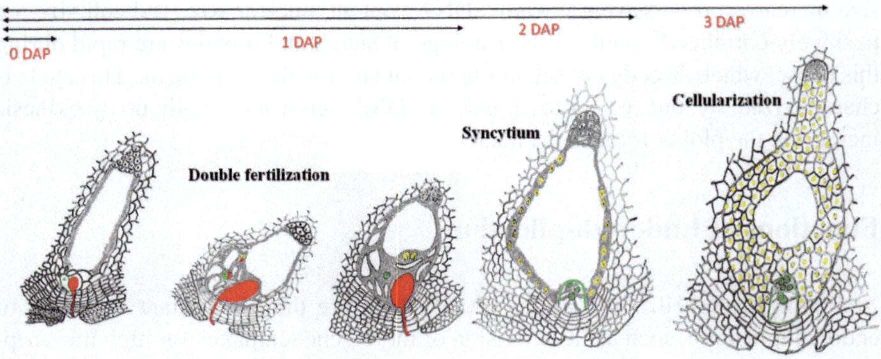

Fig. 5.3 Phases in endosperm development. Although this figure refers to maize; it is a good example of endosperm development in other grasses as well. Double fertilization, syncytium formation, and cellularization of the endosperm occur within 3- 4DAP (Days after pollination). The pollen tube and sperm nuclei are shown in red, polar nuclei in the central cell of the female gametophyte and endosperm nuclei are shown in yellow, and the egg cell nucleus and embryo nuclei are shown in green. Outlines of the multicellular endosperm and embryo are drawn in yellow and green, respectively. (Figure adopted from Kiesselbach 1999)

regulation during these three consecutive phases of endosperm development (Fig. 5.3) is crucial for seed growth. The manipulation of cell cycle regulation and programmed death of endosperm cells are believed to hold great potential for improvement of seed weight and quality in cereal crops (Sabelli et al. 2013).

Cell Cycle Regulators

The progression of cells through their division cycle is operated by the activity of a serine/threonine cyclin-dependent protein kinase (CDK) enzyme. The protein kinases are enzymes that phosphorylate proteins using ATP. There are several protein kinases showing activation during different phases of cell cycle progression. Different CDKs control the transition between various states of the cell cycle and these dynamics ensures the unidirectional flow of cell cycle steps between DNA replication and mitosis and progression of events during the transition. In precise, CDK activation is necessary for transitions from G1 to S and G2 to M phases and entry of non-dividing cells into the cell cycle. The activity of M-phase associated CDK is high in endosperm when the cells undergo mitotic division, but S-phase CDK activity peaks at the endoreduplication stage of the cells. CDKs are regulatory enzymes, activation of which depends on the regulatory subunit cyclin. There are several cyclins in eukaryotic cells. With rapid synthesis and degradation, the turnover is very fast in most of the cyclins. At a specific stage of the cell cycle, the cyclins are formed and then degraded by using ATP in the proteolytic complex called the proteasome. A small protein ubiquitin is used as a marker for fixing the target in proteasome-mediated degradation.

Endoreduplication in Rice Seed

In rice, endosperm cell division rarely increases after day 12 post-anthesis (Patel and Mohapatra 1996) indicating the time of initiation of endoreduplication. In reality, DNA duplication was found higher in large size superior grains compared to the inferior grains of rice panicle at this stage (Fig. 5.4) and it contributed positively to greater accumulation of starch and cell size of the former (Panda et al. 2016). Su'udi et al. (2012) reported that endoreduplication occurs in rice endosperm when the CDK/cyclin complex is low or inactive being targeted in ubiquitin-mediated degradation by the anaphase promoting complex nuclear protein OsCCS52A. Cyclins that regulate the transition from G1 to the S phase are called G1 cyclins, while those involved in the transition from G2 to M phase are mitotic cyclins. CDK activity can also be regulated by phosphorylation and dephosphorylation for which two tyrosine phosphorylation sites are available. One site causes activation and the other is responsible inactivation of CDK. Protein phosphatase enzymes can remove the phosphate group of CDK causing activation or inactivation depending on the site of action. Therefore addition or removal of phosphate groups from CDK is an important mechanism for control of cell cycle progression.

Besides, the role of cyclin inhibitors is also crucial for cell cycle regulation. Panda et al. (2016) reported that promotion of expression of *CDKAs* and *CYCD2:2* and inhibition of *CYCB2: 2* genes are necessary for a switch of mitotic to endoreduplication in central cells of rice endosperm increasing cell size and DNA ploidy level. Endogenous hormone level also interferes with cell cycle progression and endoreduplication. In maize endosperm, Mambelli and Setter (1998) found exogenously applied ABA depressed growth by reducing both cell division and endoreduplication, but the effect was greater on the former than the latter. Similarly, the application of ethylene action inhibitor 1-methylcyclopropene improved endoreduplication and increased cell size and starch accumulation in rice endosperm by inhibiting the expression of ethylene signal transducer genes (Panda et al. 2016).

Programmed Cell Death (PCD)

In sequence to the endoreduplication phase, endosperm enters into the terminal phase of programmed cell death. The phase is genetically programmed in which the assembly of cells is disorganized in course of cell maturation. Common to all eukaryotic plants, it is an integral part of development associated with tolerance mechanisms for biotic and abiotic stresses. The onset of programmed cell death indicates closure of cell lifetime and determines endosperm growth and grain yield. PCD initiates with permeabilization of mitochondria membrane. In the process, several caspases like proteases appear in the cells to target proteins at positions of different amino acid sequences. Chichkova et al. (2010) reported that a phytase in rice cleaved amino acid sequence of valine-glutamic acid- isoleucine- aspartic acid

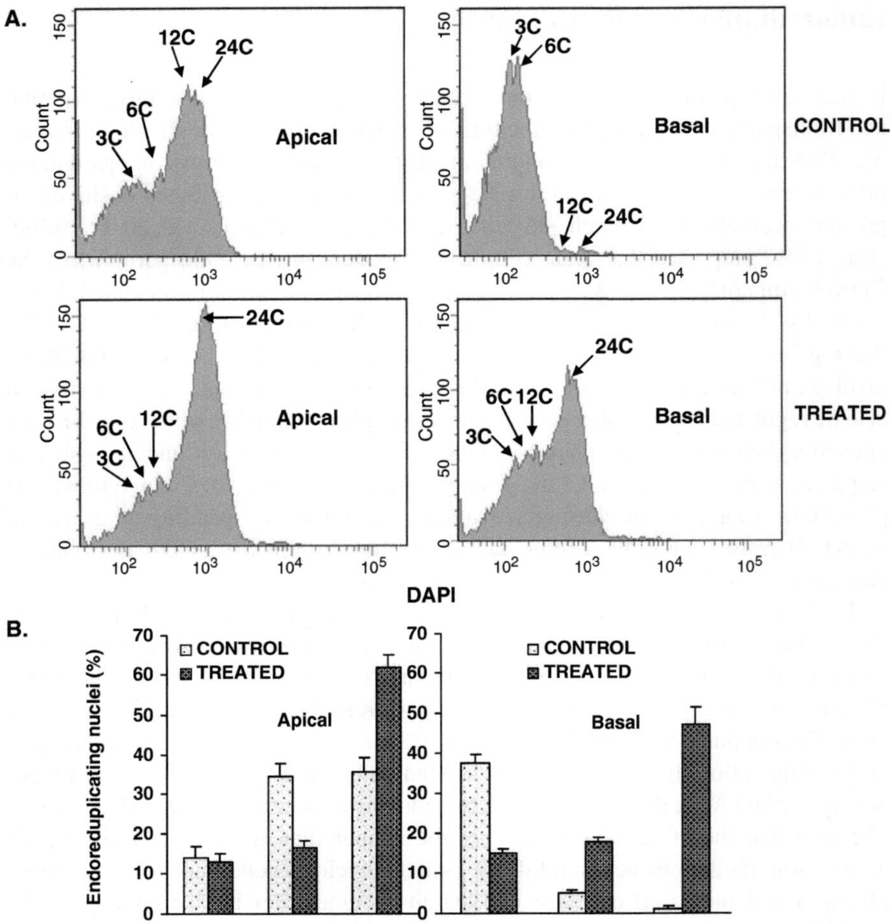

DNA class	6C	12C	24C	6C	12C	24C
Control	14.18 ±2.77	34.58 ±3.14	35.6 ±3.50	37.65 ±2.12	4.95 ±0.94	1.15 ±0.49
Treated	12.85 ±2.22	16.67 ±1.58	61.83 ±3.01	14.93 ±1.08	17.75 ±1.19	46.98 ±4.36

Fig. 5.4 The effect of 1-methylcyclopropene treatment on DNA content size classes of apical (superior type grain) and basal (inferior type grain) spikelets of rice cultivar OR-1918 on 12 days post-anthesis (Panda et al. 2016). (**A**) Paraffin-embedded endosperm cells were sectioned longitudinally in a microtome and stained with DAPI for comparison on nuclear size. (**B**) % of endoreduplicating nuclei

(VEID). Along with proteins, PCD also results in the degradation of nuclear DNA. It is also reported that PCD in rice starchy endosperm begins with nuclear degradation along with mitochondrial degeneration. Denucleation occurs asynchronously in the endosperm cells and is completed by day 13 post-anthesis (Lan et al. 2004). The endosperm has high voltage metabolism at this stage with the rapid achievement of grain filling rate and grain weight. The denucleated cells do not die and exhibit a high rate of starch synthesis by increasing activities of starch synthesizing enzymes like AGPase, starch synthase, and starch branching enzyme. At this stage initiation of PCD is marked by enhancement of catalase and superoxide dismutase (Van Breusegem and Dat 2006). Close examination of developing starchy endosperm of rice grain with fluorescence microscopy has revealed the spatial and temporal progress of programmed cell death (Kobayashi et al. 2013). Cell death is initiated in the older cells located in the center of the endosperm and spreads to the peripheral region. The spread of PCD in the endosperm is marked by mitochondrial permeabilization and activity of a caspase-like protease targeting the VEID sequence of proteins. It is surmised that with the progress of PCD the seed loses water promoting the expression of genes for synthesis of stress inductive proteins for resilience. Similarly, gradual dehydration prepares the seed for dormancy. There is some information that implicate role of hormones in PCD. Ethylene level rises and correlates positively with PCD (Young et al. 1997; Young and Gallie 2000), while ABA appears to inhibit or delay PCD (Nguyen et al. 2007). The gene expression pattern controlling signaling network for PCD is related to ethylene and ABA pathways in the endosperm (Sreenivasulu et al. 2006). However, the biochemical factors that induce stress tolerance or seed dormancy of rice seeds are not extensively studied hitherto.

References

Chichkova NV, Shaw J, Galiullina RA, Drury GE, Tuzhikov AI, Kim SH, Kalkum M, Hong TB, Gorshkova EN, Torrance L, Vartapetian AB, Taliansky M (2010) Phytaspase, a relocalisable cell death promoting plant protease with caspase specificity. EMBO J 29:1149–1161

Kiesselbach TA 1999. The structure and reproduction of corn. 50th anniversary edn. Cold Spring Harbor Laboratory Press. Cold Spring Harbor

Kobayashi H, Ikeda TM, Nagata K (2013) Spatial and temporal progress of programmed cell death in the developing starchy endosperm of rice. Planta 237:1393–1400

Lan S, Zhong F, Yang Z, Jin D, Xu Z (2004) The starchy endosperm denucleation by a process of programmed cell death during rice grain development. Shi Yan Sheng Wu Xue Bao 37:34–44

Mambelli S, Setter TL (1998) Inhibition of maize endosperm cell division and endoreduplication by exogenously applied abscisic acid. Physiol Plant 104:266–272

Mohapatra PK, Patel R, Sahu S (1993) Time of flowering affects grain quality and spikelet partitioning within rice panicle. Aust J Plant Physiol 20:231–241

Nguyen HN, Sabelli PA, Larkins BA (2007) Endoreduplication and programmed cell death in cereal endosperm. In: Olsen OA (ed) Plant cell monograph 8: Endosperm. Springer, Berlin/Heidelberg, pp 21–43

Panda B, Badoghar A, Sekhar S, Shaw B, Mohapatra PK (2016) 1-MCP treatment enhanced expression of genes controlling endosperm cell division and starch biosynthesis for improvement of grain filling in a dense-panicle rice cultivar. Plant Sci 246:11–25

Patel R, Mohapatra PK (1996) Assimilate partitioning within floret components of contrasting rice spikelets producing qualitatively different types of grains. Funct Plant Biol 23:85–92

Sabelli PA, Larkins BA (2009) The development of endosperm in grasses. Plant Physiol 149:14–26

Sabelli PA, Liu Y, Dante RA, Lizarraga LE, Nguyen HN, Brown SW, Kingler JP, Yu J, LaBrant E, Layton TM, Feldman M, Larkins BA (2013) Control of cell proliferation, endoreduplication, cell size, and cell death by the retinoblastoma-related pathway in maize endosperm. Proc Natl Acad Sci 110:E1827–E1836

Sreenivasulu N, Radchuk V, Strickert M, Miersch O, Weschke W, Wobus U (2006) Gene expression patterns reveal tissue specific signaling networks controlling programmed cell death and ABA-regulated maturation in developing barley seeds. Plant J 47:310–327

Su'udi M, Cha J-Y, Jung MH, Ermawati N, Han C, Kim MG, Woo YM, Son D (2012) Potential role of the rice OsCCS52A gene in endoreduplication. Planta 235:387–397

Van Breusegem F, Dat JF (2006) Reactive oxygen species in plant cell death. Plant Physiol 141:384–390

Young TE, Gallie DR (2000) Regulation of programmed cell death in maize endosperm by abscisic acid. Plant Mol Biol 42:397–414

Young TE, Gallie DR, DeMason DA (1997) Ethylene-mediated programmed cell death of wild type and shrunken2 genotypes. Plant Physiol 115:737–751

Chapter 6
Genetic Analyses of Floral Development on Rice Panicle

Abstract The relationship of gene activity with morphogenesis of floral organ development in rice is complex. Genetic screening of floral mutants in comparative studies and QTL analyses has been extensively performed to isolate genes involved in spikelet meristem specification. Flower organs in rice are organized in consecutive whorls for which the ABCDE model has been proposed for explanation of the molecular mechanism. Accordingly, several MADS-box genes determining meristem identity of the lemma, palea, lodicules, stamens, and megagametophyte during panicle development have been characterized and grouped into four/five classes. In exceptional cases of rice mutants, abnormalities in the expression of some genes also lead to the formation of double and triple grain rice. The genes/QTLs associated with panicle branch number, panicle length, grain number, grain size, and spikelet number have been explored.

Keywords ABCDE model · MADS-box genes · Floral mutants

Spikelet Development

Spikelet is a grass-specific structural unit of the rice panicle. Panicle structure is directly linked to spikelet/grain number and finally grain yield. Therefore, high-throughput genetic screening and QTL analyses have been extensively performed to isolate genes involved in spikelet meristem specification. Spikelet primordia develop as protuberances on the nodes of both primary and secondary branches in an acropetalous order. The molecular biology studies of spatial and temporal control of meristem identity transition during reproductive development that ultimately determines the number and shape of spikelets have been explored further (Yoshida and Nagato 2011). Inflorescence meristem initiation begins when the shoot apical meristem accumulates florigen and Hd3a and RFT1 proteins. *FLORAL ORGAN NUMBER1* (*FON1*) and *FON2* genes maintain the activity of the reproductive meristem, and their absence in mutated alleles enlarges floral meristems increasing the number of floral organs. *LONLEY GUY* (*LOG*) and *GN1* genes encode enzymes for activation and degradation of cytokinin respectively, in the meristems; mutation of the former restricts cytokinin synthesis limiting panicle size, whereas mutation of

P. K. Mohapatra, B. B. Sahu, *Panicle Architecture of Rice and its Relationship with Grain Filling*, https://doi.org/10.1007/978-3-030-67897-5_6

the latter allows cytokinin accumulation expanding panicle size. Cultivars with reduced *GN1* mRNA transcription maintain a high cytokinin level that increases the number of panicle branches.

There are some other genes reported to control panicle size by influencing meristematic activity determining branch number. Up-regulation of the *IDEAL PLANT ARCHIECTURE1* (*IPA1*) gene increases the primary branch number of panicles (Jiao et al. 2010; Miura et al. 2010). Spikelet identity is specified by the *ABERRANT PANICLE ORGANISATION1 (APO1)* gene. This gene encodes an F-box protein to promote cell proliferation of the meristems and regulation of timing for meristem transitions (Ikeda-Kawakatsu et al. 2009). A gain of expression of *APO1* increases spikelet and branch numbers, whereas loss of expression decreases their numbers (Ikeda et al. 2005, 2007) because it prolongs indeterminate meristem fate by delaying the transition of branch meristem to spikelet meristem (Kyozuka et al. 2014). This gene also improves the number of vascular bundles of the panicle for greater mobilization of assimilates to the developing sink (Terao et al. 2010). *MADS34/ PANICLE PHYTOMER2* (*PAP2*) is another gene, the expression of which controls primary branch number; this gene permits transformation of early arising spikelets into branches ultimately increasing their number (Gao et al. 2010; Kobayashi et al. 2010). Further, Kobayashi et al. (2012) identified expression of three *AP1/FUL-like (APETALA1/FRUITFUL)* genes responsible for the control of transformation of shoot apical meristem to inflorescence meristem of rice along with that of *PAP2* using microdissection and microarray. *PAP2* is a MADS-box gene belonging to the grass-specific *SEPALLATA* sub-family. Because *PAP2* and the three *AP1/FUL-like* genes jointly act in the meristem specificity for inflorescence identity, suppression or knockdown of the function of the genes delay the transition to flowering and provoke leaf primordium production. The role of transcription factors in the formation of floral organs of rice panicle is elucidated with the study of the expression of the RFL (RICE FLORICAULA/LEAFY) allele, which is a rice homolog of LFY, a transcription activator of floral differentiation in Arabidopsis. It was reported that RFL regulates two important attributes of rice panicle, i.e. transition of flowering and panicle branching; down regulation of RFL causes reduction in branch number (Rao et al. 2008) and spikelet number. The genetic loci positions of the *QTLs* have been enlisted in Table 6.1 according to Xu et al. (2020).

Floret Development

The sequence of spikelet development does not follow a pattern similar to the initiation of branch or spikelet primordia. The topmost spikelet primordium of the uppermost primary branch starts the differentiation process first with the initiation of glume initials (Xu and Vergara 1986). The primordium differentiation on each spikelet occurs in an acropetal order. The appearance of two glume initials is followed by two sterile lemma primordia and then by differentiation of lemma and palea primordia in close succession. Genetic analyses have identified sterile lemmas

Table 6.1 List of QTLs with their position and reported linked genes

Traits*	Genetic loci	Chr.	Marker interval	Reported genes/QTLs
SBN	*qSBN1*	1	H1-7–H1-8	*qSSB1-1*
	qSBN3	3	H3-36–H3-37	*OsIDS1*
	qSBN4	4	H4-5–H4-6	*qSNP-4a*
	qSBN7	7	H7-12–H7-13	*OsFOR1*
SPP	*qSPP1*	1	H1-7–H1-8	*qSSB1-1*
	qSPP3	3	H3-35–H3-36	*DST*
	qSPP4	4	H4-9–H4-10	*spp4-2*
	qSPP7	7	H7-12–H7-13	*OsFOR1*
GW	*qGW1*	1	H1-26–H1-27	
	qGW3-1	3	H3-12–H3-13	*qGW3*
	qGW3-2	3	H3-27–H3-28	*gw3.1*
	qGW6	6	H6-24–H6-25	
	qGW9	9	H9-10–H9-11	
	qGW12	12	H12-3–H12-4	
PL	*qPL2*	2	H2-27–H2-28	
	qPL3	3	H3-7–H3-8	
	qPL4-1	4	H4-13–H4-14	
	qPL4-2	4	H4-23–H4-24	
	qPL7	7	H7-7–H7-8	
LWR	*qLWR5*	5	H7-7–H7-5	*GS5, SRS32010*
	qLWR7	7	H7-13–H7-14	*GL7*
GL	*qGL6*	6	H6-18–H6-19	
	qGL7	7	H7-13–H7-14	*GL7*
PBN	*qPBN12*	12	H12-19–H12-20	

in rice as counterparts of glumes in other members of the grass family. This inference could be drawn because, in *fzp* mutant, the floral meristem is suppressed to promote spikelet meristems producing extra bracts with axillary buds (Komatsu et al. 2001, 2003). On the other hand, *FZP* encodes a transcription factor with an ERF (ethylene responsive factor) domain that prevents the formation of axillary meristems to establish the identity of floral meristems in rice spikelets (Komatsu et al. 2003; Yoshida and Nagato 2011).

The appearance of lemma and palea primordia in half alternate phyllotaxy succeeds the developmental course after initiation of the sterile lemma. The presence of lemma and palea in the outer whorl is a grass-specific floral character that distinguishes the family from eudicot flowers possessing sepals and petals instead. Comparative studies of floral mutants and information available in whole-genome sequencing of several plants in recent years have provided evidence in favor of lemma and palea being equivalents of sepals (Lombardo and Yoshida 2015). The distal tip of the lemma extends into a needle-like appendage called awn, which is considered to be a modified leaf blade (Dahlgren and Bremer 1985). Lemma of wild rice species have long awns for seed dispersal and deterrent to grazing animals, but its utility is lost in cultivated species (Furuta et al. 2015). These grass-specific perianth organs enclose the inner floral organs including two lodicules, six stamens, and a pistil with bifid and hairy stigma. The floral organs along with the lemma and palea comprise a flower unit called floret (Ikeda et al. 2004). After the palea, stamen primordia appear in a whorl as spherical swellings at the same time below the summit of the floral meristem. The lodicule primordia are formed between the palea and stamen primordia. Thereafter, the floral meristem, which is dome-shaped and larger than the stamen primordium is converted to the pistil primordium (Xu and Vergara 1986).

Genetic Mechanism of Flower Development

Flowers in eudicots consist of four different organs arranged consecutively in whorls, such as sepals (whorl 1), petals (whorl 2), stamens (whorl 3), carpel, and ovules (whorl 4). Based on genetics and molecular studies using homeotic flower mutants of model dicotyledonous plants Arabidopsis (*Arabidopsis thaliana*), petunias (*Petunia hybrida*), and snapdragons (*Antirrhinum majus*), the ABCDE (or ABCE) hypothesis was stated for the molecular mechanism of flower development (Lohmann and Weigel 2002; Prunet and Jack 2014). ABC model was established where SEPALLATA-like MADS-box A, B, C, D, and E class genes are required to confer identity to organ development either individually or in a cohort in each of the four whorls of flowers in eudicots. Individuals of the grass family *Poaceae* have floral architecture significantly different from eudicots and other monocots. Spikelet is the basal unit of the inflorescence of grasses including rice, where the specification of the whorls is exactly not similar to eudicots. Instead of distinct sepals and petals, rice bears lemma, palea, and lodicules may be considered homologous to the first two whorls of the eudicots.

MADS-Box Genes of Rice Spikelet

Because several MADS-box genes were characterized, Yoshida and Nagato (2011) proposed that the genetic mechanism of flower development in rice is comparable to the ABCDE model. The model proposed suggests that genes involved in floral organ specification and identity could be grouped into four or five classes. The organ identity is determined by genes of the same class or two different classes coordinately. Genes of class A provide identity to sepals in whorl one, A and B for petals in whorl two, B and C for stamen in whorl three and C-class gene determine carpel identity in whorl four. Reverse genetic studies extended the model to include two additional classes of homeotic genes D and E; while the former is crucial for ovule development, the latter is required for the identity of all flower organs. All these genes encode transcription factor proteins that bind with the cis-regulatory region of target genes named CArG-box to determine floral identities of the organs (Cui et al. 2010). Among all, the E-class genes exhibit the greatest diversity in terms of function and number. Wang et al. (2017) in their review suggested that five E-class *SEPALLATA-like* genes act to specify all floral organ identities and floral meristem determination, such as *OsMADS7* and *OsMADS8* for all organs, *OsMADS34/PANICLE PHYTOMER2* (*PAP2*) for branch, rudimentary glume, and sterile lemma and *OsMADS1* for lemma and palea. The role of some other genes encoding the identity of various parts of the floret organs has been established also.

Lemma and Palea Development

Development of lemma and palea directly influence grain weight and quality because they define the space for grain growth first, and secondly being photosynthetic, contributes to assimilate partitioning. Although the outer morphology of lemma and palea is similar, they have a few critical differences in some other features, such as the presence of five vascular bundles in the former compared to three in the latter. Besides, palea is awnless, unlike the lemma. For a marginal variation on morphological features of the two, there has been confusion on their equivalence and specifications and only a few genes involved in the development of lemma and palea have been explored. The presence of awn in *indica* rice is controlled by two genes, namely *DROOPING LEAF (DL)* and *OsETT2* genes, and loss of expression of the latter results in failure of awn formation in the japonica strains (Toriba and Hirano 2014). Transcriptome profiles of lemma and palea are similar, although 23 genes express differentially between them (Sato et al. 2011). Out of the 23, 13 genes encode transcription factors including *DROOPING LEAF* and *MOSAIC FLORAL ORGAN1/OsMADS6*; the former is specifically expressed in lemma and the latter is expressed in palea. *MOSAIC FLORAL ORGAN1/OsMADS6* is a member of the MADS-box gene that is reported to specify the identity of not only palea but also embraces ovule development, lodicules identity, flower patterning, and floral meristem determinacy. More recently Wang et al. (2017), revealed that the E-class gene *OsMADS1* plays an important role in the specification of the identity of lemma and palea in rice spikelets. Ectopic expression of the gene was first noticed in the floral meristem at the early stage and transgressed later into the lemma and palea (Prasad et al. 2001, 2005).

Lodicule Development

The two lodicules of rice spikelets are scale-like structures adjoining the base of palea. They are floral organs specific to grasses. They take origin just outside the stamen whorl and hence, are considered analogous to eudicot petals (Yoshida 2012).

At anthesis, they swell and force flower openings to expose the mature floral parts (Hoshikawa 1989). Because lodicules are considered petal equivalents, grass-specific B-class genes control their identity and differentiation (Nagasawa et al. 2003). In Arabidopsis, B-class gene expression is determined by a heterodimer of closely related MADS-box proteins encoded by genes of two sub-families *PISTILATA (PI)* and *APELATA3 (AP3)* (Yoshida and Nagato 2011) of which the latter type is evident in rice. *SUPERWOMEN1*, the *AP3* orthologue of rice, interacts with two homologs of *PI*, namely, *OsMADS2* and *OsMADS4* for lodicules identity and development (Yao et al. 2008), the contribution of the former is greater than the latter.

Stamen Development

ABCDE model suggests interaction in the expression of B-, C- and E-class MADS-box genes in the specification of stamens of rice. Expression of B-class genes, namely *SPW1*, *MADS2,* and *MADS4*, provide identity to both stamen and lodicules (Yoshida and Nagato 2011). Studies with genetic mutants have provided evidence in favor of two C-class gene functions in stamen determinacy of rice. They are *MADS3* and *MADS58*; the expression of the former is more important than the latter (Yamaguchi et al. 2006). Using a double-stranded RNA mediated interference technique, Cui et al. (2010) explored the expression of SEP-like genes and their interaction with the proteins encoded by them and concluded that the development of all floral organs including the stamens is regulated by E-class *OsMADS7* and *OsMADS8* genes; the loss of which leads to a homeotic transformation of lodicules, stamens, and carpels into palea/lemma like organs.

Abnormalities of Megagametophyte Development

Double- and triple-grain rice is a precious indigenous cultivar seen in several parts of western Odisha (Photo 6.1). The name varies with geographic location; Rama-Lakshman, Laba- Kusha, Do Dana Jugal and Sateen, etc. The first scientific report of collection and scientific enumeration of this natural rice mutant bearing multi-pistillate spikelets in the panicle is credited to a group of rice breeders of ICAR-National Rice Research Institute, Cuttack (Misra et al. 1992). The workers accounted for this productive multi-pistillate mutation in rice by locating the presence of 2 to 6 ovaries per spikelet distributed across the panicle. At maturity, of course, the panicle had only 3-grain types, i.e. single, double and triple grains. The other ovaries might have been lost due to competition among the fertilized ovaries for development within respective spikelets. The frequency of one grain per spikelet was higher followed by double and triple grain types (Sahu et al. 2017). It was reported that

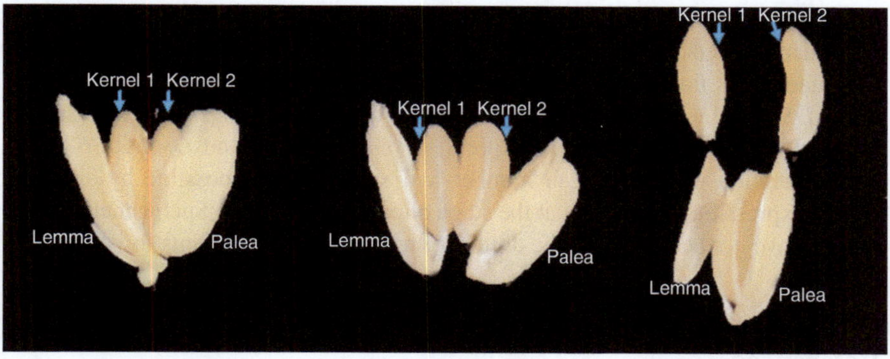

Photo 6.1 Parts of matured rice grain of multi-pistillate natural mutant indigenous rice Rama- Lakshman

these multiple traits are governed by a single recessive gene that often offsets the normal development of one grain per spikelet in a natural mutation.

The genetic mechanism controlling ontogeny of abnormal cellular proliferation of various spikelet organs has been a subject of interest recently. In rice one ovule is located within each carpel. The ovule arises from the interior meristematic tissue of the carpel and develops into a seed in the post-fertilization period. Cucinotta et al. (2020) suggested that ovule number and carpel size are inter-related through a gene regulatory network involving transcription factors and hormonal communications. However, knowledge of the genes and hormonal factors involved in determining the number of ovules and size of the carpel is scant. Against this backdrop, abnormalities in megagametophyte development in rice panicle leading to the formation of the multi-pistillate phenotype could be conjectured to the regulation of different functional classes of *MADS-box* genes. The expression of the *MADS-box* genes controls cell fate of the proliferating primordial tissues of ovule and carpel and thereby specifies identity. For example loss of function of *OsMADS13* replaces ovule with carpelloid structures (Dreni et al. 2007; Yamaki et al. 2011). *OsMADS6* and *OsMADS58* genes are associated with the regulation of stamens, carpels, and ovules (Dreni et al. 2011; Li et al. 2011). Zhang et al. (2015), using RNA in situ hybridization analysis characterized an *LBD* gene *OsIG1* (*indeterminate gametophyte1* mutation in rice) that acts as a key regulator specifying organ development in rice spikelet. This gene modulates female gametophyte development in rice in addition to its role in the control of empty glume identity and floral organ number because of its associated effects on the expression of *OsMADS1* and *OsMADS6*. *OsIG1* expression signal is strongly detected in inflorescence meristem, floral meristem, primordia of empty glume and florets, more so in the primordia of stamens, ovules, and ovary. The workers noticed several reproductive defects in *OsIG1*-RNAi hybridized plants, such as, increase in size pistil meristem, initiation of ovule primordial on both lemma and palea side of the floral meristem, the formation of fused double ovules in ovaries, development of abnormal embryo sac with anomalous polar nuclei, embryo sac with an extra egg cell, embryo sac without a female germ unit, embryo sac with abnormal nuclear migration and embryo sac of smaller size. Similar defects were also noticed in the lemma and palea development of the RNAi lines. It was concluded that floral organ identity in rice is the outcome of the interplay of *OsMADS* and the *LBD* genes because downregulation of *OsIG1* expression in *OsIG1*-RNAi hybrid led to overexpression of *OsMADS1* and *OsMADS6* genes. Thus alteration of floral organ number and size could be possible by silencing *OsIG1* expression particularly at the stage of carpel protrusion. Such genetic explorations have not been done in the double or triple grain rice mutants for so long.

References

Cucinotta M, Marzo MD, Guazzotti A, De Folter S, Kater MM, Colombo L (2020) Gynoecium size and ovule number are interconnected traits that impact seed yield. J Exp Bot 71:2479–2489

Cui R, Han J, Zhao S, Su K, Wu F, Du X, Xu Q, Chong K, Theissen G, Meng Z (2010) Functional conservation and diversification of class E floral homeotic genes in rice (*Oryza sativa*). Plant J 61:767–781

Dahlgren R, Bremer K (1985) Major clades of the angiosperms. Cladistics 1:349–368. https://doi.org/10.1111/j.1096-0031.1985.tb00433.x

Dreni L, Jacchia S, Fornara F, Fornari M, Ouwekerk PB, An G, Colombo L, Kater M (2007) The D-lineage MADS-box gene *OsMADS13* controls ovule identity in rice. Plant J 52:690–699

Dreni L, Pilatone A, Yun D, Erreni S, Pajaro A, Caporali E, Zhang D, Kater MM (2011) Functional analysis of all AGAMOUS subfamiliy members in rice reveals their roles in reproductive organ identity determination and meristem determinacy. Plant Cell 23:2850–2863

Furuta T, Komeda N, Asano K, Uehara K, Gamuyao R, Angeles-Shim RB, Nagai K, Doi K, Wang DR, Yasui H, Yoshimura A, Wu J, McCouch SR, Ashikari M (2015) Convergent loss of awn in two cultivated rice species *Oryza sativa* and *Oryza glaberrima* is caused by mutations in different loci. G3 (Bethesda) 5:2267–2274

Gao X, Liang W, Yin C, Ji S, Wang H, Su X, Guo C, Kong H, Xue H, Zhang D (2010) The SEPALLATA-like gene *OsMADS34* is required for rice inflorescence and spikelet development. Plant Physiol 153:728–740

Ikeda K, Sunohara H, Nagato Y (2004) Developmental course of inflorescence and spikelet in rice. Breed Sci 54:147–156

Ikeda K, Nagasawa N, Nagato Y (2005) ABERRANT PANICLE ORGANIZATION 1 temporally regulates meristem identity in rice. Dev Biol 282:349–360

Ikeda K, Ito M, Nagasawa N, Kyozuka J, Nagato Y (2007) Rice *ABERRANT PANICLE ORGANIZATION 1*, encoding an F-box protein, regulates meristem fate. Plant J 51:1030–1040

Ikeda-Kawakatsu K, Yasuno N, Oikawa T, Iida S, Nagato Y, Maekawa M, Kyozuka J (2009) Expression level of *ABERRANT PANICLE ORGANIZATION1* determines rice inflorescence form through control of cell proliferation in the meristem. Plant Physiol 150:736–747

Jiao Y, Wang Y, Xue D, Wang J, Yan M, Liu G, Dong G, Zeng D, Lu Z, Zhu X, Qian Q, Li J (2010) Regulation of OsSPL14 by OsmiR156 defines ideal plant architecture in rice. Nat Genet 42:541–544

Kobayashi K, Maekawa M, Miyao A, Hirochika H, Kyozuka J (2010) PANICLE PHYTOMER2 (PAP2), encoding a SEPALLATA subfamily MADS-box protein, positively controls spikelet meristem identity in rice. Plant Cell Physiol 51:47–57

Kobayashi K, Yasuno N, Sato Y, Yoda M, Yamazaki R, Kimizu M, Yoshida H, Nagamura Y, Kyozuka J (2012) Inflorescence meristem identity in rice is specified by overlapping functions of three AP1/FUL-like MADS box genes and PAP2, a SEPALLATA MADS box gene. Plant Cell 24:1848–1859

Komatsu M, Maekawa M, Shimamoto K, Kyozuka J (2001) The *LAX1* and *FRIZZY PANICLE 2* genes determine the inflorescence architecture of rice by controlling rachis-branch and spikelet development. Dev Biol 231:364–373

Komatsu M, Chujo A, Nagato Y, Shimamoto K, Kyozuka J (2003) *FRIZZY PANICLE* is required to prevent the formation of axillary meristems and to establish floral meristem identity in rice spikelets. Development 130:3841–3850

Kyozuka J, Tokunaga H, Yoshida A (2014) Control of grass inflorescence form by the fine-tuning of meristem phase change. Curr Opin Plant Biol 17:110–115

Li H, Liang W, Hu Y, Zhu L, Yin C, Xu J, Dreni L, Kater MM, Zhang D (2011) Rice *MADS6* interacts with floral homeotic genes *SUPERWOMEN1, MADS3, MADS58, MADS13,* and *DROOPING LEAF* in specifying floral organ identities and meristem fate. Plant Cell 23:2536–2552

Lohmann JU, Weigel D (2002) Building beauty: the genetic control of floral patterning. Dev Cell 18:135–142

Lombardo F, Yoshida H (2015) Interpreting lemma and palea homologies: a point of view from rice floral mutants. Front Plant Sci. https://doi.org/10.3389/fpls.2015.00061

Misra RN, De RN, Bose LK, Panigrahi A (1992) A productive multipistillate mutation in rice. Oryza 29:358–360

Miura K, Ikeda M, Matsubara A, Song X-J, Ito M, Asano K, Matsuoka M, Kitano H, Ashikari M (2010) OsSPL14 promotes panicle branching and higher grain productivity in rice. Nat Genet 42:545–549

Nagasawa N, Miyoshi M, Sano Y, Satoh H, Hirano H, Sakai H, Nagato Y (2003) *SUPERWOMAN1* and *DROOPING LEAF* genes control floral organ identity in rice. Development 130:705–718

Prasad K, Sriram P, Kumar SC, Kushalappa K, Vijayraghavan U (2001) Ectopic expression of rice *OsMADS1* reveals a role in specifying the lemma and palea, grass floral organs analogous to sepals. Dev Genes Evol 211:281–290

Prasad K, Parameswaran S, Vijayraghavan U (2005) *OsMADS1*, a rice MADS-box factor, controls differentiation of specific cell types in the lemma and palea and is an early-acting regulator of inner floral organs. Plant J 43:915–928

Prunet N, Jack TP (2014) Flower development in *Arabidopsis*: there is more to it than learning your ABCs. In: Riechmann JL, Wellmer F (eds) Flower development: methods and protocols. Springer, New York, pp 3–33

Rao NN, Prasad K, Kumar PR, Vijayraghavan U (2008) Distinct regulatory role for RFL, the rice LFY homolog, in determining flowering time and plant architecture. Proc Natl Acad Sci 105:3646–3651

Sahu GR, Sarawgi AK, Tiwari JK (2017) New genotypes with multiple pistils and inheritance pattern of this strain in rice. Electron J Plant Breed 8:998–1000

Sato Y, Antonio B, Namiki N, Motoyama R, Sugimoto K, Takehisa H, Minami H, Kamatsuki K, Kusaba M, Hirochika H, Nagamura Y (2011) Field transcriptome revealed critical developmental and physiological transitions involved in the expression of growth potential in japonica rice. BMC Plant Biol 11:10. http://www.biomedcentral.com/1471-2229/11/10

Terao T, Nagata K, Morino K, Hirose T (2010) A gene controlling the number of primary rachis branches also controls the vascular bundle formation and hence is responsible to increase the harvest index and grain yield in rice. Theor Appl Genet 120:875–893

Toriba T, Hirano H-Y (2014) The *DROOPING LEAF* and *OsETTIN2* genes promote awn development in rice. Plant J 77:616–626

Wang L, Zeng X-Q, Zhuang H, Shen Y-L, Chen H, Wang Z-W, Long J-C, Ling Y-H, He G-H, Li Y-F (2017) Ectopic expression of *OsMADS1* caused dwarfism and spikelet alteration in rice. Plant Growth Regul 81:433–442

Xu X, Vergara BS (1986) Morphological changes in rice panicle development: a review of literature, IRRI research paper series no 117. International Rice Research Institute, Manila

Xu X, Zhang M, Xu Q, Feng Y, Yuan X, Yu H, Wang Y, Wei X, Yang Y (2020) Quantitative trait loci identification and genetic diversity analysis of panicle structure and grain shape in rice. Plant Growth Regul 90:89–100

Yamaguchi T, Lee DY, Miyao A, Hirochika H, An G, Hirano H-Y (2006) Functional diversification of the two C-class MADS box genes *OSMADS3* and *OSMADS58* in Oryza sativa. Plant Cell 18:15–28

Yamaki S, Nagato Y, Kurata N, Nonomura K (2011) Ovule is a lateral organ finally differentiated from the terminating floral meristem in rice. Dev Biol 351:208–216

Yao S-G, Ohmori S, Kimizu M, Yoshida H (2008) Unequal genetic redundancy of rice *PISTILLATA* orthologs, *OsMADS2* and *OsMADS4*, in lodicule and stamen development. Plant Cell Physiol 49:853–857

Yoshida H (2012) Is the lodicule a petal: molecular evidence? Plant Sci 184:121–128

Yoshida H, Nagato Y (2011) Flower development in rice. J Exp Bot 62:4719–4730

Zhang J, Tang W, Huang Y, Niu X, Zhao Y, Han Y, Liu Y (2015) Down-regulation of a *LBD*-like gene, *OsIG1*, leads to occurrence of unusal double ovules and developmental abnormalities of various floral organs and megagametophyte in rice. J Exp Bot 66:99–112

Chapter 7
Diversity of Panicle Architecture and Traits Influencing Grain Filling

Abstract In the long history of domestication through cultivation and selection under diverse habitats; the plant acquired adaptability for growth matching with the physico-chemical behavior of the environment, which is most prominently reflected in the phenology of panicle. Panicle morphology varies significantly more than any other phenotypic attribute of the plant and it is both cultivar and environment-specific. The architectural traits, including branching, panicle length, and spikelet number, are established during the reproductive phase from panicle initiation to heading depending on activities rachis and branch meristems. Termination of rachis meristem activity fixes primary branch number, whereas, the transition of the primary or secondary meristem into spikelet limits spikelet number; the earlier the closure of the events less are the numbers. Many genes, either small effects or major effects, have been identified that are associated with panicle architecture and grain yield. They are distributed across the 12 rice chromosomes.

Keywords Domestication · Panicle phenology · Rachis meristem · Branch meristem

The Uniqueness of Rice Biodiversity

The success of rice as a long-lasting human food crop has relied heavily on its wide biodiversity unparalleled in any other crop of agricultural importance. In the long history of domestication through cultivation and selection under diverse habitats; the plant acquired adaptability for growth matching with the physico-chemical behavior of the environment, which is most prominently reflected on the phenology of 108,000 rice accessions preserved in the gene bank of IRRI (Jackson and Lettington 2003) and hundreds of rice strains more in CGIAR, WARDA and International Institute of Tropical Agriculture. Environment-induced expression of phenotypic traits is displayed in growth patterns of all plant parts, but more particularly the flowering organs, which represent the climax of development. Crowell et al. (2014) considered floral morphology and complex panicle architectures of rice evolutionary embellishments on reproduction and gene transfer, and hence, variation in inflorescence development is an important target of crop selection. Besides, rice possesses a uniqueness of flower structure that has greatly diverged from other

P. K. Mohapatra, B. B. Sahu, *Panicle Architecture of Rice and its Relationship with Grain Filling*, https://doi.org/10.1007/978-3-030-67897-5_7

model eudicots like *Arabidopsis*, *Antirrhinum*, or *Petunia*, and hence considered, the most fascinating model species of monocots for the study of flower development (Yoshida and Nagato 2011). Cultivar-specific variations that exist in size, shape, number and distribution of spikelets on primary and secondary branches and individual grain weight of panicle far outnumber the typical differences in the characteristics of vegetative organs like leaf, stem, and roots. Additionally, the biodiversity of a cultivar traditionally and markedly depends on grain texture, colour, aroma and quality, not only for crop agronomy, but also for commercial purposes.

Diversity of Panicle Architecture

Rice has a wide range of panicle architecture (Fig. 7.1). Several morphological characteristics determine panicle architecture (FRA 2007); prominent among them are size, branch and grain number, the curvature of the rachis, spikelet, and secondary branch density, and attitude of branches of the panicle. The color, texture, size, and design of the caryopsis also contribute to the morphological diversity of panicle. Hence, panicle architecture has been a major agronomic factor critically linked to grain yield and grain quality. Although panicle architecture is determined by a plethora of characters, the most prominent among them is the branching pattern. Hoshikawa (1989) observed that the number of primary branches and spikelets produced per panicle is cultivar specific. The architectural traits, including branching, panicle length, and spikelet number, are established during the reproductive phase from panicle initiation to heading (Adriani et al. 2016).

Determination of Branch and Spikelet Number in Panicle

Branch number varies among cultivars, because of the differences in patterns of rachis and primary branch meristem formation and specification of their identities. Inflorescence structure is a result of activities of various meristems in course of ontogeny. After ontogeny, the fate of a meristem changes sequentially during development. The diversity in panicle architecture of rice cultivars is incumbent on differences in the functional plasticity of these meristems, terminal activity of which lead to the fixation of numbers of primary branches and spikelets of the panicle. As reported earlier, the termination of the rachis meristem fixes the primary branch number while consumption of the primary branch meristem into a spikelet signals the end of its activity determining the final spikelet number. The branch meristem activity is indeterminate, but its transition to floral meristem, i.e. spikelet meristem makes it determinate, indicating closure of meristem activity. A transition early during the meristematic phase results in the production of fewer and shorter branches with less number of spikelets, whereas delaying transition favors more branches and larger panicles (Kyozuka et al. 2014) (Figs.7.2 and 7.3).

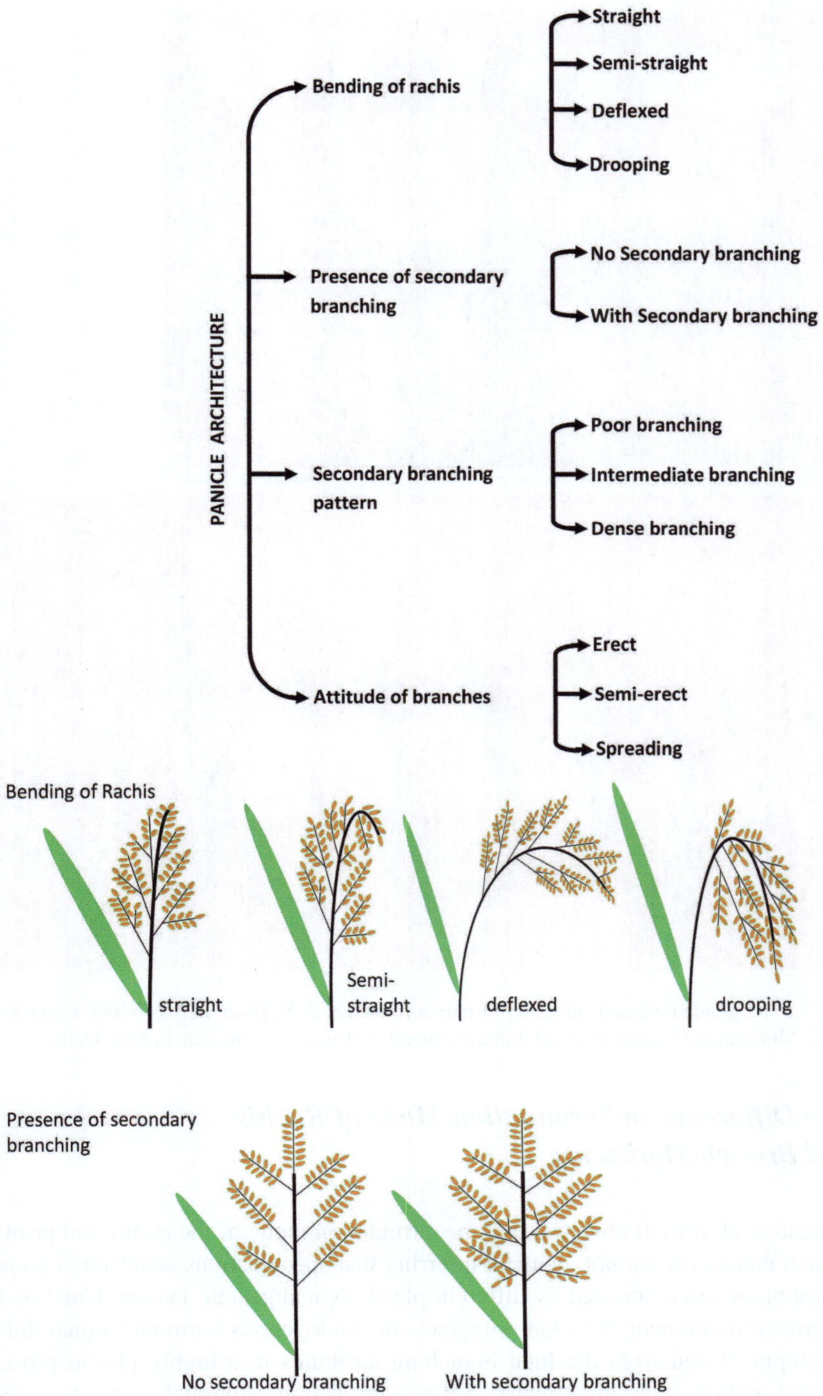

Fig. 7.1 Diversity of panicle architecture in rice

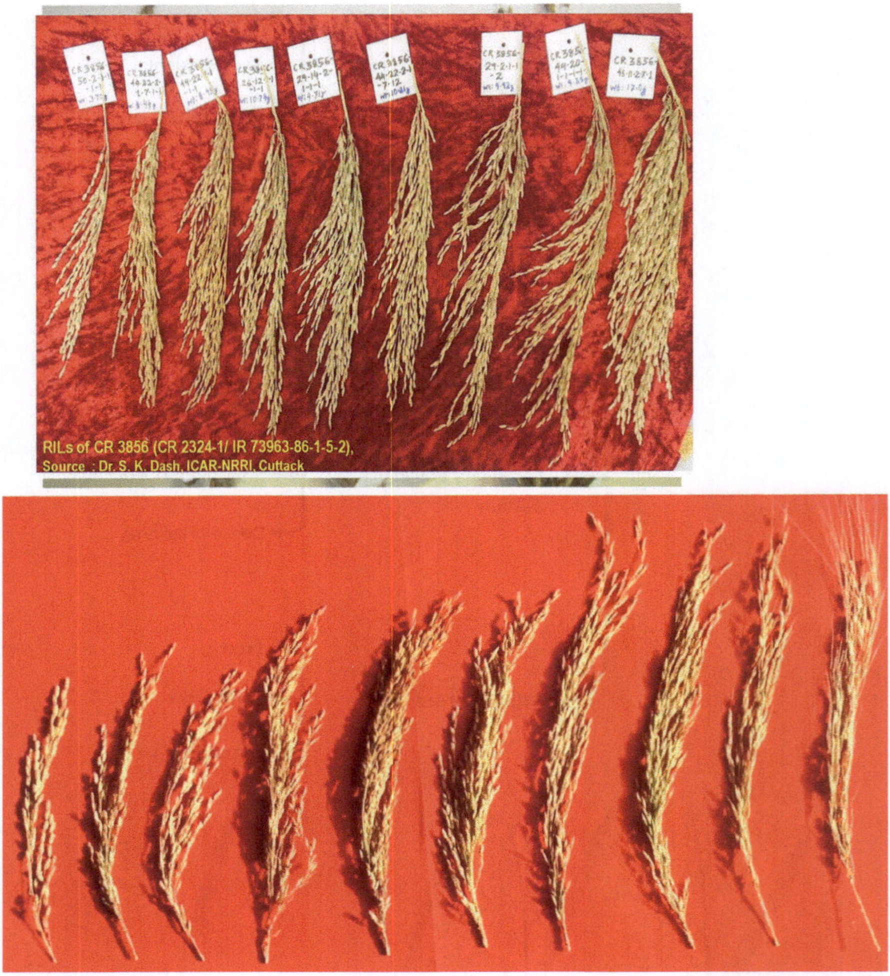

Fig. 7.1 (continued) Genetic diversity of rice in India. Dr. S. K. Dash, ICAR- NRRI, Cuttack and Dr. T. Mohapatra, Director General, Indian Council of Agricultural Research, New Delhi

The Difference in Termination Mode of Rachis and Branch Meristems

Yamaki et al. (2010) observed that the termination mode of the rachis and primary branch meristems are not identical inferring that specifications achieved in panicle morphology are controlled by different physiological/genetic factors. Cued by the external environment, the plant integrates the endogenous hormonal signal during development and fixes the final branching attributes in a highly plastic process. Several authors have investigated differences in transcriptional programs, which revealed temporal and spatial control of meristem transition identities during reproductive development of rice panicle. A delay in transition can increase branch

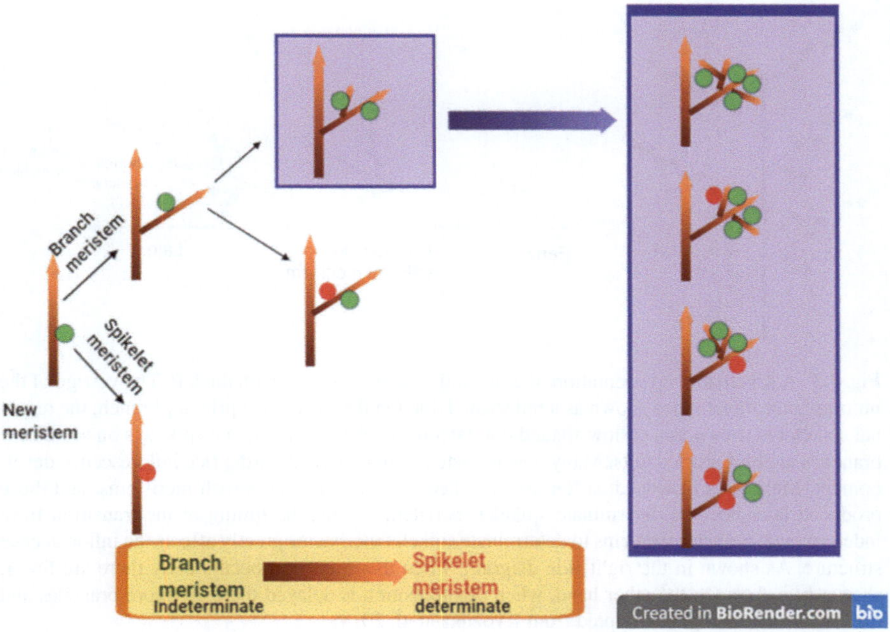

Fig. 7.2 Concept of meristem identity specification. A new meristem (green shaded circle), which then acquires the identity of a branch meristem (BM, brown arrow) or a spikelet meristem (SM, red circle). While the SM has a determinate developmental fate and grows as a spikelet, the BM is indeterminate and further produces new meristems that acquire either SM or BM identities. (Adopted from Kyozuka et al. 2014)

number and *vice versa*. Because rice cultivars differ considerably in the primary branch number of panicle, the mode of termination of rachis meristem activity and its transition into a degenerative structure in them are also controlled by different genetic factors. Similarly, the number of spikelets produced on the primary branches is determined by genes controlling the timing of phase transition of the branch meristem to acquire a floral identity. It was observed that some transgenic rice plants exhibited highly branched and denser panicle morphology because of the delay in the phase change from the branch meristem from shoot to floral meristem state. The delay in change over promotes overexpression of *TERMINAL FLOWER 1 (TFL1)/ CENTRORADIALIS (CEN)*-like genes leading to the generation of more numbers of branches in the panicle (Nakagawa et al. 2002). Hence, the specification in panicle morphology achieved in a cultivar is determined by the expression of genes, which might have been influenced by temporal and spatial controls of the regulatory system. Collectively, the regulatory system that controls the course of inflorescence development and final panicle architecture is divided into three major factors, such as, (a) timing of rachis meristem abortion, (b) transition of branch meristem to the terminal spikelet meristem, and (c) programme shift of lateral meristem identity on a rachis branch in acropetal fashion (Ikeda-Kawakatsu et al. 2009).

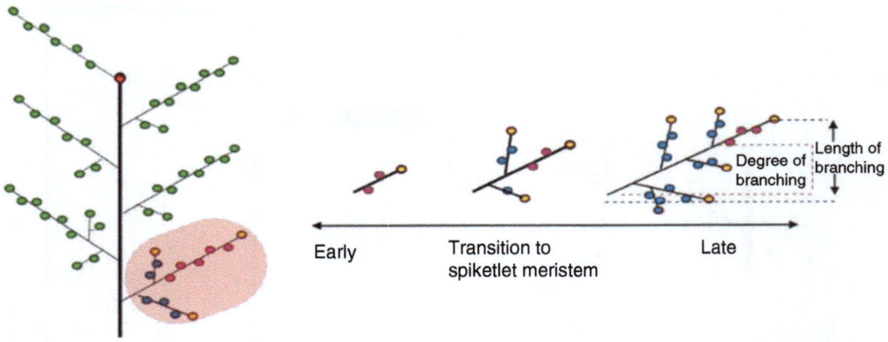

Fig. 7.3 A schematic representation of a rice inflorescence is shown on the left. The vestige of the inflorescence meristem is shown as a red shaded dot. On the lowermost primary branch, the terminal spikelet is shown as a yellow shaded dot, lateral spikelets are pink, and spikelets on secondary branches are blue shaded dots. Many new meristems are produced during rice inflorescence development. Meristems produced earlier are specified as indeterminate branch meristems, and those produced later become determinate spikelet meristems. Thus, the timing of the transition from indeterminate branch meristems to determinate spikelet meristems greatly affects the inflorescence structure. As shown in the right-side diagram, when this transition occurs early there are fewer, shorter branches. On the other hand, when this transition is delayed there are more branches and the branches are longer. (Adopted from Kyozuka et al. 2014)

Kato (2017) identified a new short tip panicle (STP) type in the japonica cultivar Gimbozou which is characterized by short primary branches bearing few spikelets only at the top two nodes of the panicle in comparison to the dense spikelet proximal branches. The cultivar is unique for the expression of the STP phenotype that is controlled by a dominant allele *Stp* at the STP locus. *Stp* expression is inhibited by an incompletely dominant allele *I- Stp* at another independent I-STP locus. Because of decreased spikelet number, this novel phenotype does not contribute to high yielding ability directly but holds promise for future studies on molecular biology and its application in panicle morphogenesis.

Identification of QTLs for Grain Yield

Rice yield is a measurable phenotype being controlled by cumulative actions of a group of genes with minute individual effects known as QTL (quantitative trait loci). These minor effects genes/loci (QTLs) are distributed in the 12 paired chromosomes (http://www.gramene.org). Because rice breeding is cumbersome and time taking, and hence increasingly difficult, breeders have a preference for molecular design breeding. Identification of the QTLs for yield-related traits in rice of course has been stupendous for the fact that diversity is high and various alleles in different lines cause different phenotypes. The QTL mapping experiments are performed by a molecular geneticist to identify DNA marker-trait association that helps plant breeders for appropriate selection of cultivar in a segregating population.

The segregating population, derived from recombinant inbred lines, is usually used in the experiments for screening certain traits of interest (Table 7.1). DNA extracted from individual plants of the population is genotyped and DNA marker information is used to generate a linkage map. The information available with a linkage map is used to locate specific regions of the genome associated with the trait by using statistical methods, such as interval mapping. The positions of the QTLs controlling the traits are identified relative to the markers used in the construction of the map. The contribution of the QTL is assessed by a logarithm of odds (LOD) score, whereas the R^2 value represents the effect of the QTL on the trait. A high LOD score

Table 7.1 List of the SSR genetic markers tightly associated with agronomic traits of rice

Trait[a]	Chr.	Locus	Known-loci
PH	1	RM6466	
TGW	6	RM1370	TGW6
TN	1	RM3425	
	10	RM6144	OsCesA7
AC	3	RM411	
GT	3	RM1022	
GC	1	RM5310	
	10	RM331	
GW	3	RGS1	GS3
	3	RM6283	GL3.2/Qgw-3
	5	RM6645	
	5	RM3777	GW5
	5	RM1089	GW5
	5	RMw-513	GW5
	7	RM351	GL7/GW7
	8	RM1345	
GL	3	RGS1	GS3
	3	RM6283	GL3.2/Qgw-3
	4	RM7187	D11
L/W	3	RGS1	GS3
	3	RM6283	GL3.2/Qgw-3
	4	RM7187	D11
	5	RM6645	
	6	RM204	
	7	RM351	GL7/GW7
	8	RM1345	
	9	RM3700	SG1
PL	1	RM3425	
TPC	1	RM495	

[a]Traits in the table: *PH* Plant height, *TGW* thousand-grain weight, *TN* tiller number, *AC* amylose content, *GT* gelatinization temperature, *GC* gel consistency, *GW* grain width, *GL* grain length, *L/W* grain length-to-width ratio, *PL* panicle length, *TPC* total protein content (Wang et al. 2016)

signifies a greater probability of the presence of a trait in the genomic region, whereas a large R^2 value accounts for a greater effect of QTL in terms of contribution to the phenotype.

Cumulative Genes

Because QTL effects to phenotype are determined by statistical methods only, the amount of contribution made by the individual gene effects cannot be verified through Mendelian methods. The genes embodied on QTLs are non-allelic and called polygenes or cumulative genes. The QTLs' effects are strongly influenced by environmental conditions. Unlike qualitative traits exhibiting discontinuous variations and forming distinct phenotypic classes with single gene effects, quantitative traits show continuous variation and give a measure of polygenic control through estimation of population parameters like the mean and standard deviation. Because of the complexity of the polygenic system of inheritance, identification and elucidation of QTLs for rice grain yield have been mired with mystery. In recent years using the database of the International Rice Genome Sequencing Project (IRGSP-2005) and http://www.gigasciencejournal.com/content/3/1/7,2014, much different rice QTLs have been identified and characterized. Information on rice QTLs is available in http://www.gramene.org/qtl. Altogether more than 80,000 genes of the rice genome have been identified in the IRGSP and each gene has an unknown number of alleles. The wide spectrum of genetic diversity provides a good choice for rice breeders for management rice genome for yield benefits under challenging environments.

QTLs for Yield Components

Grain yield in rice is a complex trait, multiplicatively determined three important components. The components of rice grain yield are panicle number, grain number, and weight per panicle (Yoshida 1983; Sakamoto and Matsuoka 2008) and they correlate positively with final yield (Akinwale et al. 2011). Grain yield has been considered one of the most important indexes in rice breeding. It is controlled by the expression of multiple quantitative trait loci (QTLs) subject to environmental changes (Fig. 7.4) (Bai et al. 2012; Seo et al. 2019). Wang et al. (2016) conducted genome-wide analyses of the association between SSR markers and seven yield traits and four-grain quality traits of 277 primal parental lines of hybrid rice and revealed a correlation of 29 loci with the 11 agronomic characters; one for plant height, two for tiller number, one for panicle length and the others for grain size and grain quality. In another review, the number of increasing yield genes has been listed to 27 (Xu et al. 2016b). Assessing a population of 386 recombinant sister inbred lines, Zhang et al. (2020) intended mapping and confirming QTLs for

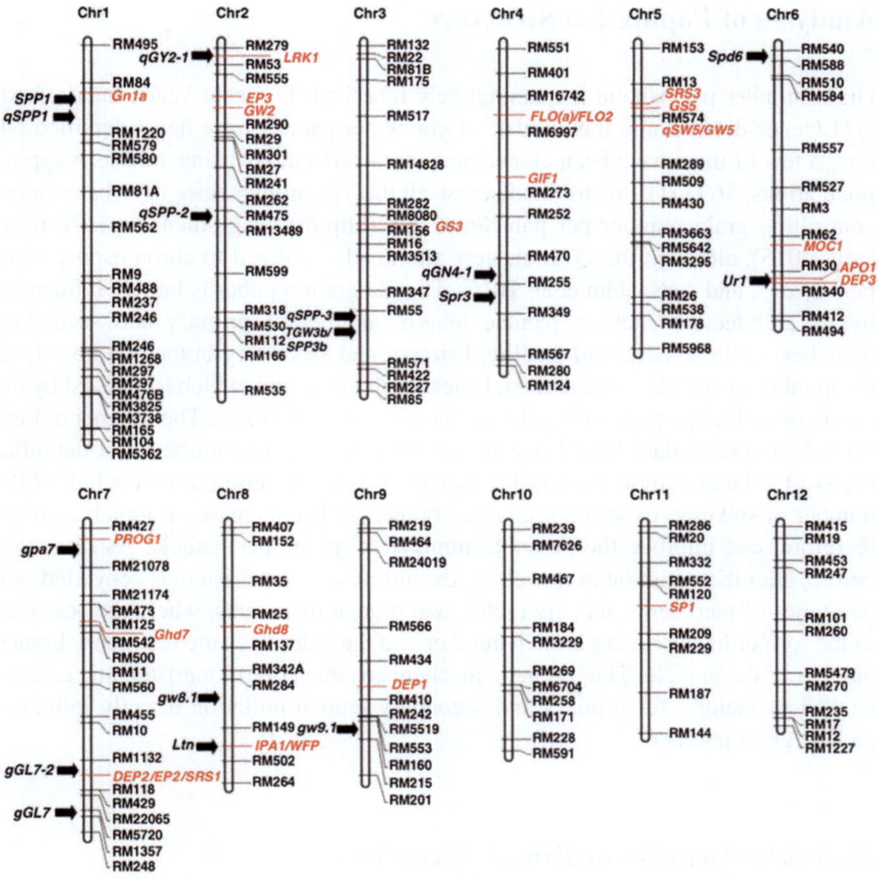

Fig. 7.4 The QTLs/genes on the genetic map in rice Chromosomes for yield potential (red color). (Adopted from Bai et al. 2012)

yield-related traits of rice and highlighted the significance of panicles per plant, grain numbers per panicle, and 1000-grain weight for overall grain yield; using whole genome re-sequencing the Chinese workers identified 43 yield-related QTLs distributed across all 12 chromosomes. Out of the three yields related traits, grain number per panicle exhibits the highest variation under prevailing environments. Thus, grains per panicle are the most important factor that determines the grain yield of rice, and an increase in grain number has been a hallmark of domestication. With the wealth of publicly available genomic resources in rice, QTL mapping experiments have been performed to identify association of markers for various yield-determining traits. The yield related traits are quantitative traits determined by many minor effect genes distributed on different chromosomes, which have been identified with SSR markers as given in review (Table 7.1) (Wang et al. 2016).

Analyses of Population Structure

Grain number per panicle is quantitatively inherited. In recent years, hundreds of QTL/genes determining the number of grains per panicle have been identified but only a few of them have been cloned and characterized. By using various mapping populations, 369 QTLs distributed across all the 12 chromosomes have been found controlling grain number per panicle of rice (http://www.gramene.org) (Noroozi et al. 2015), although the QTL clusters are mostly confined to chromosome numbers 1, 2, 3, and 4 (Neelam et al. 2016). Panicle grain number is largely influenced by its architecture such as panicle length, number of primary and secondary branches, grain density, and fertility. Primary and secondary branch number fixes the number of spikelets, the potential sites for grain setting, which is realized by the nature of assimilate partitioning during the post-anthesis phase. The number of both primary and secondary branches accounts for panicle grain numbers, but the influence of the latter is more discernible than the former because more than half of the number of spikelets exists on secondary branches. The increase of branch number, therefore, can improve the potential number of grains per panicle. As discussed before, after the initiation of flowering, the inflorescence meristem is converted into two types of meristems, namely rachis, and branch meristems, where the activities of the former fixes primary branch number and the latter fixes the secondary branch number of the panicle. Thus, genetic mechanisms that control meristematic activity as well as identity for primary and secondary branch initiation directly influence panicle grain number.

Molecular Genetics of Panicle Meristems

The molecular genetic framework that is crucial for meristem activity and identity has been explored by isolating mutants with defective panicle architecture in addition to population mapping (Wang and Li 2005; Furutani et al. 2006). The QTL mapping for the primary branch number of courses has been explored more than that of the secondary branches. Yamagishi et al. (2002) identified five and three *QTLs* responsible for the numbers of primary branches and secondary branches per panicle respectively. In contrast, Fujita et al. (2009) used several introgression lines to map chromosomal regions associated with important agronomic traits. In the process, six QTLs were identified for total spikelet number, five on Chromosome 4, namely *qTSN4.1*- *qTSN4.5* (Fujita et al. 2012) and one on chromosome 7, called *qTSN7.1* (Koide et al. 2013). They visualized that some more QTLs for total spikelet numbers could be present on other chromosomes. In recent work, Sasaki et al. (2017) detected two QTLs, namely *qTSN12.1* and *qTSN12.2* in the long arm of chromosome 12, overexpression of which in the panicle formation stage increases both primary and secondary branch number for high spikelet number. Singh et al. (2018) have separated genetic control of spikelet number from grain number. By using biparental recombinant inbred lines populations of NPT *indica* rice genotype

Pusa 126, they reported mapping of a major QTL for grain number per panicle, i.e. *qGN4.1*. In related work, they revealed the presence of a major QTL for spikelet number in the same chromosomal region by three different names, such as *SPIKE*, *GPS*, or *LSCHL4*, and have implicated overexpression of the *Nal1* gene as the causal factor for the generation of high spikelet number.

Regulation of Panicle Architecture Using Genetic Mutants

APO, DEP, TAW, and LAX Genes

In addition to the population structure analyses, using various sub-groups of *indica* and *japonica* rice, several workers have cloned yield-related traits. It was found that rice *apo1* mutant exhibits precocious transition of inflorescence meristem to spikelet meristem identity and possesses a smaller number of primary branches and spikelets (Ikeda et al. 2005). The gain of function causes high *APO1* mRNA expression for an F-box protein increasing numbers of both branches and grains per panicle and finally grain yield (Terao et al. 2010). *APO1*positively regulates spikelet number by closing precocious conversion of inflorescence meristem into spikelet meristem. Mutants with a greater expression level of *APO1* produce more spikelets through an increase of cell proliferation in the meristem. But an over-expression results in the development of abnormal panicles (Ikeda et al. 2007) because of additional cell proliferation in the meristem (Huang et al. 2009). Similarly, the expression of another gene *DEP1* (*DENSE AND ERECT PANICLE 1*) increases rachis branch number and grain number (Huang et al. 2009). This dominant allele is a gain of function mutation, where it encodes a G-protein γ-subunit that regulates erect panicle, grain number per panicle, nitrogen uptake, and stress tolerance through the G-protein signal pathway (Xu et al. 2016a) (Table 7.2). The gene enhances the spikelet number of panicle by increasing meristem activity and cell proliferation at the cost of reduced internode length of the inflorescence. Xu et al. (2016a) stated that the erect panicle architecture controlled by *DEP1* has been exploited in rice

Table 7.2 Change in morphological features of rice panicle caused by *dep1* allele

Genetic background	Subspecies	Grain number		Primary branch		Secondary branch		Yield increasing
		DEP1	dep1	DEP1	dep1	DEP1	dep1	
ZF802	*Indica*	60	290	–	–	–	–	–
Shao314	*Japonica*	130	251	11	16	18	49	40.90%
Zhonghua 11	*Japonica*	104	150	–	–	–	–	–
RIL-ipa1	*Japonica*	–	–	15	15	63	60	Not significant
Toyonishiki	*Japonica*	108	146	11	13	16	25	10.16%
R6547	*Indica*	303	288	17	17	62	57	−17.20%

Adopted from Xu et al. (2016a)

breeding for over a century for its high yield, lodging tolerance with strong stem, reasonable population structure, and nitrogen use efficiency. With variation to *DEP1*, Li et al. (2010) identified a *dense and erect panicle 2* (*dep2*) mutant that exhibited dense and erect panicle phenotype. *DEP2* is highly expressed in young panicles, which controls panicle erectness. Its mutation restricts elongation of branches but not primordium initiation resulting in a compact arrangement of the organs. The mutation induces a defect in cell proliferation to reduce panicle elongation. The mutation did not cause any difference in grain production between the wild type and *dep2* mutant. However, the alleles regulating grain number are later found different from that of branch number (Terao et al. 2010). Yoshida et al. (2013) identified *TAWAWA1 (TAW1)* gene encoding a nuclear protein showing high expression both in the inflorescence and branch meristems of rice during meristem phase transition. The gene regulates spikelet number by suppression of the transition from branch meristem to spikelet meristem. Thereby *TAW1* promotes branch meristem activity at the cost of change over to floral meristem identity. In the dominant gain of function extension of inflorescence meristem activity was prolonged and spikelet meristem specification was delayed increasing branch as well as spikelet number. In contrast, suppression of *TAW1* led to formation of a small inflorescence with less number of branches. Komatsu et al. (2001) also reported that *LAX1* (*LAX PANICLE1*) gene is necessary for the initiation and activity of lateral meristems of the rachis for development of primary branches, secondary branches and lateral spikelets of panicle. Mutant allele lax1 severely restricts initiation and/or development of rachis branches, lateral spikelets and terminal spikelets. *LAX* encodes a basic helix-loop-helix (bHLH) transcription factor that expresses between the summit of the shoot apex and region of the new meristem formation. This factor is necessary for formation of all types of lateral meristems of rice panicle. The loss of function of gene in *lax* mutants', *lax1* exhibits defects in formation of branches and spikelets by abrogating initiation of axillary meristem initiation. It has been shown that meristematic cells are specified in *lax1*, but they fail to develop axillary meristem (Oikawa and Kyozuka 2009). It has been shown that the transcription factor gene *RFL* promotes panicle branching by modulating expression of *LAX1* gene (Deshpande et al. 2015). Further it was found that *SPA* (*SMALL PANICLE*) gene like *LAX1* is another main regulator of rice panicle architecture (Komatsu et al. 2003a). Similar to *lax1*, branches and spikelets of the panicle are severely reduced in *spa* mutant.

FZP Gene

Komatsu et al. (2001; 2003b) reported that both *FRIZZY PANICLE2 (FZP2)* and *LAX1* genes determine panicle architecture by controlling rachis branch and spikelet development in rice through functions in genetically independent pathways, which could be additive. According to the rice database, *Oryzabase* gene details (21 April 2019) frizzy plants produce no seeds being propagated as heterozygotes. *FZP* is located on chromosome 7 encodes a potential transcription factor with an ethylene-responsive element- binding factor (ERF) which establishes floral meristem

Table 7.3 Morphological variation of different attributes of the mutant *fzp-11*, wild type (Dj), and *FZP*-over-expressing OX-*FZP*-(*fzp-11*) rice plants

Traits	*fzp-11* (M ± SD)	Wt (DJ) (M ± SD)	OX-FZP-(*fzp-11*) (M ± SD)
No. of roots	51 ± 15	48 ± 13	34 ± 11**
Panicle length (cm)	18 ± 3	18.3 ± 2.7	8.5 ± 1.1**
No. of primary branch	9.9 ± 2.2	10.3 ± 2.0	9.8 ± 2.1
No. of secondary branch	38 ± 9.1	23.3 ± 5.6	5.2 ± 2.7**
Spikelets per panicle	0	100 ± 6	53 ± 19**

Adopted from Bai et al. (2016)

** Significantly different between the plants of OX-*FZP* (*fzp*-11) and wild-type (DJ), and between OX-*FZP* (*fzp*-11) and *fzp*-11 at $P < 0.01$ (Student's *t*-test)

identity by suppression of axillary meristem formation within the spikelet meristem. There are several mutant alleles of *FZP*, the most recent one identified is *fzp12* (Ren et al. 2018). The loss of function mutation of *FZP* promotes the formation of supernumerary axillary meristems. Thus, *FZP* represses panicle branching and positively influences floral meristem identity in the reproductive stage to promote spikelet number. The *fzp* mutation replaces the formation of florets with branches as spikelet meristem produces extra rudimentary glumes with axillary buds (Xing and Zhang 2010). Bai et al. (2016) compared phenotypes of spikelets and floral organs in mutant *fzp-11*, wild type, and *FZP*-overexpressing plants. The vegetative features and number of primary branches of panicle did not change between the wild and *fzp-11* mutant, while the number of spikelets was nil and secondary branches were higher significantly in the former compared to the latter. Over-expression of *FZP* produced a miniature plant bearing a small panicle with less number of spikelets and secondary branches (Table 7.3). The results obtained identified *FZP* as a negative regulator of *RFL/APO2* and is responsible for transition of panicle branching to spikelet formation. The over-expression of *FZP* limits meristematic activities that not only reduce vegetative growth but also restrict secondary branch and spikelet number significantly. *FZP* overexpression regulates the expression of *OsMADS* box class B *AGL6* (*OsMADS6*, *OsMADS17*) and class E (*OsMADS1*, *OsMADS7*, *OsMADS 8*) genes. Most recently, Fujishiro et al. (2018) have shown that the *FZP* allele associated with the quantitative locus *qSrn7* increases higher-order branching in the upper part of the panicle and final grain yield.

Cytokinin Oxidase/Dehydrogenase Gene

Ashikari et al. (2005) cloned a major QTL *GN1a (GRAIN NUMBER1a)*, located on chromosome 1 that encodes cytokinin oxidase/dehydrogenase (OsCKX2), the enzyme responsible for the degradation of cytokinin. Loss of function of *OsCKX2* expression increases cytokinin accumulation in inflorescence meristems for promotion of grain number and yield. This QTL increases grain number by increasing the number of secondary branches of the panicle, but it also has an adverse effect in impacting grain ripening (Nagata et al. 2002). In contrast, *lonely guy* (*log*) mutation has a defect in cytokinin synthesis and produces a panicle much smaller than the wild type. Similar to *Gn1a*, Li et al. (2011) found two allelic larger panicle-(*lp-1*, *lp-2*) mutants, where the mutant panicle possesses more primary branches and more grains with increased grain weight compared to the wild type. The *LARGER PANICLE (LP)* gene encodes a kelch repeat-containing F-box protein localized in the endoplasmic reticulum-associated with protein degradation. In the mutants, cytokinin oxidase/dehydrogenase expression is down-regulated for benefit of the panicle assuming that LP might be involved in modulating cytokinin level in the branch primordial region. In contrast to this type of gene regulation improving panicle size, overexpression of *OsDof12* leads to smaller panicle size of rice with reduced numbers of primary and secondary branches, and such effect has resulted from hyposensitivity of brassinosteroids (Wu et al. 2015). *Dof* stands for DNA binding with one finger, which is a transcription factor protein.

Wealthy Farmer's Panicle

Similarly, Miura et al. (2010) showed that the QTL *WFP* (*WEALTHY FARMERS PANICLE*) encodes OsSPL14 (SQUAMOSA PROMOTER LIKE-14) protein, and its higher expression during the reproductive phase of rice increases panicle branching and grain yield. *SPL* genes represent a family of plant-specific transcription factors that control various aspects of development. In the rice genome, there are 19 *SPL* genes, of which *OsSPL14* is predominantly expressed in the shoot apex during both vegetative and reproductive stages controlling meristematic activity for organ development. Similar to *WFP*, *OsSPL14* is also known as *IPA1(IDEAL PLANT ARCHITECTURE1)*; it is responsible for bearing phenotypes accommodating gain in yield potential of new plant type rice over the semi-dwarf high yielding rice (Jiao et al. 2010). It was found that increased expression of *IPA1* in the panicle conferred ideal panicle architecture in rice embodying fewer tillers, stronger culms larger panicles, and higher grain weight. Wang et al. (2017) identified an IPA1 interacting ring finger E3 ligase protein (IPI1) that promotes IPA1 degradation in panicles but increases its stability in the shoot apex. The regulatory role of *IPA1* is antagonistic to that of *IPI1*, where *IPI1* controls tiller number and panicle size by modulating IPA1 protein levels in a tissue- specific manner (Wang and Wang 2017). Additionally, *OsSPL14*-overexpression transgenic rice plants confers higher ABA and gibberellin acid 3 (GA3) levels at seedling and tillering stages (Lian et al. 2020). Hashida et al. (2013) suggested that a gene encoding sucrose phosphate synthase *OsSPS1* introduced into japonica rice Koshihikari from *indica* rice Kasalath improved spikelet number by 38–47% in transgenic rice because the higher activity of the enzyme increased sucrose synthesis of source leaf and improved partitioning of dry matter for panicle growth that increased secondary branch number. More recently, Chen et al. (2015) showed an evolutionarily conserved gene *FUWA* controls panicle architecture and grain shape and weight in rice; the gene encodes an NHL domain protein that interferes with excessive cell division of roots, shoot and inflorescence meristems and knockdown of the gene expression with RNA interference makes the panicle erect accommodating larger size grains. Some of the QTLs have pleiotropic effects with other characters, such as *Ghd7* not only controls grain number but also heading date and plant height.

Genes Controlling Panicle Length

Panicle length is considered an important yield-related trait for rice. Panicle length is associated with seed setting rate, spikelet number and density, and grain plumpness that ultimately reflect upon grain number and final yield. Panicle length is inherited quantitatively and regulated by both major and minor QTLs and they are commonly associated with the expression of QTLs for spikelet density (Liu et al. 2011) and heading dates, such as *Hd1, EHD4, Hd6, DTH7*, and *DTH8*. Approximately the presence of 253 QTLs for panicle length has been detected in all 12 chromosomes of rice. Liu et al. (2016) delimited *LONG PANICLE1* (*LP1*) locus to a 90 kb region on the long arm of chromosome 9 with help of linkage and association mapping and predicted it to be a major regulator of panicle length in rice. Spikelet density is lax and panicle is longer in *indica* subspecies rice compared to japonica subspecies. The genetic variation in panicle length within the former is greater than the latter. Inheritance of panicle length is not simply being determined quantitatively, for which there are several major and minor QTLs. Li et al. (2009) identified a novel rice mutant, *short panicle 1* (*sp1*); the mutant bears a short panicle because of defects in branch elongation, but does not affect floral meristem initiation or formation. *SP1* gene is highly expressed in the phloem of panicle branches and encodes a peptide transporter protein SP1 with 12 transmembrane domains localized in the plasma membrane.

QTLs for Panicles Per Plant

In rice panicle number per plant is considered an important agronomic trait for the determination of final grain yield. Shoot architecture exhibiting the number of tillers per plant, tiller angle, and plant height determine panicle number. High tillering is desired for successful competition with weeds and an increase in the number of effective tillers for better yield. However, there is a negative correlation between panicle number and spikelet number per panicle. Because primary branching from the main rachis and secondary branching of primary branches and spikelet formation on both types of branches determine grain number and yield per panicle, more attention has been given to the study of the molecular biology of panicle branching and tillering in rice recently (Liang et al. 2014; Bai et al. 2016). Xu et al. (2001) mapped two panicle number QTLs close to the related traits for spikelet number per panicle. Phenotyping for effective tillers of the rice plant is usually done at the heading stage (Duan et al. 2015). The identification of genes/QTLs controlling panicle number has an important bearing in breeding for high rice yield. Zhu et al. (2011)

reported a major QTL *qPN1* location on the long arm of chromosome 1 and expression of this QTL was negatively related to panicle number. *qPN1* expression also affects panicle length, grain number, plant height, and yield as well.

QTLs for Tiller Number

There is no consensus on the number of QTLs for effective tillers, however. Using a set of recombinant inbred lines, Pinson and Jia (2016) mapped five QTLs for early tiller production of which only three had a stronger association with effective panicle number. They are considered panicle number QTLs as well, and markers linked to them can be used for marker-assisted selections in breeding programs. According to Xing and Zhang (2010), there are 21 distinct QTLs in rice for the number of tillers per plant and they are located on chromosome 7. With composite interval mapping, Nagabhusana et al. (2006a, b) detected four QTLs for tiller numbers residing each on chromosome numbers 3, 6, 8, and 10. Arite et al. (2007) identified the *D10* (*DWARF10*) gene that is involved in the synthesis of an inhibitor for shoot branching of rice. Another gene *IDEAL PLANT ARCHITECTURE1* (*IPA1*), was found regulating both tiller and panicle branch number (Lu et al. 2013). *IPA1* suppresses tiller bud outgrowth by directly interacting with the promoter of *TEOSINITE BRANCHED1* (*TB1*). *OsTB1* in rice is a negative regulator for tiller development and has been a direct target for positive regulation of *IPA1* action. *OsTB1* expresses itself in axillary buds, deficiency of which produces a semi-dwarf plant with a large number of tillers. In contrast, over-expression of *OsTB1* in *ipa1* plants could produce a reduced tiller phenotype. More recently single recessive mutant *lt1*of rice with few tillers and panicle branches was identified (Yu et al. 2017), where transcript levels of tillering inhibitor genes *D3 (DWARF3)*, *D10*, *HTD1 (HIGH TILLERING DWARF1),* and *HTD2 (HIGH TILLERING DWARF2)* increased considerably to the detriment of tiller production. On the other hand, the relative expressions of positive regulators of tillering genes like *MOC1*, *MOC2*, and *MOC3* decreased significantly.

QTLs for Grain Weight

An increase in grain weight of matured kernels is a method of improving rice yield (Matsushima and Tsunoda 1957). Grain weight depends on grain size, which is fixed by the hull space between lemma and palea, which is a stable genetic character (Matsushima 1970). It has been used in rice to distinguish the population structure of cultivars with enormous biodiversity. Grain size is defined by its three dimensions, length, breadth, and thickness. These three traits are quantitatively inherited by several genes. The combination of expression of genes controlling various seed characteristics finally determines seed size. The cultivar diversity of grain weight and size in rice is very wide. In general, wild-type alleles have small size grains in comparison to cultivar alleles. Within the cultivated species, there is a large diversity of grain size between the sub-populations. The varietal stability of the constitution precludes genetic manipulation of grain size. In recent years, QTL mappings have revealed the association of several yield-related traits including grain weight in one part of the genome indicating linkage or pleiotropic effects of genes. The complimentary linkage of beneficial traits in association with heritability attribute produces higher yields and allows simultaneous improvement of the traits by using molecular markers. Similar to weight, grain quality traits are also yield-related traits inherited quantitatively and some of them are pleiotropically related to other agronomic characters.

Population structure analyses using SSR markers have identified several loci, three for grain length, eight for grain width, eight for grain length/width ratio, one for thousand-grain weight, one for amylose content, two for gel consistency, one for gelatinization temperature, and one for total protein content (Wang et al. 2016). Xing and Zhang (2010) identified 24 distinct grain weight QTLs and found them located on chromosomes 1, 3, 5, and 7. Luo et al. (2013) identified the existence of strong linkage of 2 QTLs, *qSPP5* for spikelet number per panicle and *qTGW5* for 1000 grain weight on chromosome number 5 of rice; the allele for an increase of SPP however, reduced TGW. Several other genes are having pleiotropic effects. *GS3* is the major effect controller QTL for grain weight and length, but minor effect controller QTL for grain width and plumpness (Mao et al. 2010; Wang et al. 2011). *GS3* gene has been delimited to a DNA fragment approximately of 7.9 bp in length that has five exons and encodes 232 amino acids for a PEBP domain in the N-terminus, a trans-membrane region and a TNFR/NGFR family cysteine rich domain in the middle part and a VWFC module in the C-terminus (Fan et al. 2006). *GS3* functions as a negative regulator for grain size, a mutation in the second exon causing truncation of 178 amino acids in the C-terminus area, increases grain size in large grain varieties. Xu et al. (2016b) have observed a close association of *GS3* marker with grain length and width in widely distributed rice cultivars. *GW2* gene determining grain weight and width is in association with panicle number, heading date whereas *GL7* controls both grain length and width (Wang et al. 2015). *GL7* encodes a LONGIFOLIA type protein, homologous to *Arabidopsis*, up-regulation in expression of which, leads to elongation of grain length and changes in gene expression levels at this locus contributes to grain size diversity (Wang et al. 2015).

QTLs for Grain Size

Genes/QTLs controlling different aspects of grain size have been cloned and characterized by several workers. Xia et al. (2017) found three QTLs *qGW1*, *qGS7* and *qGS3* have large effects on grain size, and where the first two regulate grain size and the last one determines grain length. Besides, Xu et al. (2015) reported that differential expression of *GS5* regulates the grain size of rice. Song et al. (2007) identified *GW2* QTL encoding Ring type protein with E3 ubiquitin ligase activity for controlling kernel size and weight. *GW2* negatively controls cell division by targeting substrates to the proteasome for proteolysis and loss of its function augurs increase of cell division for widening spikelet hull capacity and kernel size. Ubiquitin-associated domain protein, encoded by *HGW* gene is found in cytoplasm and nucleus; and it functions as a key upstream regulator to promote expression of heading date and grain weight-related genes (Li et al. 2012). Similarly, Shomura et al. (2008) cloned a QTL *qSW5* on chromosome number 5 deletion of which results in enhanced cell division in the outer glumes covering the hull cavity for accommodating a larger size kernel. On the contrary, *GLW7* QTL encodes a transcription factor OsSPL13 high expression of which increases cell size of the hull and grain length (Si et al. 2016). The large grain allele *GLW7* in tropical japonica might have been introgressed from *indica* cultivars during artificial selection. *GW8* is another quantitative trait locus, synonymous with *OsSPL16*, which encodes a positive regulator protein for cell proliferation; enhanced expression of the QTL increases grain width, and mutation in its promoter region makes the grain more slender for the betterment of quality (Wang et al. 2012). Ishimaru et al. (2013) reported that loss of function of IAA glucose hydrolase gene *TGW6* enhances grain weight and yield of rice. Grain filling is an important trait that contributes to the grain weight of rice. By screening for mutants with grain filling defects, Wang et al. (2008) reported the isolation and functional analyses of the *GIF1* (*Grain incomplete filling1*) gene that encodes a cell wall invertase required for carbon partitioning during the early grain filling stage. *GIF1* is necessary for the unloading of sucrose in the ovular and stylar vascular tissues for the starch synthesis of the endosperm. A *gif1* mutation shows slower grain filling and exhibits increased grain chalkiness on account of abnormally developed and loosely packed starch granules. The latest report on the molecular biology of grain size comes with the identification of a brassinosteroid-deficient mutant *ltbsg 1* showing longer top branches in panicle and shorter grain (Qin et al. 2018). In the wild *LTBSG1* located on chromosome 10 is constitutively expressed and the proteins are found in the chloroplast, cell membrane and nucleus. Conversely, in the mutant a nucleotide substitution at the exon 2 of *LOC_Os10g25780* caused amino acid variation (Glycine- Arginine) in the FAD-binding domain of delta (24)-sterol reductase, which is involved in brassinosteroid synthesis.

References

Adriani DE, Dingkuhn M, Dardou A, Adam H, Luquet D, Lafarge T (2016) Rice panicle plasticity in near isogenic lines carrying a QTL for larger panicle is genotype and environment dependent. Rice 9(1). https://doi.org/10.1186/s.12284-016-0101-x

Akinwale M, Gregorio G, Nwilene F, Akinyele B, Ogunbayo S, Odiyi A (2011) Heritability and correlation coefficient analysis for yield and its components in rice (Oryza sativa L.). Afr J Plant Sci 5:207–212

Arite T, Iwata H, Ohshima K, Maekawa M, Nakajim M, Kojiam M, Sakakibara H, Kyozuka J (2007) DWARF10, an RMS1/MAX4/DAD1 ortholog, controls lateral bud outgrowth in rice. Plant J 51:1019–1029

Ashikari M, Sakakibara H, Lin S, Yamamoto T, Takeshi T, Nishimura A, Angeles ER, Qian Q, Kitano H, Matsuoka M (2005) Cytokinin oxidase regulates rice grain production. Science 309:741–745

Bai X, Wu B, Xing Y (2012) Yield-related QTLs and their applications in rice genetic improvement. J Integr Plant Biol 54:300–311

Bai X, Huang Y, Mao D, Wen M, Zhang L, Xing Y (2016) Regulatory role of FZP in the determination of panicle branching and spikelet formation in rice. Sci Rep 6:19022. https://doi.org/10.1038/srep19022

Chen J, Gao H, Zheng X-M, Jin M, Weng JF, Ma J, Ren Y, Zhou K, Wang Q, Wang J, Gao H, Zheng XM (2015) An evolutionarily conserved gene, FUWA, plays a role in determining panicle architecture, grain shape and grain weight in rice. Plant J 83:427–438

Crowell S, Falcão AX, Shah A, Wilson Z, Greenberg AJ, Mccouch SR (2014) High-resolution fluorescence phenotyping using a novel mage-analysis pipeline, PANorama. Plant Physiol 165:479–495

Deshpande GM, Ramakrishna K, Chongloi GL, Vijayraghavan U (2015) Functions for rice RFL in vegetative axillary meristem specification and out growth. J Exp Bot 66:2773–2784

Duan L, Huang C, Chen G, Xiong L, Liu Q, Yang W (2015) Determination of rice panicle numbers during heading by multi-angle imaging. Crop J 3:211–219

Fan C, Xing Y, Mao H, Xu C, Li X, Zhang Q (2006) *GS3*, a major QTL for grain length and weight and minor QTL for grain width and thickness in rice, encodes a putative trans-membrane protein. Theor Appl Genet 112:1164–1171

FRA P (2007) Guidelines for the conduct of test for distinctiveness, uniformity and stability on (*Oryza sativa* L.). Plant Var J India 1

Fujishiro Y, Agata A, Ota S, Ishihara R, Takeda Y, Kunishima T, Ikeda M, Kyozuka J, Hobo T, Kitano H (2018) Comprehensive panicle phenotyping reveals that qSrn7/FZP influences higher-order branching. Sci Rep 8:12511. https://doi.org/10.1038/s41598-018-30395-9

Fujita D, Santos RE, Ebron LA, Telebanco-Yanoria MJ, Kato H, Kobayashi S, Uga Y, Araki E, Takai T, Tsunematsu H, Imbe T, Khush GS, Brar DS, Fukuta Y, Kobayashi N (2009) Development of introgression lines of an Indica-type rice variety, IR64, for unique agronomic traits and detection of the responsible chromosomal regions. Field Crop Res 114:244–254

Fujita D, Tagle AG, Ebron LA, Fukuta Y, Kobayashi N (2012) Characterization of near-isogenic lines carrying QTL for high spikelet number with the genetic background of an indica rice variety IR64 (Oryza sativa L.). Breed Sci 62:18–26

Furutani I, Sukegawa S, Kyozuka J (2006) Genome-wide analysis of spatial and temporal gene expression in rice panicle development. Plant J 46:503–511

Hashida Y, Aoki N, Kawanishi H, Okamura M, Ebitani T, Hirose T, Yamagishi T, Ohsugi R (2013) A near isogenic line of rice carrying chromosome segments containing OsSPS1 of Kasalath in the genetic background of Koshihikari produces an increased spikelet number per panicle. Field Crop Res 149:56–62

Hoshikawa K (1989) The growing rice plant. An anatomical monograph. Nosan Gyoson Bunka Kyokai:199–205

Huang X, Qian Q, Liu Z, Sun Z, He S, Luo G, Xia CC, Fu X (2009) Natural variation at the DEP1 locus enhances grain yield in rice. Nat Genet 41:494–497

Ikeda K, Nagasawa N, Nagato Y (2005) ABERRANT PANICLE ORGANIZATION 1 temporally regulates meristem identity in rice. Dev Biol 282:349–360

Ikeda K, Ito M, Nagasawa N, Kyozuka J, Nagato Y (2007) Rice ABERRANT PANICLE ORGANIZATION 1, encoding an F-box protein, regulates meristem fate. Plant J 51:1030–1040

Ikeda-Kawakatsu K, Yasuno N, Oikawa T, Iida S (2009) Expression level of ABERRANT PANICLE ORGANISATION1 determines rice inflorescence form through control of cell proliferation in the meristem. Plant Physiol 150:736–747

Ishimaru K, Hirotsu N, Madoka Y, Murakami N, Hara N, Onodera H, Keshiwagi T, Ujiie K, Shimizu B-I, Onishi A (2013) Loss of function of the IAA-glucose hydrolase gene TGW6 enhances rice grain weight and increases yield. Nat Genet 45:707–711

Jackson MT, Lettington RJL (2003) Conservation and use of rice germplasm, an evolving paradigm under the international treaty on plant genetic resources for food and agriculture. In: Sustainable rice production for food security. Proceedings of the 20th session of the international Rice Commission. Bangkok, Thailand 23–26 July 2002, pp 75–88

Jiao Y, Wang Y, Xue D, Wang J, Yan M, Liu G, Dong G, Zeng D, Lu Z, Zhu X, Qian Q, Li J (2010) Regulation of OsSPL14 by OsmiR156 defines ideal plant architecture in rice. Nat Genet 42:541–544

Kato T (2017) Inheritance of a novel panicle type, 'short tip panicle', found in a rice cultivar Glimbozu. J Crop Res 62:41–45

Koide Y, Fujita D, Tagle AG, Sasaki K, Ishimaru T, Fukuta Y, Kobayashi N (2013) QTL for spikelet number from a high-yielding rice variety, Hoshiaoba, detected in an introgression line with the genetic background of an indica rice variety, IR64. Euphytica 192:97–106

Komatsu M, Maekawa M, Shimamoto K, Kyozuka J (2001) The LAX1 and FRIZZY PANICLE 2 genes determine the inflorescence architecture of rice by controlling rachis-branch and spikelet development. Dev Biol 231:364–373

Komatsu K, Maekawa M, Ujiie S et al (2003a) LAX and SPA: major regulators of shoot branching in rice. Proc Natl Acad Sci U S A 100:11765–11770

Komatsu M, Chujo A, Nagato Y, Shimamoto K, Kyozuka J (2003b) FRIZZY PANICLE is required to prevent the formation of axillary meristems and to establish floral meristem identity in rice spikelets. Development 130:3841–3850

Kyozuka J, Tokunaga H, Yoshida A (2014) Control of grass inflorescence form by the fine-tuning of meristem phase change. Curr Opin Plant Biol 17:110–115

Li S, Qian Q, Fu Z, Zeng D, Meng X, Kyozuka J, Maekawa M, Zhu X, Zhang J, Li J, Wang Y (2009) Short panicle1 encodes a putative PTR family transporter and determines rice panicle size. Plant J 58:592–605

Li F, Liu W, Tang J, Chen J, Tong H, Hu B, Li C, Fang J, Chen M, Chu C (2010) Rice DENSE AND ERECT PANICLE 2 is essential for determining panicle outgrowth and elongation. Cell Res 20:838–849

Li M, Tang D, Wang K, Wu X, Lu L, Yu H, Gu M, Yan C (2011) Mutations in the F-box gene LARGER PANICLE improve the panicle architecture and enhance the grain yield in rice. Plant Biotechnol J 9:1002–1013

Li J, Chu H, Zhang Y, Mou T, Wu C, Zhang Q, Wu J (2012) The rice HGW gene encodes a ubiquitin-associated (UBA) domain protein that regulates heading date and grain weight. PLoS One 7:e34231. https://doi.org/10.1371/journal.pone.0034231

Lian L, Xu H, Zhang H, He W, Cai Q, Lin Y, Wei L, Pan L, Xie X, Zheng Y, Wei Y, Zhu Y, Xie H, Zhang J (2020) Overexpression of *OsSPL14* results in transcriptome and physiology changes in indica rice 'MH86'. Plant Growth Regul 90:265–278

Liang W-H, Shang F, Lin Q-T, Lou C, Zhang J (2014) Tillering and panicle branching genes in rice. Gene 537:1–5

Liu T, Li L, Zhang Y, Xu C, Li X, Xing Y (2011) Comparison of quantitative trait loci for rice yield, panicle length and spikelet density across three connected populations. J Genet 90:377–382. https://doi.org/10.1007/s12041-011-0083-9

Liu E, Liu Y, Wu G, Zeng S, Thu GTT, Liang L, Liang Y, Dong Z, She D, Wang H, Zaid IU, Hong D (2016) Identification of a candidate gene for panicle length in rice (Oryza sativa L.) via association and linkage analysis. Fron Plant Sci 7:596. https://doi.org/10.3389/fpls.2016.00596

Lu Z, Yu H, Xiong G, Wang J, Jiao Y, Liu G, Jing Y, Meng X, Hu X, Qian Q, Fu X, Wang Y, Li J (2013) Genome-wide binding analysis of the transcription activator IDEAL PLANT ARCHITECTURE1 reveals a complex network regulating rice plant architecture. Plant Cell 25:3743–3759

Luo X, Ji SD, Yuan PR, Lee HS, Kim DM, Balkunde S, Kang JW, Ahn SN (2013) QTL mapping reveals a tight linkage between QTLs for grain weight and panicle spikelet number in rice. Rice (New York) 6:Article number 33

Mao H, Sun S, Yao J, Wang C, Yu S, Xu C, Liu X, Zhang Q (2010) Linking differential domain functions of the GS3 protein to natural variation of grain size in rice. Proc Natl Acad Sci U S A 107:19579–19584

Matsushima S (1970) Crop science in rice. Fuji Publishing Co Ltd., Tokyo, 365 pp

Matsushima S, Tsunoda K (1957) Analysis of developmental factors determining yield and yield prediction in lowland rice: XLII. On the prediction of the number of spikelets per panicle and that of spikelets per hill. Jpn J Crop Sci 26:87–88

Miura K, Ikeda M, Matsubara A, Song X-J, Ito M, Asano K, Matsuoka M, Ktano H, Ashikari M (2010) *OsSPL14* promotes panicle branching and higher grain productivity in rice. Nat Genet 42:545–549. https://doi.org/10.1038/ng.592

Nagabhusana K, Vaishali MG, Hittalmani S (2006a) Temporal expression of QTLs for plant height and tiller number in rice (*Oryza sativa* L.). Indian J Genet 66:177–181

Nagabhusana K, Mane SP, Hittalmani S (2006b) Comparative studies on QTL mapping by simple interval mapping and composite interval mapping models for selected growth and yield traits in rice (*Oryza sativa* L.). Indian J Crop Sci 1:97–101

Nagata K, Fukuta Y, Shimizu H, Yagi T, Terao T (2002) Quantitative trait loci for sink size and ripening traits in rice (Oryza sativa L.). Breed Sci 52:259–273

Nakagawa M, Shimamoto K, Kyozuka J (2002) Overexpression of RCN1 and RCN2, rice TERMINAL FLOWER 1/CENTRORADIALIS homologs, confers delay of phase transition and altered panicle morphology in rice. Plant J 29:743–750

Neelam K, Kumar K, Dhaliwal HS, Singh K (2016) Introgression and exploitation of QTL for yield and yield components from related wild species in rice cultivars. In: Rajpal VR, Rao SR, Raina SN (eds) Molecular breeding for sustainable crop improvement, vol 2. Springer, Cham, pp 171–202. https://doi.org/10.1007/978-3-319-27090-6

Noroozi M, Sattari A, A-fakheri B (2015) Detection of QTLs on panicle and grain number in rice (*Oryza sativa* L.): a review. Int J Farm Allied Sci 4:314–317

Oikawa T, Kyozuka J (2009) Two-step regulation of LAX PANICLE1 protein accumulation in axillary meristem formation in rice. Plant Cell 21:1095–1108

Pinson SRM, Jia Y (2016) QTLs for early tiller production and relationships with rapid seedling growth and increased panicle number in rice. Crop Sci 56:505–519

Qin R, Zeng D, Yang C, Akhter D, Alamin M, Jin X, Shi C (2018) LTBSG1, a new allele of BRD2, regulates panicle and grain development in rice by brassinosteroid biosynthetic pathway. Genes 9:292. https://doi.org/10.3390/genes9060292

Ren D, Hu J, Xu Q, Cui Y, Zhang Y, Zhou T, Rao Y, Xue D, Zeng G, Zhang G, Gao Z, Zhu L, Shen L, Chen G, Guo L, Qian Q (2018) *FZP* determines grain size and sterile lemma fate in rice. J Exp Bot 69:4853–4866

Sakamoto T, Matsuoka M (2008) Identifying and exploiting grain yield genes in rice. Curr Opin Plant Biol 11:209–214

Sasaki K, Fujita D, Koide Y, Lumanglas PD, Gannaban RB, Tagle AG, Obara M, Fukuta Y, Kobayashi N, Ishimaru T (2017) Fine mapping of a quantitative trait locus for spikelet number per panicle in a new plant type rice and evaluation of a near-isogenic line for grain productivity. J Exp Bot 68:2693–2702

Seo J, Lee S-M, Han J-H, Shin N-H, Koh H-J, Chin JH (2019) Identification of yield and yield-related quantitative trait loci for the field high temperature condition in backcross populations of rice (*Oryza sativa* L.). Plant Breed Biotechnol 7:415–426

Shomura A, Izawa T, Ebana K, Ebitani T, Kanegae H, Konishi S, Yano M (2008) Deletion in a gene associated with grain size increased yields during rice domestication. Nat Genet 40:1023–1028

Si L, Chen J, Huang X, Gong H, Luo J, Hou Q, Zhou T, Lu T, Zhu J, Shangguan Y, Chen E, Gong C, Zhao Q, Jing Y, Zhao Y, Li Y, Cui L, Fan D, Lu Y, Weng Q, Wang Y, Zhan Q, Liu K, Wei X, An K, An G, Han B (2016) OsSPL13 controls grain size in cultivated rice. Nat Genet 48:447–456

Singh VK, Ellur RK, Singh AK, Nagarajan M, Singh BD, Singh NK (2018) Effect of qGN4.1 QTL for grain number per panicle in genetic backgrounds of twelve different mega varieties of rice. Rice (New York) 11:8

Song X-J, Huang W, Shi M, Zhu M-Z, Lin H-X (2007) A QTL for rice grain width and weight encodes a previously unknown RING-type E3 ubiquitin ligase. Nat Genet 39:623–630

Terao T, Nagata K, Morino K, Hirose T (2010) A gene controlling the number of primary rachis branches also controls the vascular bundle formation and hence is responsible to increase the harvest index and grain yield in rice. Theor Appl Genet 120:875–893

Wang Y, Li J (2005) The plant architecture of rice (Oryza sativa). Plant Mol Biol 59:75–84

Wang B, Wang H (2017) IPA1: a new "Green revolution" gene? Mol Plant 10:779–781

Wang E, Wang J, Zhu X, Hao W, Wang L, Li Q et al (2008) Control of rice grain-filling and yield by a gene with a potential signature of domestication. Nat Genet 40:1370–1374
Wang C, Chen S, Yu S (2011) Functional markers developed from multiple loci in GS3 for fine marker-assisted selection of grain length in rice. Theor Appl Genet 122:905–913
Wang S, Wu K, Yuan Q, Liu X, Liu Z, Lin X, Zeng R, Zhu H, Dong G, Qian Q, Zhang G, Fu X (2012) Control of grain size, shape and quality by OsSPL16 in rice. Nat Genet 44:950–954
Wang Y, Xiong G, Hu J, Jiang L, Yu H, Xu J, Fang Y, Zeng L, Xu E, Xu J, Ye W, Meng X, Liu R, Chen H, Jing Y, Wang Y, Zhu X, Li J, Qian Q (2015) Copy number variation at the GL7 locus contributes to grain size diversity in rice. Nat Genet 47:944–948
Wang Y, Zheng Y, Cai Q, Liao C, Mao X, Xie H, Zhu Y, Lian L, Luo X, Xie H, Zhang J (2016) Population structure and association analysis of yield and grain quality traits in hybrid rice primal parental lines. Euphytica 212:261–273
Wang J, Yu H, Xiong G, Lu Z, Jiao Y, Meng X, Liu G, Chen X, Wang Y, Li J (2017) Tissue-specific ubiquitination by IPA1 INTERACTING PROTEIN 1 modulates IPA1 protein levels to regulate plant architecture in rice. Plant Cell. https://doi.org/10.1105/tpc.16.00879
Wu Q, Li D, Li D, Liu X, Zhao X, Li X, Li S, Zhu L (2015) Overexpression of *OsDof12* affects plant architecture in rice (*Oryza sativa* L.). Front Plant Sci 6:833. https://doi.org/10.3389/fpls.2015.00833
Xia D, Zhou H, Qiu L, Jiang H, Zhang Q, Gao G, He Y (2017) Mapping and verification of grain shape QTLs based on an advanced backcross population in rice. PLoS One 12(11):e0187553. https://doi.org/10.1371/journal.pone.0187553
Xing Y, Zhang Q (2010) Genetic and molecular bases of rice yield. Annu Rev Plant Biol 61:421–442
Xu J, Xue Q, Luo L, Li Z (2001) QTL dissection of panicle number per plant and spikelet number per panicle in rice (Oryza sativa L.). Yi chuan xue bao, Acta Genetica Sinica 28:752–759
Xu C, Liu Y, Li Y, Xu X, Xu C, Li X, Xiao J, Zhang Q (2015) Differentia expression of GS5 regulates grain size in rice. J Exp Bot 66:2611–2623
Xu H, Zhao M, Zhang Q, Xu Z, Xu Q (2016a) The DENSE AND ERECT PANICLE 1 (DEP1) gene offering the potential in the breeding of high-yielding rice. Breed Sci 66:659–667
Xu F, Jin L, Huang Y, Tong C, Chen Y, Bao J (2016b) Association mapping of quantitative trait loci for yield-related agronomic traits in rice (Oryza sativa L.). J Integr Agric 15:2192–2202
Yamagishi M, Takeuchi Y, Kono I, Yano M (2002) QTL analysis for panicle characteristics in temperate *japonica* rice. Euphytica 128:219–224
Yamaki S, Miyabayashi T, Eiguchi M, Kitano H, Nonomura K-I, Kurata N (2010) Diversity of panicle branching patterns in wild relatives of rice. Breed Sci 60:586–596
Yoshida S (1983) Rice. In: Smith WH, Banta SJ (eds) Potential productivity of field crops under different environments. International Rice Research Institute Los Baños Laguna, Los Baños, pp 103–127
Yoshida H, Nagato Y (2011) Flower development in rice. J Exp Bot 62:4719–4730
Yoshida A, Sasao M, Yasuno N, Tagaki K, Daimon Y, Chen R, Yamazaki R, Tokunaga H, Kitaguchi Y, Sato Y, Nagamura Y, Ushijima T, Kumamaru T, Iida S, Maekawa M, Kyozuka J (2013) TAWAWA1, a regulator of rice inflorescence architecture, functions through the suppression of meristem phase transition. Proc Natl Acad Sci U S A 110:767–772
Yu H, Qiu Z, Xu Q, Wang Z, Zeng D, Hu J, Zhang G, Zhu L, Gao Z, Chen G, Guo L, Qian Q, Ren D (2017) Fine mapping of LOW TILLER 1, a gene controlling tillering and panicle branching in rice. Plant Growth Regul 83:93–104
Zhang M, Zhou Z, Chen Y, Cao Y, Deng C, Xue P, Zhan X, Cheng S, Cao L, Zhang Y (2020) Finding new addictive QTL for yield traits based on a high-density genetic map in hybrid rice. Plant Growth Regul 93:105–115
Zhu J, Zhou Y, Liu Y, Wang Z, Tan Z, Yi C, Tang S, Gu M, Liang G (2011) Fine mapping of a major QTL controlling panicle number in rice. Mol Breed 27:171–180

Chapter 8
Change of Panicle Architecture During Domestication

Abstract The 12,000 years of rice antiquity embodies evolutionary history beholden with significant changes in panicle architecture. In the process of domestication, rice might have diffused into a gradation of habitats permitting phenotypic plasticity for changes in panicle shape as well as speciation by artificial selection. In tandem with evolution, lax and spreading type panicle shape changed into a compact type amenable for the accommodation of larger numbers of spikelets and higher grain yield. However, crowding of a large number of spikelets into crammed space of inflorescence in large panicle cultivars widened inter-grain apical dominance to the detriment of grain filling in the spikelets on the basal part. Panicle branches produced relatively more on the apical part of a panicle of newer cultivars tended to establish a cylindrical pattern in place of the triangular-shaped panicle of vintage rice, and thereby increased grain number and grain filling.

Keywords Evolution · Panicle shape · Lax orientation · Compact type

Domestication

Because of the economic considerations, panicle architecture has always been a major target for crop domestication and improvement. The 12000-year-old rice antiquity embodies evolutionary history beholden with the changes in cultivation, domestication, dispersal, and diversification. Khush (1997) suggested that the genus *Oryza* originated in Gondwana land and species differentiation occurred with the breakup of its landmass later. The wild species of the genus were the original inhabitants. They might have diffused into a gradation of habitats permitting phenotypic plasticity for speciation followed by artificial selection (Figs. 8.1 repeat of Fig. 1.1A). Chang (2003) proposed that the prehistoric selection of rice in primitive agriculture followed a sequence of gathering, cultivation, and domestication. The diversifications that have accompanied the adaptation through its diffusion into different agro-climatic conditions were initiated with the origin of cultivated rice *Oryza sativa* from its wild ancestor *O. rufipogon* (Chang 1976; Cook and Evans 1983). The line of descent followed by the cultivated rice was from wild perennial to wild annual before moving onto annual cultigens. This evolution might have been

P. K. Mohapatra, B. B. Sahu, *Panicle Architecture of Rice and its Relationship with Grain Filling*, https://doi.org/10.1007/978-3-030-67897-5_8

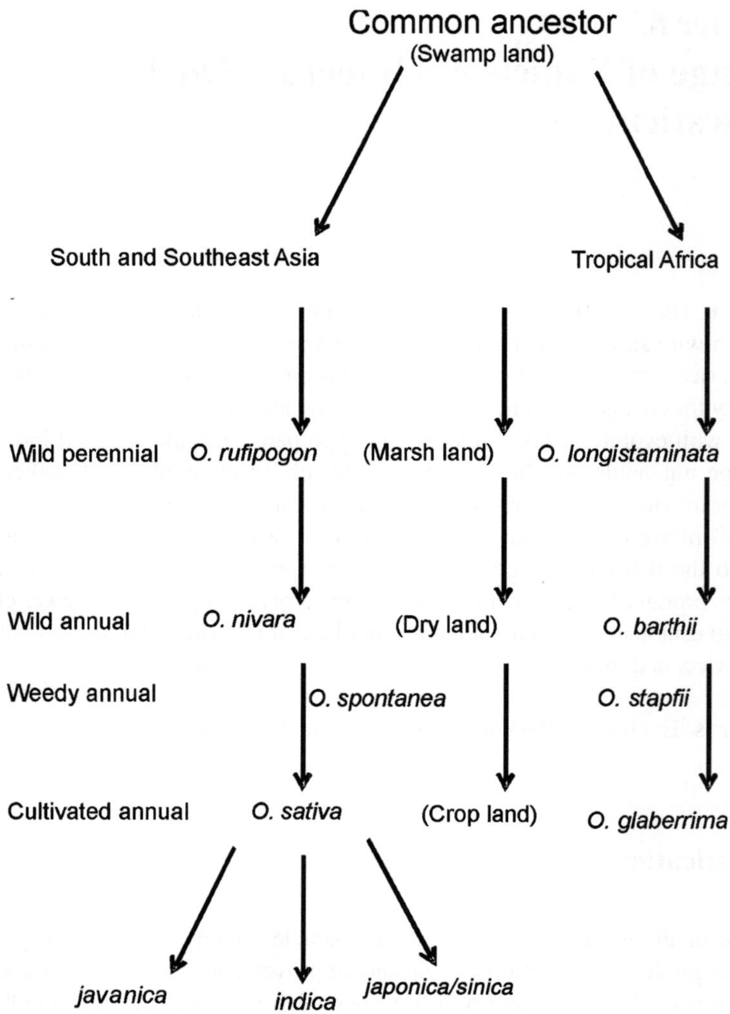

Fig. 8.1 Origin of cultivated rice from wild ancestors (According to Cook and Evans 1983)

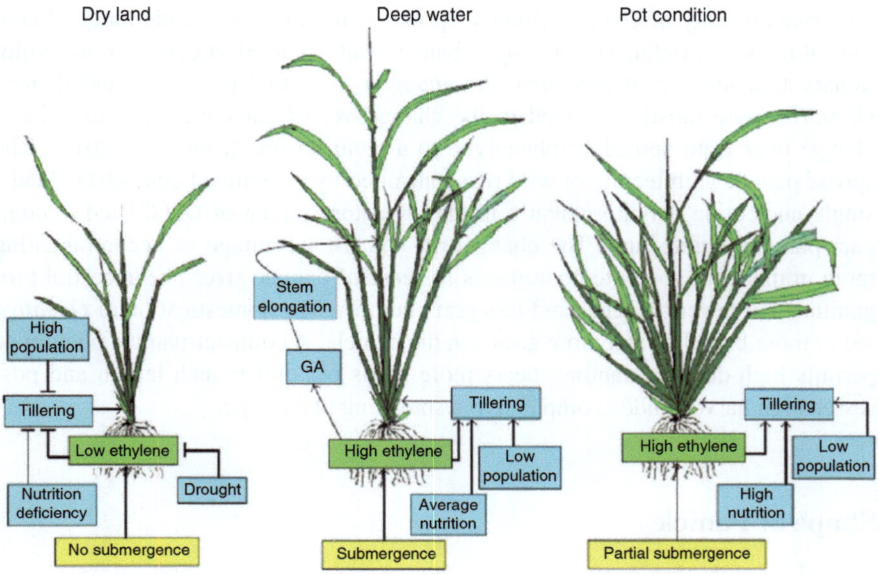

Fig. 8.2 Schematic of the factors controlling tiller dynamics of *Oryza rufipogon* Griff. growing in dryland (natural) and deep water (natural) conditions and cemented pots (manipulated) in the campus of Sambalpur University, India (Mohapatra et al. 2011a)

prompted by an ecological shift from persistently wet to seasonally dry habitats (Grillo et al. 2009; Sandamal et al. 2018). *O. rufipogon* does possess the phenotypic plasticity for such an adaptation. The plant becomes mono tillering under the parched dry land situation with limited resources for water and minerals; multi-tillered in swampy situation with ample water in growth medium, and profuse-tillering when both water and minerals are non-limiting (Fig. 8.2)

Domestication Syndromes

Domestication occurred with the appearance and fixation of syndrome traits typical to the process. A change over to tough rachis phenotype in which grains do not shatter to the ground due to failure of rachis to become brittle with progression to maturity (Allaby 2010) might be the beginning. In course of this evolution, panicle structure and floral morphology modified to embellish change over from cross to self-fertilization, and this process of domestication was controlled by *OTLs* of small to moderate effects. In the process of artificial selection of cultigens, the choice was given for heavier panicles with heavy non- shedding grains. In contrast, natural selection in wild rice might have a preference for spreading type of panicle with shedding grains and cross-pollination habit for easy and wide dispersion and maintenance of high genetic diversity (Vaughan et al. 2003). Singh et al. (2018) detected wild rice diversity of *Oryza rufipogon* species complex from a wide range of ecological niches in India. The adaptive changes that occurred in course of an evolutionary transition from perennial *O. rufipogon* to annual *O. nivara* and then to *O. sativa* were mostly reflected in the changeover of panicle architecture, i.e. a change over from spreading phenotype to a compact one (Zhu et al. 2013). The spread panicle architecture of wild rice controlled by a dominant gene *OsLG1*and a single nucleotide polymorphism-6 in the regulatory region of *OsLG1* led to compact panicle architecture. The changeover had the advantage of accommodating more grains with no contemporaneous increase of panicle size. The ancestral progenitors had fewer branches and less grain number. The domestication of *O. sativa* led to more branches and more grains in the panicle. A compact-panicle phenotype permits high-density planting, bears more seeds per unit branch length and possesses high harvest index compared to a spreading phenotype.

Shape of Panicle

Because of the short and sturdy attitude, the panicle remains mostly erect during grain maturation to avoid shading effects on photosynthetic leaves underneath (Setter et al. 1995). One more significant change that occurred in morphology during domestication was that of the shape of the panicle. Wild rice distantly related to cultivated rice had primary branches with an acropetally receding number of secondary branching organs (branches and spikelets) towards the apex from the base presenting a conical/pyramidal pattern. This pattern changed to a truncated conical pattern in closely related species by an increase of secondary organ formation in the distal primary branches of the conical-shaped panicles. A moderate reduction of primary branch meristem activity along the rachis axis promoted secondary branching of distal primary branches. This arrangement was followed by a cylindrical pattern with an increase of secondary branching structure number on distal primary branches. Yamaki et al. (2010) reported that in truncated conical shaped panicles secondary spikelet formation declined acropetally in the primary branches from value seven in proximal position to two in the distal position. The advanced cylindrical pattern was noticed in the panicles of Kasalath and Nipponbare with very little variation of secondary spikelet number between the primary branches distributed across the panicle rachis. The range of secondary spikelet numbers per primary branch was 6 to 8 for Nipponbare and 9 to 11 in Kasalath. The evolutionary significance of truncated or cylindrical shaped panicle structure was to inhibit the production of secondary organs in proximal part and enhance their number in the distal part, and thereby improve well-filled grain number. (Fig. 8.3 reference Yamaki et al. 2010).

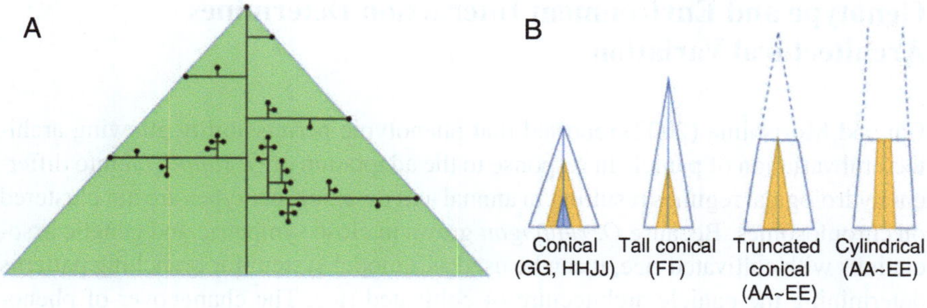

Fig. 8.3 Schemes of panicle branching patterns in the genus *Oryza* (Yamaki et al. 2010 – adopted with the permission of the author)
A) A branching pattern of the rice panicle in the case that all panicle meristems demonstrate their full branching abilities. Spikelet number on each primary branch declines acropetally and the same order exists for secondary branches allowing the conical shape of the panicle. B) Model of the evolutionary course of panicle branching pattern in the genus *Oryza*. The areas within the solid lines indicate the panicles and vertical lines indicate the rachises. The height represents the branching ability of the rachis meristem (the primary branch number per panicle) and the width represents the branching ability of the primary branch meristem. The yellowish area represents secondary branch formation on primary branches. The bluish area represents tertiary branch formation on secondary branches. The conical panicle-like that in the GG and HHJJ species seems to be plesiomorphic. In this type of panicle, the branching ability of the primary branch meristem is at first strong, resulting in tertiary branch formation at the proximal region of the panicle, but it abruptly declines acropetally. In the tall conical panicle, like that in the FF species, the branching ability of primary branch meristem is not strong at first which results in a lack of tertiary branch formation, but it is maintained to form a lot of primary organs. The common ancestor of *O. sativa* and *O. officinalis* complex (AA~EE) species increased the branching ability of rachis and primary branch meristems and acquired rachis meristem abortion to form a truncated conical panicle. Among these species, some species and accessions again have moderated the reduction in the branching ability of the primary branch meristem, resulting in the formation of a cylindrical panicle

Relationship of Panicle Architecture with Grain Filling Characteristics

In the present context of rising demand for rice production, an investigation embracing the genetic basis of the evolutionary mechanism of panicle architecture in course of diffusion into different agro-climatic conditions worldwide holds potential for yield manipulation. In a recent survey of rice research for top yields across different environments in the tropics, Dingkuhn et al. (2015) found no evidence of yield ceiling exceeding 12 t ha^{-1}. The yield dynamics strongly correlated with radiation use efficiency and best of the tropical elite cultivars had no yield advantage over varieties dating 30–40 years back. Aptly it is suggested to articulate new concepts in rice breeding for improvement of yield formation. In this context, designer rice with a regulated branching pattern of panicle amenable for high grain filling deserves foremost attention (Mohapatra 2013).

Genotype and Environment Interaction Determines Architectural Variation

Cai and Morishima (2002) reported that phenotypic permissibility allowing architectural variation of panicle in response to the adaptation of *O. rufipogon* into different hydrological regimes resulting in annual and perennial ecotypes are not clustered on chromosomes. Because *O. rufipogon* grows in close sympatric and genetic associations with cultivated rice, it can be used as a model system for branching patterns determining the panicle architecture of cultivated rice. The changeover of phenotypic attributes that not only anointed one tiller in the annual dryland ecotype *O. nivara* out of its profuse tillering perennial counterpart *O. rufipogon*, but also made the panicle architecture compact by increasing the density of spikelets. The compromise of inter grain distance on the panicle branches might be simply epigenetic cued by a strategy for resilience under perturbed environments (Mohapatra et al. 2011a). However, Wu et al. (2015) reported that plant morphology with reduced plant height, shortened leaf length, and low-yielding smaller size panicle are phenotypic traits of transgenic rice over-expressing the *OsDof12* gene, and the smaller panicle size resulted from low primary and secondary branch number of the panicle. Adriani et al. (2016) concurred with the assumption by opining that architectural variation of a panicle of rice genotypes is prone to genotype and environmental interactions. Zhu et al. (2013) have shown that the dominant gene *OsLG1* is an SBP-domain transcription factor responsible for ligule development genetic modification of which might have been responsible for an increase of panicle compactness during domestication. An increase in spikelet density of panicle compromises with inter-grain apical dominance and freedom of space for development specifying changeover in spikelet distribution from a heterogeneous to a homogeneous pattern. The heterogeneous architecture of panicle is best suited in conditions where the plant encounters several forms of stress. Because inclement weather at a sensitive stage like anthesis impacts seed filling, a staggered development permits the plant to sacrifice some spikelets, while preserving the rest to enable some grains to fill to provide seeds for the next generation (Mohapatra et al. 2011b) (Fig. 8.4). The advantage for survival inbuilt in a heterogeneous architecture, however, has a negative influence on grain filling thanks to strong inter-grain apical dominance; grains mostly located on the secondary branches of basal primary branches are discriminated and decimated for grain development (Mohapatra and Sahu 1991).

Modification of Panicle Architecture for Improved Grain Filling

Unlike the exclusive rain-fed agro-ecosystems of vintage agriculture, rice presently is grown under relatively stable irrigated conditions with the application of modern farm management techniques. The method of cultivation mostly precludes the

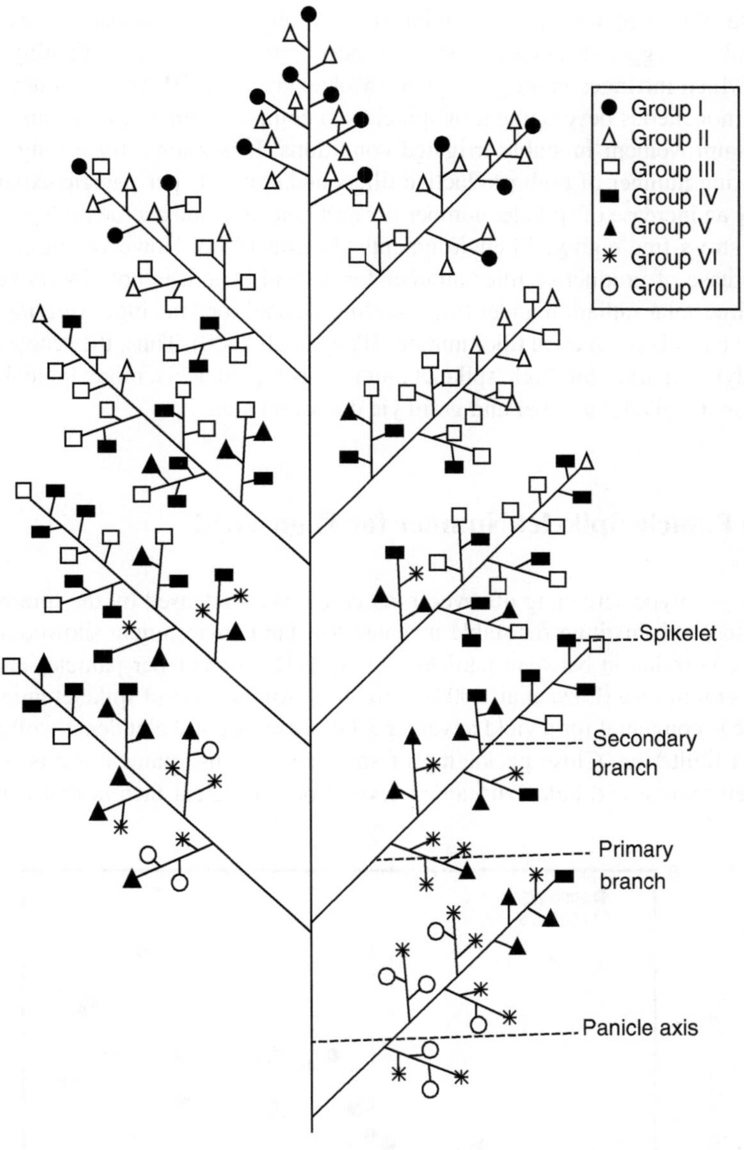

Fig. 8.4 Schematics of rice panicle showing the location of developmentally different types of spikelets on the primary and secondary branches. The spikelets flowered each day for seven successive days (Group I on day 1 to group VII on day 7). Groups I-III are considered superior spikelets producing quality grains compared to the inferior spikelets of groups IV to VII (From (Mohapatra et al. 2011b)

adverse effects of weather uncertainties. With abysmal evolutionary pressure for survival; a staggered pattern of spikelet development becomes a liability for the plant which infringes upon grain yield (Mohapatra et al. 2011b). In contrast, with the homogeneous development of spikelets, a compact panicle trait is amenable to yield improvement in stable irrigated conditions. It is known for a long that the increasing number of both productive tillers and spikelets per panicle expand sink size by an increase of spikelet number per unit land area and accrue high grain yield under stress-free high yield environments (Takeda 1984). However, the option for an increase of productive tiller number for a yield benefit is not always possible, when low solar radiation under tropical cloudy conditions infringes tillering capacity and curtails productive tiller number (Peng et al. 2000). Thus, the second option embodying a trait for high spikelet number per panicle is more logical for an increase of spikelet number and grain yield per unit area.

High Panicle Spikelet Number for High Yield

In fact genotype screening of several *indica* cultivars released by the International Rice Research Institute for stable irrigated low land environment showed a strong positive correlation between panicle size (spikelet number per panicle) and grain yield per unit area (Laza et al. 2004) – Fig. 8.5). An increase of spikelet number per panicle is congenial for a yield advantage, but increased spikelet density often forebears a limitation. Close packaging of spikelets in dense panicle limits hull size between lemma and palea and alters several physiological factors that determine

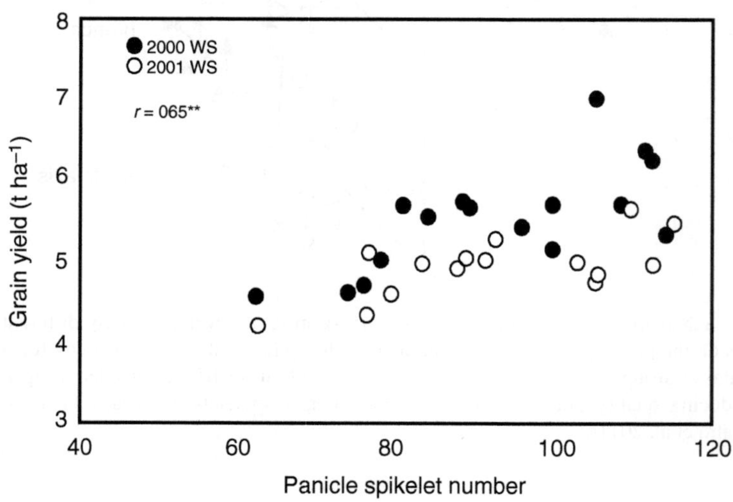

Fig. 8.5 The effect of panicle size on grain yield of IRRI-released rice cultivars grown at IRRI farm 2000 (closed circles) and 2001 (open circles) wet seasons. ** Correlation coefficient value ® was significant at the 1% probability level (Laza et al. 2004). With permission of the author

assimilate partitioning, apical dominance, and hormonal balance during grain filling undermining yield potential of the crop (Peng et al. 2000; Panda et al. 2009; Yang and Zhang 2009; Sekhar et al. 2015). Despite the shortcomings, the tendency to increase spikelet number per panicle and accommodate the spikelets densely in the structure without a corresponding increase of panicle size has been the best option for breeders to improve the yield potential of rice. The other components of grain yield like individual grain weight and panicle number per plant appear to be genetically determined and stable for decades (Evans et al. 1984). In contrast, seed number per panicle is believed to be amenable for genetic modification because of plasticity observed under a range of agro- climatic conditions in which rice is grown (Kropff et al. 1994). Using an empirical model Sheehy et al. (2001) predicted two-times increase of yield potential, if all spikelets arising on panicle would fill.

Relationship of Panicle Architecture with Grain Filling

The relationship of grain filling with panicle architecture has been a subject of intensive research in the development of new plant type (NPT) rice. High seed numbers for increased sink capacity along with extended period of grain filling were two important traits used for improving the semidwarf plant type and developing a new ideotype (Dingkuhn et al. 1991). Instead of the exclusive *indica* type, NPTs used tropical *japonica* germplasm to increase panicle size (Peng et al. 2000). Similarly, exploitation of heterosis by intercrossing of indica and *japonica* cultivars also increased panicle size larger than conventional inbred rice. However, an increase in spikelet number does not always transform into yield benefit at maturity because the degree of grain filling depends on the position of spikelets within a panicle (Kato 2006). In a heterogeneous architecture, spikelets placed on the distal parts and primary branches of panicle possess attributes for grain filling better than that of spikelets located on basal part and secondary branches; the former is called superior and the latter inferior spikelets (Mohapatra et al. 1993; Ishimaru et al. 2003). In traditional rice panicle, architecture is simple. The number of spikelets borne directly on primary branches is high, and spikelet development not impeded by space restrictions ensures good occurrence. The lax-panicle cultivars produce grains with larger biomass (Panda et al. 2009). But overall yield becomes poor because of the low spikelet number. Spikelet number per panicle has increased significantly in some newly developed cultivars, but the additional spikelets are placed on secondary branches (Kudo 1991; Yamamoto et al. 1991). These spikelets do not fill as efficiently as the spikelets located on primary branches and the objective of achieving high yield potential is not realized (Kato 1993) (Fig. 8.6).

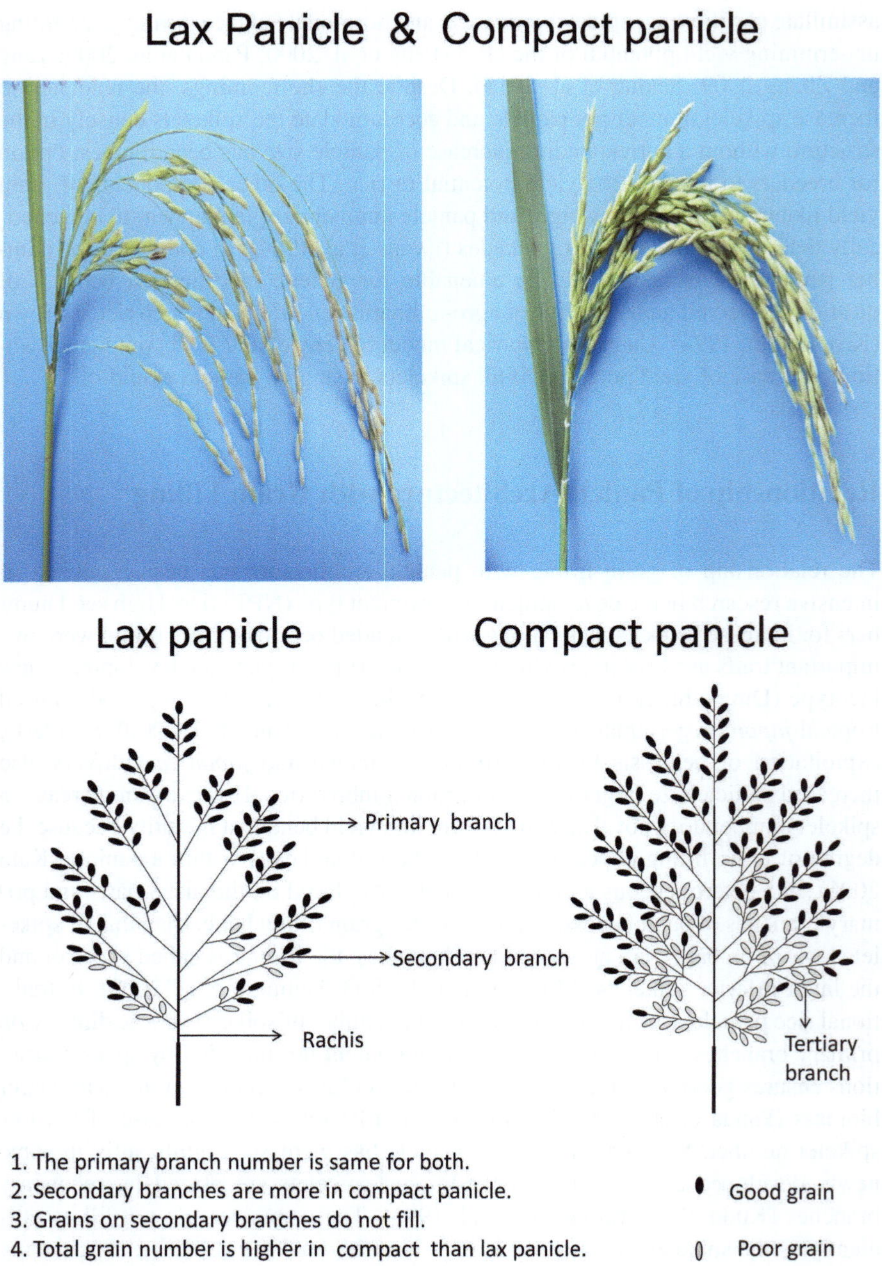

Fig. 8.6 Panicle structure of lax- and compact-panicle rice

High Spikelet Laxity Associated with Yield Loss

Comparison of *indica* inbred (low spikelet number) with new plant type (high spikelet number) confirmed that the number of filled grains of panicle did not improve with a corresponding increase of spikelet number (Khush 1995; Yang et al. 2000). The extended sink capacity of the large-panicle cultivars could not be realized because the increase in spikelet number per panicle depended on an increase in the number of inferior spikelets located mostly on secondary branches of panicle (Kato 2010; Kato et al. 2010). Nonetheless, increased grain compactness has been accepted in modern rice cultivars, because high laxity is generally associated with loss of grain yield (Moldenhauer and Gibbons 2003). Besides the increase of panicle size to accommodate more grains in the structure could have a detrimental effect on light interception and the rate of photosynthesis of source leaves which are positioned below the panicle and supply assimilates to grains during development (Xu et al. 1990; Setter et al. 1996). Therefore breeders of the heavy panicle type cultivars had a preference for more spikelets per unit length of panicle to increase spikelet number per panicle without any appreciable increase in panicle length. Additionally, not all large-panicle rice cultivars fill poorly as genetic variation exists in them for the degree of grain filling (Kato 1999, 2006; Meng et al. 2016). Grain sink strength is a factor that is related to poor grain filling of the inferior spikelets (Liang et al. 2001). Grain sink strength of large-panicle rice cultivars improves when they have genes *AS2-2* and *SUT1-2*, alleles at *OsAGPS2* (*AS2*) on chromosome 8 and *OsSUT1* (*SUT1*) on chromosome 3 respectively (Kato and Horibata 2011). *AS2* codes for the small subunit of starch synthesizing enzyme ADP-glucose pyrophosphorylase while *SUT1* controls the synthesis of sucrose transporter protein in the endosperm cells. These alleles for good grain filling are mostly present in *indica* cultivars in contrast to the *japonica* and their introgression into large-panicle rice can augur a solution to break yield frontier of modern rice. The tendency to lay more secondary branches on the upper part of panicle or have more spikelets positioned directly on primary branches can also improve sink efficiency for higher grain filling. It is proposed that a change over from the compact and determinate architecture to a simple and indeterminate type inflorescence, as seen in barley, would improve placement of spikelets more on primary than on secondary branches for the benefit of grain filling in rice (Mohapatra 2013). Laxity in spikelet orientation, i.e. conducive for grain filling, will also increase rice yield in this type of panicle.

References

Adriani DE, Dingkuhn M, Dardou A, Adam H, Luquet D, Lafarge T (2016) Rice panicle plasticity in Near Isogenic Lines carrying a QTL for larger panicle is genotype and environment-dependent. Rice 9(28):1–15. https://doi.org/10.1186/s.12284-016-0101-x

Allaby R (2010) Integrating the processes in the evolutionary system of domestication. J Exp Bot 61:935–944

Cai H, Morishima H (2002) QTL clusters reflect character associations in wild and cultivated rice. Theor Appl Genet 104:1217–1228

Chang T-T (1976) The origin, evolution, cultivation, dissemination, and diversification of Asian and African rices. Euphytica 25:425–441

Chang T-T (2003) Origin, domestication, and diversification. Chapter 1, Rice: origin, history, technology, and production. Edited by C Wayne Smith, John Wiley & Sons, Inc. ISBN 0-471-34516-4

Cook M, Evans LT (1983) Some physiological aspects of the domestication and improvement of rice (Oryza spp.). Field Crop Res 6:219–238

Dingkuhn M, de Vries FP, De Datta SK, Van Laar H (1991) Concepts for a new plant type for direct-seeded flooded tropical rice. Direct seeded flooded rice in the tropics. International Rice Research Institute, Los Banos, Philippines, pp 17–38

Dingkuhn M, Laza MRC, Kumar U, Mendez KS, Collard B, Jagadish K, Singh RK, Padolina T, Malabayabas M, Torres E (2015) Improving yield potential of tropical rice: achieved levels and perspectives through improved ideotypes. Field Crop Res 182:43–59

Evans L, Visperas R, Vergara BS (1984) Morphological and physiological changes among rice varieties used in the Philippines over the last seventy years. Field Crop Res 8:105–124

Grillo MA, Li C, Fowlkes AM, Briggeman TM, Zhou A, Schemske DW, Sang T (2009) Genetic architecture for the adaptive origin of annual wild rice, Oryza nivara. Evolution: Int J Organic Evol 63:870–883

Ishimaru T, Matsuda T, Ohsugi R, Yamagishi T (2003) Morphological development of rice caryopses located at the different positions in a panicle from early to middle stage of grain filling. Funct Plant Biol 30:1139–1149

Kato T (1993) Variation in grain-filling process among grain positions within a panicle of rice (Oryza sativa L.). SABRAO J 25:1–10

Kato T (1999) Grain filling characteristics of rice cultivars with numerous spikelets in a panicle. In Proc. Int. Symp. "World Food Security and Crop Production Technologies for Tomorrow", 1999 (Graduate School of Agriculture, Kyoto University), pp. 336–337

Kato T (2006) Panicle centroid index: an index to represent the distribution of spikelets in a rice panicle. Kinki J Crop Sci Breed 51:31–36

Kato T (2010) Variation and association of the traits related to grain filling in several extra-heavy panicle type rice under different environments. Plant Prod Sci 13:185–192

Kato T, Horibata A (2011) Non-random distribution of the alleles for good grain filling at OsAGPS2 and OsSUT1 among a wide range of rice (Oryza sativa L.) cultivars. Breed Sci 61:217–220

Kato T, Taniguchi A, Horibata A (2010) Effects of the alleles at *OsAGPS2* and *OsSUT1* on the grain filling in extra- heavy panicle type of rice. Crop Sci 50:2448–2456

Khush GS (1995) Breaking the yield frontier of rice. GeoJournal 35:329–332

Khush GS (1997) Origin, dispersal, cultivation, and variation of rice. *Plant Mol Biol* **35**:25–34

Kropff MCassman K, Peng S, Matthews R, Setter T (1994) Quantitative understanding of yield potential. Breaking the yield barrier. Proceedings of a workshop on rice yield potential in favorable environments- Edited Cassman KG. IRRI, Philippines, 21–38

Kudo K (1991) Studies on the ripening of rice [Oryza sativa] plants. 2: influence on the density of panicles on the ripening in rice varieties. Report Tohoku breeding. Crop Sci Soc Japan 34:109–112

Laza MRC, Peng S, Akita S, Saka H (2004) Effect of panicle size on grain yield of IRRI-released *indica* rice cultivars in the wet season. Plant Prod Sci 7:271–276

Liang J, Zhang J, Cao X (2001) Grain sink strength may be related to the poor grain filling of *indica-japonica* rice (Oryza sativa) hybrids. Physiol Plant 112:470–477

Meng T, Wei H, Li C, Dai Q, Xu K, Huo Z, Wei H, Guo B, Zhang H (2016) Morphological and physiological traits of large-panicle rice varieties with high filled-grain percentage. J Integr Agric 15:1751–1762

Mohapatra PK (2013) A physiological approach to design panicle architecture in the quest for high grain yield. In: Muralidharan K, Siddiq EA (eds) International dialogue on perception and prospects of designer rice. Society for Advancement of Rice Research, Directorate of Rice Research, Hyderabad, pp 131–142

Mohapatra PK, Sahu SK (1991) Heterogeneity of primary branch development and spikelet survival in rice panicle in relation to assimilates of primary branches. J Exp Bot 42:871–879

Mohapatra PK, Panda B, Ekamber K (2011a) Plasticity of tiller dynamics in wild rice Oryza rufipogon Griff.: a strategy for resilience in suboptimal environments. *Int J Agronom* 2011. 9 pages

Mohapatra PK, Panigrahi R, Turner NC (2011b) Chapter five – physiology of spikelet development on the rice panicle: is the manipulation of apical dominance crucial for grain yield improvement? Adv Agron 110:333–359

MohapatraPK PR, Sahu SK (1993) Time of flowering affects grain quality and spikelet partitioning within the rice panicle. Funct Plant Biol 20:231–241

Moldenhauer K, Gibbons JH (2003) Rice morphology and development. In: Smith C, Dilday R (eds) Rice origin, history, technology, and production. John Wiley and Sons Inc, New Jersey, pp 103–125

Panda BB, Kariali E, Panigrahi R, Mohapatra PK (2009) High ethylene production slackens seed filling in compact panicled rice cultivar. Plant Growth Regul 58:141–151

Peng S, Laza R, Visperas R, Sanico A, Cassman KG, Khush GS (2000) Grain yield of rice cultivars and lines developed in the Philippines since 1966. Crop Sci 39:1552–1559

Sandamal S, Tennakoon A, Meng Q-L, Marambe B, Ratnasekera D, Melo A, Ge S (2018) Population genetics and evolutionary history of the wild rice species Oryza rufipogon and O. nivara in Sri Lanka. Ecol Evol 8:12056–12065

Sekhar S, Panda BB, Mohapatra T, Das K, Shaw BP, Kariali E, Mohapatra PK (2015) Spikelet-specific variation in ethylene production and constitutive expression of ethylene receptors and signal transducers during grain filling of compact-and lax-panicle rice (Oryza sativa) cultivars. J Plant Physiol 179:21–34

Setter TL, Conocono EA, Egdane JA, Kropff MJ (1995) Possibility of increasing yield potential of rice by reducing panicle height in the canopy. 1. Effects of panicles on light interception and photosynthesis. Aust J Plant Physiol 22:441–451

Setter T, Conocono E, Egdane J (1996) Possibility of increasing yield potential of rice by reducing panicle height in the canopy. II. Canopy photosynthesis and yield of isogenic lines. Funct Plant Biol 23:161–169

Sheehy JE, Dionora MJA, Mitchell PL (2001) Spikelet numbers, sink size, and potential yield in rice. Field Crop Res 71:77–85

Singh B, Singh N, Mishra S, Tripathi K, Singh BP, Rai V, Singh AK, Singh NK (2018) Morphological and molecular data reveal three distinct populations of Indian wild rice Oryza rufipogon Griff species complex. *Front Plant Sci* 7:2018. https://doi.org/10.3389/fpls.2018.00123

Takeda T (1984) Physiological and ecological characteristics of high-yielding varieties of lowland rice. In Proceedings of International Crop Science Symposium, Fukuoka, Japan, pp. 17–20

Vaughan DA, Morishima H, Kadowaki K (2003) Diversity in the Oryza genus. Curr Opin Plant Biol 6:139–146

Wu Q, Li D, Li D, Liu X, Zhao X, Li X, Li S, Zhu L (2015) Overexpression of OsDof12 affects plant architecture in rice (Oryza sativa L.). Front Plant Sci 6:833. https://doi.org/10.3389/fpls.2015.00833

Xu ZJ, Chen WF, Zhang LB, Yang SR (1990) Comparative study on light distribution in rice canopies of different panicle types. *Scientia Agricultura Sinica* **23**:6–10

Yamaki S, Miyabayashi T, Eiguchi M, Kitano H, Nonomura K-I, Kurata N (2010) Diversity of panicle branching patterns in wild relatives of rice. Breed Sci 60:586–596

Yamamoto Y, Yoshida T, Enomoto T, Yoshikawa G (1991) Characteristics for the efficiency of spikelet production and the ripening in high-yielding *japonica-incdica* hybrid and semidwarf *indica* rice varieties. Japanes J Crop Sci 60:365–372

Yang J, Zhang J (2009) Grain-filling problem in 'super' rice. J Exp Bot 61:1–5

Yang J, Peng S, Visperas RM, Sanico AL, Zhu Q, Gu S (2000) Grain filling pattern and cytokinin content in the grains and roots of rice plants. Plant Growth Regul 30:261–270

Zhu Z, Tan L, Fu Y, Liu F, Cai H, Xie D, Wu F, Wu J, Matsumoto T, Sun C (2013) Genetic control of inflorescence architecture during rice domestication. Nat Commun 4:2200

Chapter 9
Physiology of Variation in Individual Grain Weight of Rice Panicle

Abstract Each spikelet borne on a panicle node is genetically competent to bear a well-developed high- density grain at maturity. But there is a wide margin between the numbers of spikelets differentiated at ontogeny and that come to maturity in the panicle because not all developing spikelets fill efficiently. Position-specific grain occurrence leading to heterogeneous grain development has been a major impediment in breaking the genetic yield ceiling of rice. Comparison of two extreme types of spikelets of panicle, superior and inferior, delineates the physiological factors determining kernel growth, endosperm cell division, starch synthesis, and assimilates partitioning between the spikelets and highlights a concept for inter-spikelet dominance for regulation of grain filling. Grain growth sustains on assimilates partitioning from source organs by mass flow through phloem sieve tube to pericarp of caryopsis. Sucrose, the dominant sugar of mass flow, is transported symplastically in the vascular bundles of the pericarp up to the upper terminal part of the caryopsis. Phloem sap moves symplastically from sieve elements and companion cell complex through pigment strand on the dorsal side of caryopsis and circumferentially via nucellus before being released into the apoplast between nucellus and aleurone layer. Sucrose actively unloaded from nucellus into the embryonic apoplasm is loaded energetically into aleurone layer cells. The onward movement of sucrose into the starchy endosperm cells is symplastic. Poor sink strength of inferior caryopsis slackens endosperm growth leading to accumulation of assimilates.

Keywords Grain filling · Assimilates · Inferior and superior spikelets · Phloem solutes · Transport pathway

P. K. Mohapatra, B. B. Sahu, *Panicle Architecture of Rice and its Relationship with Grain Filling*, https://doi.org/10.1007/978-3-030-67897-5_9

Grain Filling in Rice Spikelet

Depending on the variety and environmental conditions rice plants take around 3–6 months for completion of the life cycle and reach maturity. In course of the process, they use to pass through 3 different stages, such as vegetative, reproductive, and ripening. A tropical rice cultivar with 120 days life span spends around 60 days for vegetative and 30 days each for reproductive and ripening stages. The fertile floret of rice spikelet is hermaphrodite and self- pollinated for fertilization. The ripening phase starts with fertilization. This stage is the final stage of the rice life cycle at which the fertilized ovary develops into caryopsis. The initial stage of caryopsis growth after fertilization involves rapid cell division of the endosperm with a feeble gain of dry matter. Thereafter rate of dry matter accumulation increases exponentially before slowing down in the final phase of grain maturation. The filling rate and duration determine the final grain weight and quality. Comparable to *japonica*, the filling rate of *indica* rice caryopsis is faster, but the duration for growth is smaller. The growth duration of *japonica* rice grain could be as long as 65 days in the cool temperate climate. There are four steps for the ripening process. The terms used are based on the color and texture of the ripening grains.

(a) Milk stage
Starch synthesis primarily determines grain filling. At this stage the starch granules are soft and the interior of the kernel is filled with a white liquid similar to milk.
(b) Soft dough stage
At this stage the starch granules begin to harden and firm, but still not very hard.
(c) Hard dough stage
The whole grain is firm with moisture content above 22%. The lemma and palea surrounding the developing caryopsis change color from greenness to yellow.
(d) Maturity
The grain is hard. Moisture content drops below 20%. Lemma and palea are yellow.

Heterogeneity of Spikelet Development

Each spikelet borne on a panicle node is genetically competent to bear a well- developed high-density grain at maturity. But there is a wide margin between the numbers of spikelets differentiated at ontogeny and that come to maturity in the panicle because not all developing spikelets fill efficiently (Fig. 9.1). The physiological significance of the process of initiation of a large number of spikelets before flowering and elimination or partial success of occurrence of many of them post-flowering are not known. This happens especially in the heavy panicle type rice cultivars, which have not received adequate attention from scientists. The heterogeneous architecture of rice panicle has been a major factor for the disparity in grain filling of individual spikelets of the panicle. In contrast, a homogeneous architecture with synchronous

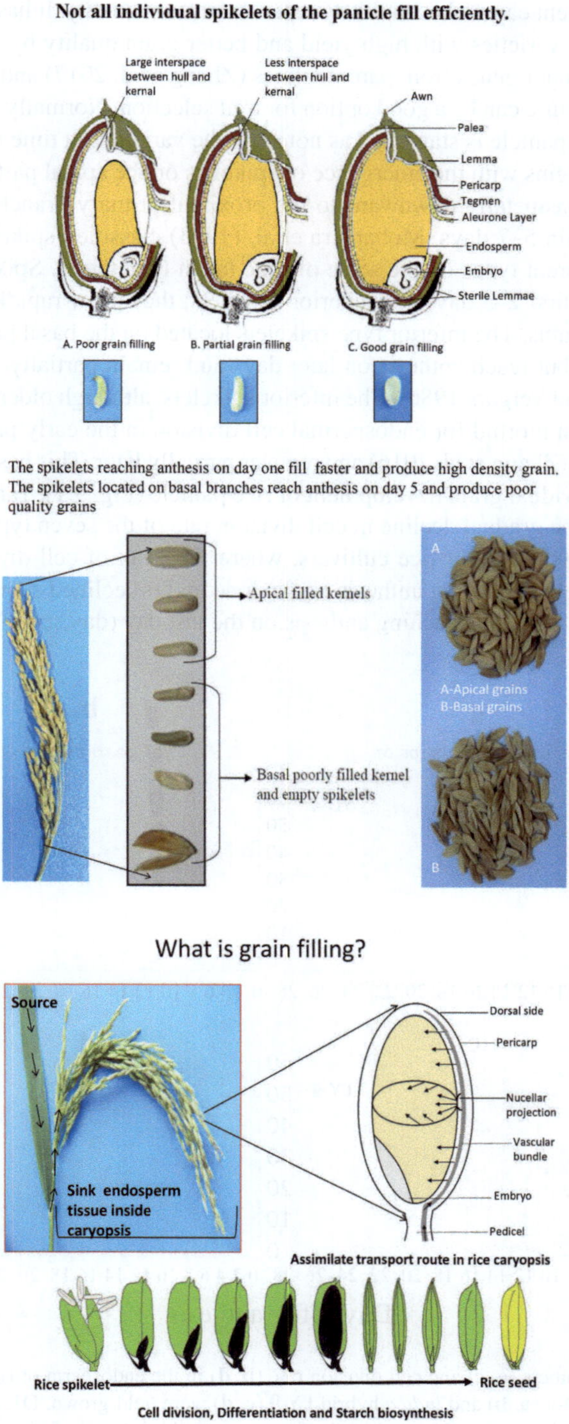

Fig. 9.1 Grain filling in different types of spikelets of rice panicle

grain development can make a cultivar superior in grain quality. It has been possible to develop elite varieties with high yield and better grain quality by selective pyramidisation of major genes from parental lines (Zheng et al. 2017) and homogeneous panicle architecture can be a good option for trait selection. Normally spikelet development on rice panicle is staggered as noted by the variation in time taken to anthesis. Anthesis begins with the emergence of spikelets on the apical part of the panicle and proceeds basipetally downward to the proximal primary branch. The event is completed within 5–7 days. Mohapatra et al. (1993) classified spikelets of panicle into seven different types on the scale of time taken to anthesis. Spikelets reaching anthesis in the first 2–3 days are superior spikelets; they grow rapidly and produce high-density grains. The inferior type spikelets located on the basal part of the panicle arise early, but reach anthesis on later days and remain partially- or un-filled at maturity (Xu and Vergara 1986). The inferior spikelets, although older than the superior ones remain morbid for endospermal cell division in the early part of the post-anthesis period (Zhang et al. 2016) and recover partially later. This has been a unique feature for individual grain development of rice panicle (Fig. 9.1). Yang et al. (2006) has described the gradual decline in cell division rate of the seven types of spikelets of panicle in two different rice cultivars, where the peak of cell division rate was highest in spikelets reaching anthesis on day one and it declined sequentially to the lowest value in spikelets reaching anthesis on the last day (day seven) (Fig. 9.2).

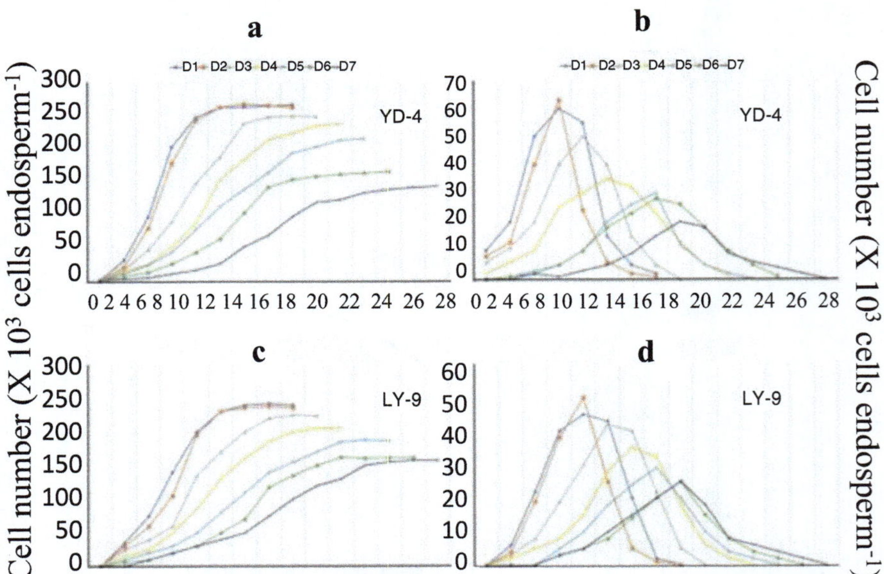

Fig. 9.2 Cell numbers (**a**, **c**) and cell division rate (**b**, **d**) in the endosperm of rice spikelets. The *indica* cultivar YD-4 (**a**, **b**) and *indica* hybrid LY-9 (**c**, **d**) were field-grown. D1, D2, D3, D4, D5, D6, and D7 are spikelets that flowered on the first, second, third, fourth, fifth, sixth, and seventh days, respectively, on a panicle (Yang et al. 2006 with permission of the author)

The Difference in the Growth of Superior and Inferior Spikelets

Delay in active cell division, at the early stage of development, ultimately lowers final grain weight significantly in late-developing inferior spikelets compared to the fast- developing superior spikelets. In the post-anthesis stage, dry matter content and starch deposition increase until a week before maturity and stabilize thereafter. The growth rate increases rapidly until the middle stage of grain maturation and declines to naught at maturity. The rate of rise and fall in growth is slower and continues for longer in the inferior spikelet than the superior spikelet (Jeng et al. 2003; Fig. 9.3).

Yang et al. (2002b) have compared the superior versus inferior spikelets of six different rice genotypes for growth during the post-anthesis period (Table 9.1). The active period of endosperm cell division was short and initiated early in superior spikelets and they gained more numbers of cells, greater cell division rate, and higher biomass. In comparison, inferior spikelets initiated an active period of cell division late and the activity continued for a longer time. However, the long span of cell division could not remedy equality in endosperm cell number or the final grain weight of the superior spikelets.

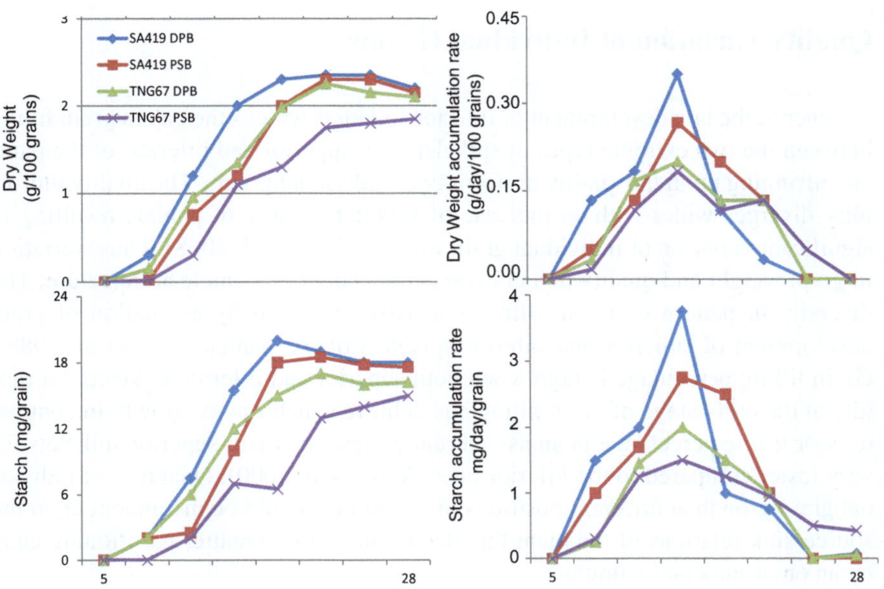

Fig. 9.3 Changes of dry weight, dry matter accumulation rate, starch content, and starch accumulation rate in growing rice grains, collected from distal primary branches (DPB) and proximal secondary branches (PSB) of panicle, in cultivar Tainung 67 and mutant SA419 (modified from Jeng et al. 2003)

Table 9.1 The rate and active period of endosperm cell division, maximum cell number, cell weight, endosperm weight, and grain weight of six different rice cultivars

Genotype	Types of spikelet	Mean Cell division rate (X 10^3 cells per endosperm d^{-1})	Maximum Cell number (X 10^3 cells per endosperm)	Active division period (DPA)	Cell weight (ng)	Endosperm weight (mg)	Grain weight (mg)
PC/Zao	Superior	21.2	212	2–10	92.0	19.5	25.4
	Inferior	11.8	171	6–18	80.1	13.7	18.1
JW-8/ IR36	Superior	21.2	236	2–11	93.6	22.1	27.8
	Inferior	11.5	191	6–20	92.2	17.6	23.3
Sujing 1	Superior	27.5	275	2–10	90.2	24.8	30.8
	Inferior	24.2	242	4–12	93.8	22.7	28.2
Yangdao 4	Superior	25.7	257	2–10	92.6	23.8	29.2
	Inferior	13.5	195	6–18	94.4	18.4	24.7
Yanjing 2	Superior	25.2	252	2–10	92.4	23.3	29.0
	Inferior	18.0	220	4–14	95.0	20.9	26.1
IR72	Superior	25.9	231	2–9	94.8	21.9	26.8
	Inferior	18.0	200	3–12	94.0	18.8	24.5

Adopted from Yang et al. 2002b with permission of the author

Quality Variation of Individual Grains

In general, the late development of inferior spikelets widens the gap in grain filling between the two extreme types of spikelets, i.e. superior and inferior, of the panicle infringing upon the quality uniformity of individual grains. The quality uniformity diverges wider with an increase of spikelet number of panicle resulting in significant variation of individual grain weight (Wang et al. 2008). Thus, variation in grain weight and quality in a cultivar is dependent on panicle architecture. The diversity in pattern of grain filling is usually measured by estimation of grain development of superior and inferior spikelets of the panicle (Zhu et al. 1988). Grain filling percentage is high when both superior and inferior spikelets fill rapidly at the early stage of grain filling and achieve synchrony of growth in contrast to poor filling percentage in an asynchronous system where superior spikelets fill very faster compared to the inferior ones (Yang et al. 2000). Quantum of individual grain growth at different positions of the panicle could be incumbent upon the source-sink relations of the plant that determines carbon matter partitioning cued by an environmental stimulus.

Assimilate Partitioning

Several authors have assessed the physiological factors controlling assimilate partitioning by examining metabolic and hormonal regulation mechanisms of spikelet development. A rice plant under stable growth conditions holds potential for increased yield because it possesses excess photosynthetic capacity that could be used for grain filling subject to a greater demand for assimilates. Grain growth during the maturation phase is sustained by assimilates concurrently supplied by leaf photosynthesis and stem and leaf sheath remobilization (Yoshida 1972). Carbon assimilates produced more than the demand of young seedlings in the vegetative and early reproductive stages are stored in the leaf sheath and stem in the form of non-structural carbohydrates that move to the developing grains in the maturation phase (Li et al. 2017). The apparent contribution of non-structural carbohydrates has been estimated at 10–40% of final grain; this contribution becomes crucial when either challenging environments infringe upon current photosynthesis or yield potential is based on high biomass production (Yang and Zhang 2010). Incidentally, panicle development coincides with stem growth and heavy panicle rice may be source limited for assimilates to sustain the occurrence of a large number of spikelets (Yang et al. 2002a; Adriani et al. 2016). Competition for resources between the organs being exacerbated, the inferior spikelets would be discriminated more as against the superior spikelets.

Assimilate Pool Is High in Inferior Spikelets

However, Takai et al. (2005) suggested otherwise; they identified the QTLs involved in partitioning of non-structural carbohydrates (NSC) from stem and sheath to panicle and linked the NSC dynamics during the grain filling period to variation in sink size of the panicle. It was reported that poor partitioning of assimilates from the source leaves and stem deprive grain development of inferior type spikelets borne on secondary branches of panicle (Kato 2004). These spikelets make unsuccessful competition for a limited pool of assimilates, which results in poor seed setting. However, the concept of source limitation of grain growth has been contradicted. Vergara (1988) reported that many spikelets of rice panicle do not produce high-density grains despite the presence of enough assimilates. Starch synthesis in endosperm cells of inferior grain is poor (Umemoto et al. 1994) and assimilates partitioned to them remain unutilized so that grain sink may be limiting transport and storage of assimilates. Because residual assimilates accumulate in the inferior spikelets, they exhibit callusing ability in anther culture greater than the superior ones (Afza et al. 2000). Conversely, spikelets of upper primary branches exhibit metabolic dominance in starch biosynthesis and produce better quality grains than

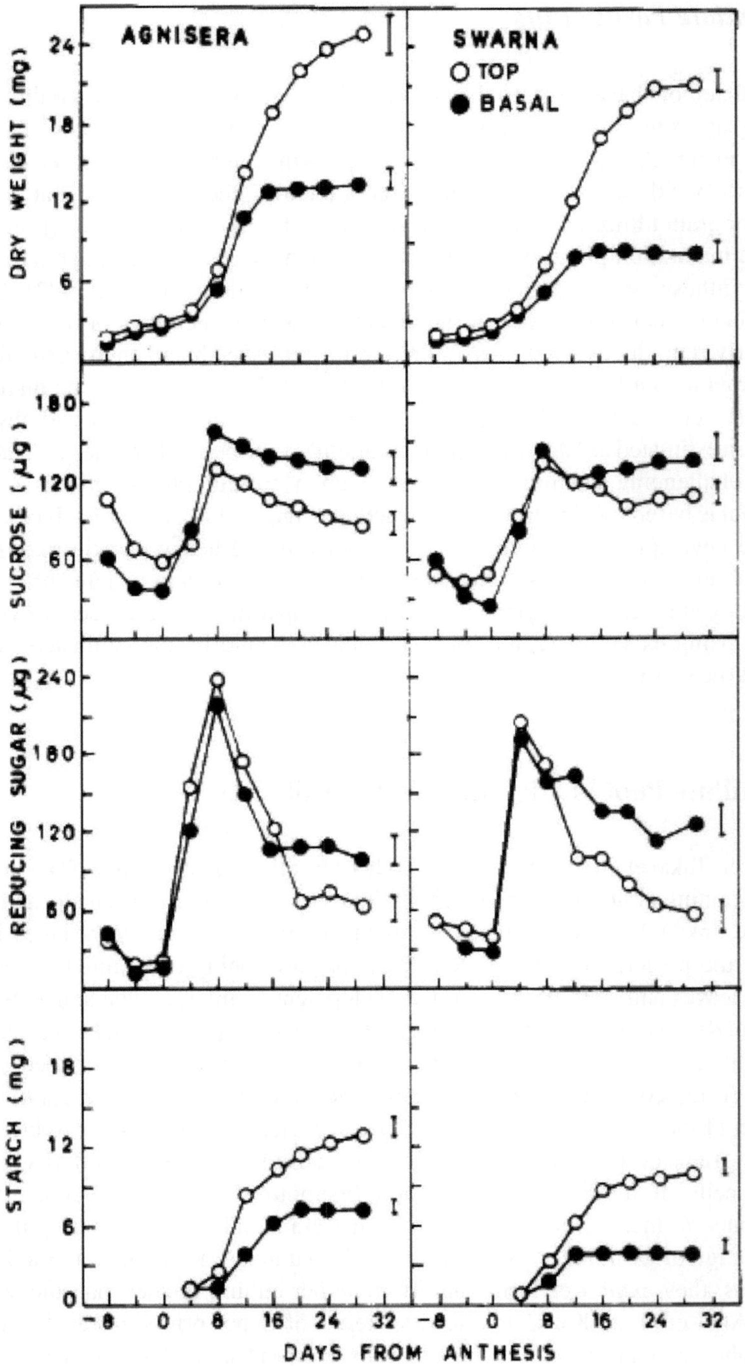

Fig. 9.4 The dry weight, starch, sucrose, and reducing sugar contents of the fertile top (superior) and basal (inferior) spikelets of rice varieties Agnisera (traditional tall *indica*) and Swarna (semi-dwarf high yielding) during the period of grain development. The vertical bars represent sd values at $p \leq 0.05$ (Patel and Mohapatra 1996)

the late flowering inferior ones on secondary branches (Mohapatra et al. 1993; Ishimaru et al. 2003; Yang et al. 2006). Poor starch synthesis in inferior basal spikelets of panicle in the post-anthesis stage of grain development results in the accumulation of unused soluble carbohydrates. The gradient in accumulation between superior and inferior kernels becomes larger in compact panicle rice cultivars because the grain development of inferior spikelets in these cultivars is much poorer. Irrespective of differences in cultural type and plant phenotype, assimilate partitioning to developing kernels of inferior spikelets is always sub-optimal, as shown in the comparison between a traditional rice cultivar Agnisera with a high yielding semi-dwarf cultivar Swarna (Fig. 9.4 (Patel and Mohapatra 1996). Similarly greater discrimination in assimilate partitioning during grain development of compact panicle cultivars Mahalaxmi and CR-29 compared to their respective contemporaneous lax-panicle cultivars Upahar and CR- 45 has been shown in (Panda et al. 2015) - Fig. 9.5.

Inter-Spikelet Dominance in Grain Filling

Mohapatra et al. (2011) proposed heterogeneous panicle architecture leading to inter- grain apical dominance an important strategy for plant survival under a pernicious environment. During panicle development, grains located on apical branches exhibit strong metabolic dominance for assimilate partitioning over their basal counterparts. The magnitude of difference in grain growth between the two types of spikelets is larger in compact than that of lax-panicle cultivars. Hence, a large panicle rice cultivar with increased sink capacity would conceptually encourage greater source activity and biomass production. But in reality, reduced sink strength of the failing spikelets of panicle does not augur such an effect. The intricate relationship between source and sink in dry matter partitioning has been worked out by partial excision of competing sinks of developing panicles and comparing ^{13}C distribution pattern between traditional low yielding cultivar Nakateshinsenbon and high yielding cultivar Takanari (Mohapatra et al. 2004). Because the demand for assimilates for spikelet development is high in the latter, the source area for photosynthate supply is sustained by the upper three leaves, whereas the source activity of flag leaf only supports spikelet growth in the former. Further, partial detachment of some branches of panicle improved assimilates partitioning in favor of the spikelets of residual branches. Similarly thinning of dominant upper and middle spikelets of panicle improved setting rate and grain filling of inferior basal spikelets (Xu et al. 2013). The treatment up-regulates the expression of genes involved in the conversion of sucrose to starch metabolism and promotes sink efficiency of the disadvantaged spikelets by removing physiological dominance imposed by the superior spikelets (You et al. 2016). Hence, inferior spikelets that do not produce quality grains are amenable to grain filling when intrinsic physiological constraints are removed, the magnitude of which is beholden with an increase of panicle size.

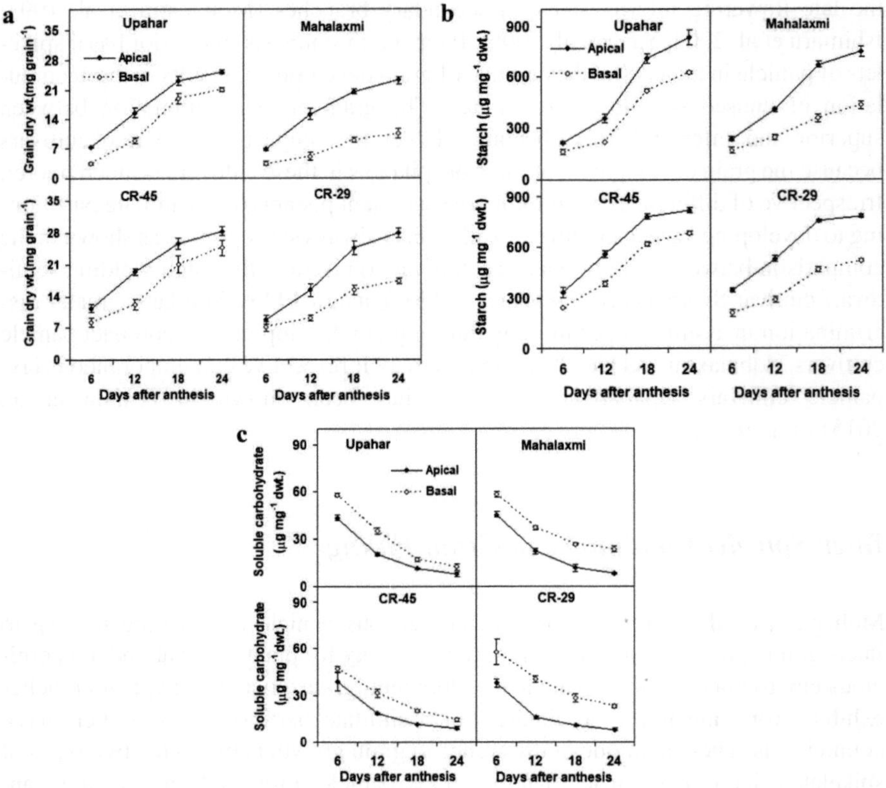

Fig. 9.5 (**a**) Dry weight and (**b**) starch and (**c**) soluble carbohydrate concentration of grains of apical and basal spikelets of panicle on the main shoot of compact-panicle (Mahalaxmi and CR-29) and lax- panicle-panicle (Upahar and CR-45) rice cultivars on different days after anthesis. The vertical bars represent ±s.d. (n = 3). (Panda et al. 2015)

Pathway of Assimilate Transport in Rice Caryopsis

In matured unpolished rice grain starch contributes 80–90% of the final dry weight. Sucrose is the major phloem solute transported to the developing seeds, wherein it is converted to starch and stored in endosperm cells. In the recently developed large panicle rice bearing numerous spikelets and high sink capacity, grain yield potential is undermined for poor inferior spikelet filling (Chen et al. 2019). It was proposed that efficient sugar transport can satisfy high sink strength and strong sink activity and thereby improve sugar unloading of disadvantaged spikelets. Assimilates necessary for endosperm and embryo development and storage in the endosperm are transported from the vegetative organs to the growing panicle. The path of sugar transport to rice kernel is not simple and involves both physical and physiological factors. At the source end, sucrose synthesized in mesophyll cells is loaded into the sieve cell-companion cell complex; the process of loading is either apoplastic by H^+- sucrose symporters or symplastic through plasmodesmatal connections. In recent years the importance of sucrose transporters has been elucidated. It was reported that gene expression levels of sucrose transporter OsSUTs are more efficient in the functional leaves of cultivar possessing a large number of superior spikelets compared to cultivar with less number of such spikelets (Chen et al. 2019).

After long-distance transport, sucrose is finally unloaded at the sink end after traversing through vascular bundles of the leaf sheath, stem, and rachis sequentially. A rachis branch vascular bundle reaches the base of spikelet and forms more branches to supply to the rudimentary glumes, two sterile glumes, lemma and palea, and filaments of six stamens. Finally, the main vascular bundle reaches the base of the ovary. At the ovary base, the main vascular bundle gives rise to three thin branches of which, one goes to the ventral side while the other two move to the lateral side. The lateral branches pass through the style and reach stigma. The main vascular bundle rises from the base of the ovary to the top of the ovule on the dorsal side and it is responsible for the major conduction of assimilates. As the ovary elongates in the early stage of grain development, the vascular bundle also elongates on the dorsal side. Close contact of vascular bundles of pericarp with growing endosperm ensures transfer of phloem assimilates from the former to the latter. On the dorsal side, the vascular bundles have their terminals at different levels and become thinner gradually as they reach the top of the ovule. The vascular bundles do not have direct contact with endosperm for the transfer of assimilates. The phloem sap of the pericarp moves through the nucellar projections laterally, before entering the endosperm.

Movement of Solutes

Phloem solutes are released into the maternal tissues of the pericarp that has sieve element-companion cell complex, vascular parenchyma; pigment strand, and nucellus (inner layer of pericarp) bound together in symplastic connections. In rice seed, phloem sap is transported symplastically in the vascular bundles of the pericarp on the dorsal side of caryopsis up to the terminal part of the grain, where it is released into companion cells, vascular parenchyma, pigment strand, and nucellar cells through plasmodesmata. The nucellus extends more than halfway through seed length. Sucrose moving out of the sieve element-companion cell complex laterally traverses through cells of the pigment strand and vascular parenchyma before reaching nucellus, just below the vascular bundles where the pigment strand is connected symplastically to the nucellus. At this point, the cuticle surrounding the endosperm is missing facilitating contact of the nucellar projection with the aleurone layers of

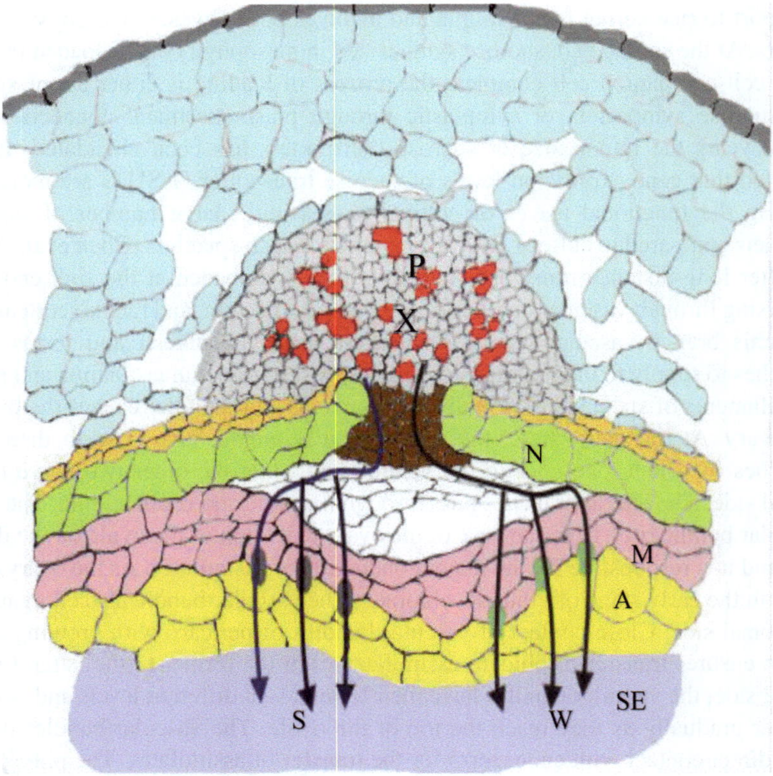

Fig. 9.6 A figure on vascular bundles of the ovule. The arrowheads indicate the direction of assimilate movement. (Adopted from Wu et al. 2016). P = phloem, X = xylem, A = aleurone, W = water, S = sugar, Se = starchy endosperm, N = nuclear tissue, M = boundary between maternal tissue and filial tissue

the endosperm. The companion cells are bigger compared to the sieve cells. The inner walls of pigment strand cells have a thick deposition of suberin-like materials that act as permeability barriers between apoplast and symplast of pigment strand. The sugars travel symplastically from pigment strand cells into the nucellus through high-frequency plasmodesmata connections at the junction between the two types of tissues only at the dorsal side of the caryopsis. This connection is not available on the ventral side. On the dorsal side, the cuticle surrounding the endosperm is absent and the pericarp maintains contact with the nucellus for easy unloading of solutes (Fig. 9.6). However, there are neither plasmodesmata between nucellus and aleurone layer and also nor between nucellus and seed coat (Oparka and Gates 1981; Furbank et al. 2001). The presence of a cuticle separates the residual part of the nucellus from the endosperm. Similar to nucellus, cells of the aleurone layers are also symplastically connected.

Movement of Assimilates and Water in Developing Caryopsis

The movement of solutes in the sieve tube from source organs to the pericarp of the developing kernel occurs by an osmotically driven hydrostatic pressure gradient-induced mass flow. Assimilates moving into the kernel in the first few days after anthesis (pre-storage phase) have more hexoses than sucrose, but the ratio between hexose and sucrose lowers rapidly with the onset of starch synthesis on day five (storage phase) (Ishimaru et al. 2005). Sucrose, the dominant sugar of mass flow, is transported symplastically in the vascular bundles of the pericarp up to the upper terminal part of the caryopsis. Unlike other cereals like wheat and barley, the unloading of sucrose in rice caryopsis is not limited to a specific part of the maternal filial interface. The unloading in rice grain is proposed to occur throughout the nucellus/endosperm interface and not confined to the tissue nearby of vascular bundles as seen in wheat and barley. The transport occurs across the entire nucellus/endosperm interface of the dorsal part of the caryopsis, rather than in the tissues close to vascular bundle/nucellar projection. During grain filling, sucrose enters into the endosperm because of the mass flow of solutes in an aqueous medium. Entry of the solutes into endosperm, however, prompts exit of water. Oparka and Gates (1981) suggested that water both enters and leaves the caryopsis through the isolated apoplast of the pigment strand, with no entry into the symplastic system. The inner walls of pigment strand cells are heavily suberised. The presence of suberin isolates the apoplast from symplast, which thereby facilitates opposing fluxes of assimilates and water to occur in the dorsal part of the caryopsis (Figs. 9.7 and 9.8) Water exiting from the developing caryopsis finds its way into the xylem vessels of vascular bundles of the pericarp. Because rice grain possesses no embryonic cavity, the movement of the solute from the maternal pericarp to the filial endosperm tissues is circumferential. The symplastic movement of phloem sap ends in the nucellar tissue and it is delivered into the apoplasm between nucellus and aleurone through active unloading. The nucellar projection/nucellus bears small densely packed cytoplasmic cells near the dorsal vascular bundle region adjacent to the multilayered aleurone/sub-aleurone of the filial endosperm. Both aleurone and nucellus are thin and single-layered on the non-vascular ventral side of the endosperm. The absence of plasmodesmata between nucellus and aleurone layer brings a change over from symplastic to apoplastic solute transport (Oparka and Gates 1981), sucrose actively unloaded from nucellus into the embryonic apoplasm is loaded energetically into aleurone layer cells. The onward movement of sucrose into the starchy endosperm cells is symplastic.

The Function of the Pericarp in Assimilate Transport

Thus, pericarp functions are crucial for grain filling. Its development is closely knitted with endosperm development. Endosperm cellularisation begins after cell multiplication stops in pericarp tissue. It functions as a temporary storage tissue for assimilates before grain filling. In rice fruit caryopsis, the pericarp, the presumptive

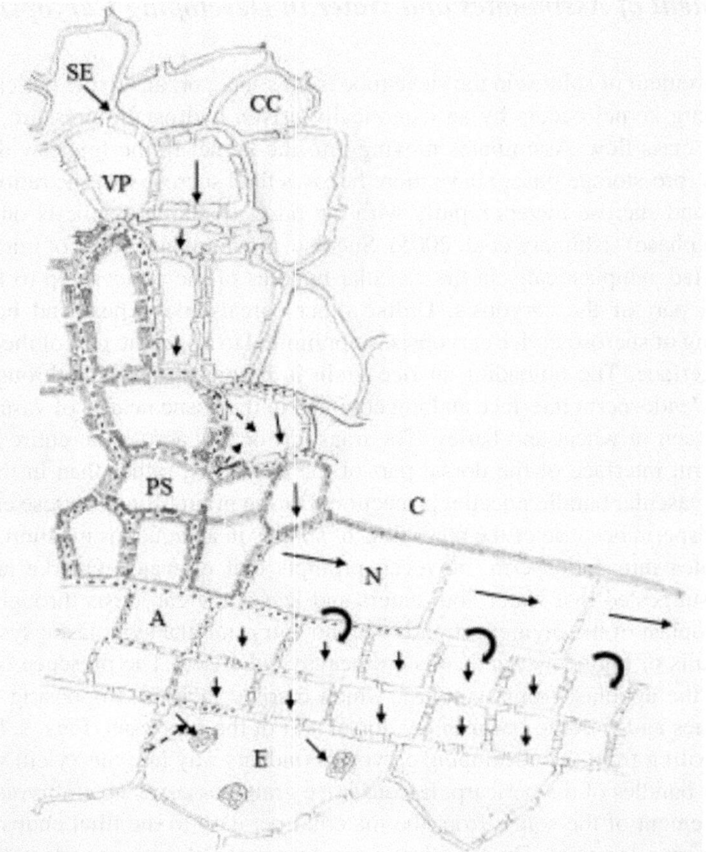

Fig. 9.7 Schematic representation of the movement of water and assimilates in the dorsal region of rice caryopsis. Assimilates move sympastically (darts) from sieve element (SE) - companion cell (CC) complexes, through the pigment strand (PS) and circumferentially via the nucellus (N). Transport across the apoplast between the nucellus and aleurone layer (A) is represented by a curved arrow. Water (broken lines) leaves the caryopsis via isolated cell walls of the pigment strand. E = endosperm, C = cuticular layer, VP = vascular parenchyma, X = xylem (Adopted from Oparka and Gates 1981)

seed coat, fuses with the endosperm at a later stage of grain development; the functions of seed coat are rendered by the external bracts lemma and palea. At the time of grain maturation, these bracts remain green and remobilize assimilates for grain filling (Zhang et al. 2016). Although the contribution to carbon fixation is insignificant, the presence of chlorophylls in the pericarp recapitulates its seed coat identity. Similar to the spikelets, distribution in time and space also influences pericarp development; pericarp growth becomes slower in the juvenile phase of seed development, but degradation is faster in the senescent phase in a basally placed late-developing inferior kernel compared to a superior kernel. A poorly-developed pericarp slackens grain filling by disturbing the delivery of assimilates passing through its tissues (Mohapatra and Mohapatra 2006).

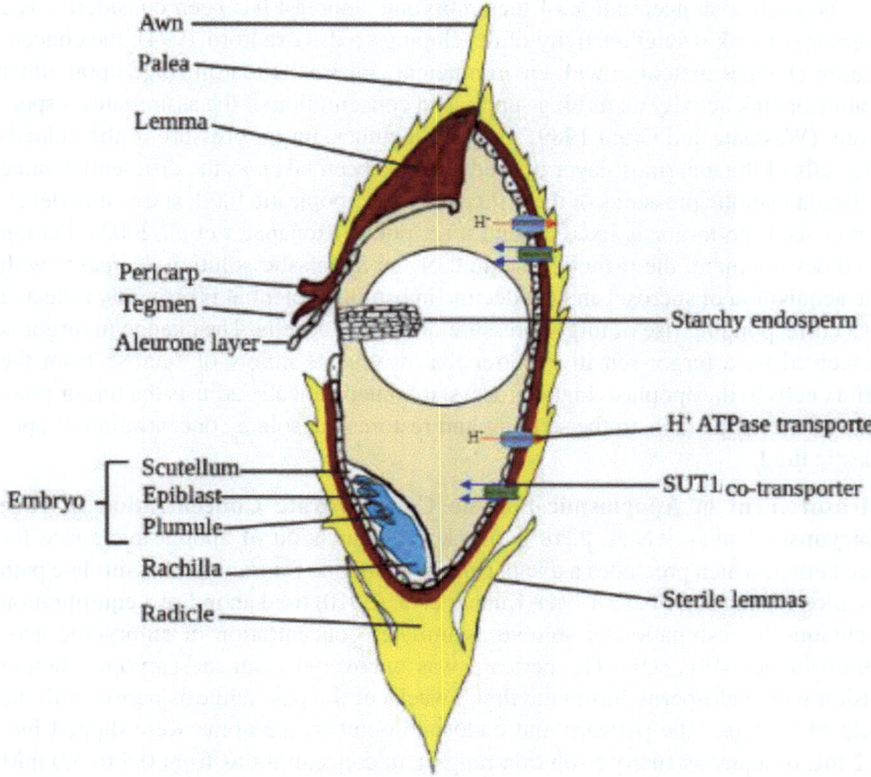

Fig. 9.8 Sucrose unloading into endosperm by SUT transporters in rice. The SUT1 keeps expressing throughout the 21 days of maturation whereas SUT2 through 5 expresses during the first week after anthesis. H^+-ATPase provides a driving force (H^+ ions) for sucrose co transporting into the endosperm

Bulk Flow of Solutes Irrespective of the Concentration Gradient

Bradford (1994) proposed the concept that assimilates movement into developing seed is a function of water relations and it is compatible with the Munch theory of the bulk flow of solutes in the phloem due to the generation of an osmotic driven pressure gradient between sources and sink organs. The seed pericarp is connected by three vascular bundles, a major one on the dorsal side and two minor ones on the lateral sides. The dorsal side vascular bundle plays a prominent role in the delivery of water, carbon assimilates, and nutrients to the developing caryopsis. During seed filling assimilates from the source organs are normally transported in the phloem as sucrose to the developing seed (Thorne 1985). It is recognized as the major translocatable solute in the phloem sap, the exchange of which between maternal and filial tissues determines the nature of carbon partitioning between the two tissues. Thus, sucrose concentration determines the osmotic potential of the sap for the generation of turgor motivating force for the movement and final unloading. There are other minor solutes like monosaccharides, amino acids and minerals and they contribute to the osmotic potential of phloem sap. However, they move along with the major solute in bulk flow, irrespective of their concentration gradient.

The sucrose concentration of the embryonic apoplast has been considered as a measure of sink strength/activity of developing seeds (Bradford 1994); the concentration changes matching with environmental constraints that infringe upon either source or sink activity disturbing supply and consumption of the assimilates respectively (Westgate and Grant 1989). In grain-legumes, turgor pressure of the unloading cells of the innermost layer of seed coat has been taken as the differential value between osmotic pressures of the efflux cells and apoplastic fluid, and that in developing seed the turgor is fixed around a set point (Mohapatra et al. 2003). During seed development, the osmotic potential of the apoplastic solution decreases with the acquisition of sucrose, and the decline in osmotic potential is promptly reflected in a corresponding rise of turgor pressure of the efflux cells. The change in turgor is perceived by a turgor-sensitive porter that drives the inflow of sucrose from the efflux cells to the apoplast. In the process, it homeostatically adjusts the turgor pressure of the efflux cells to the setpoint and restores the solute concentration of apoplastic fluid.

Measurement of Apoplasmic Soluble Carbohydrate Concentration in Rice Caryopsis Unlike wheat, there is no clear demarcation of apoplasmic space for rice kernel, which precludes a dye perfusion technique for tracing of assimilate path location (Wang and Fisher 1994). Kuanar et al. (2010) used an indirect equilibration technique for estimation of soluble assimilates concentration of embryonic apoplasm in rice (Fig. 9.9). The pericarp was uncovered from the caryopsis before fusion with endosperm during the first 2 weeks of the post-anthesis period with the help of forceps. The pericarp and endosperm-embryo combine were dipped into 0.2 mL of aqueous sucrose solution ranging in concentrations from 0.5 to 5.0 mM in small tubes for 5 seconds and shaken well allowing diffusion of solutes in the medium. The short duration of the wash is intended only to capture the apoplasmic sugars adhering to the outer layer of endosperm/embryo and the inner layer of the pericarp, and exclude the possibility of facilitated diffusion of solutes (Wang and Fisher 1995) into and from the dipped structures.

The seed components were removed and 0.1 mL of the solution was taken for estimation of soluble assimilates by the colorimetric method after the development of color in the phenol-sulphuric acid method. A control set of sucrose solutions like the previous one was prepared without the addition of the seed components and processed similarly for the colorimetric analysis. The percentage of change in absorbance in each concentration level that occurred with the addition of seed components was calculated; taking the percent of change in absorbance versus sucrose concentration, a graph was plotted. When the concentration of solutes in the surrounding solution is low, it gains assimilates from the caryopsis components reflecting a sharp increase in the percentage of rise in absorbance. The gain in assimilates concentration and consequential rise in absorbance percentage decrease progressively with the increase of sucrose concentration in the surrounding medium. Using the null hypothesis, the concentration of assimilates in the caryopsis apoplasm becomes isotonic with that of the external medium, where no change in percentage of absorbance occurs with the addition of the caryopsis components.

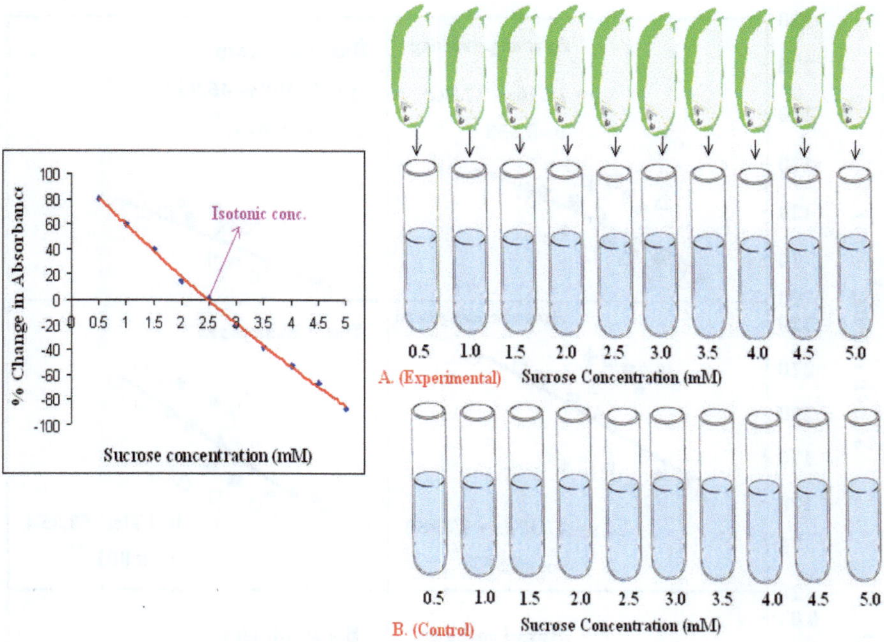

Fig. 9.9 Using the null hypothesis, the concentration of assimilates in the caryopsis apoplasm was estimated in developing rice caryopsis by Kuanar et al. 2010

Sucrose Transporter

The presence of a turgor-sensitive homeostat, similar to grain legumes, for passive unloading of sucrose in rice caryopsis, has not been evident so long. Notwithstanding the mechanism of unloading, the concentration of solutes in the apoplastic fluid, however, bears important implications determining the nature of grain filling in rice caryopsis. It was shown that the concentration of soluble carbohydrates in the fluid directly correlates with cell number, weight, and starch concentration of the developing rice caryopsis (Kuanar et al. 2010; Fig. 9.10). It implies that grain development post-anthesis depends on the assimilates unloaded into this compartment and fast-developing superior spikelets use them up more efficiently than the slow-growing inferior spikelets.

In rice caryopsis, the onset of grain development initiates carrier-mediated efflux of sucrose from the innermost cell layer of the nucellus into the apoplastic space, from which the solute enters into cells of the outermost aleurone by the carrier-mediated influx. There are five types of transporters in rice (OsSUTs) that handle sucrose loading and unloading (Lim et al. 2006) of which *OsSUT1* is expressed immediately after anthesis and maintains abundance until 25 days post-anthesis (Furbank et al. 2001; Aoki et al. 2003). In comparison *OsSUT2*, *OsSUT3*, *OsSUT4*,

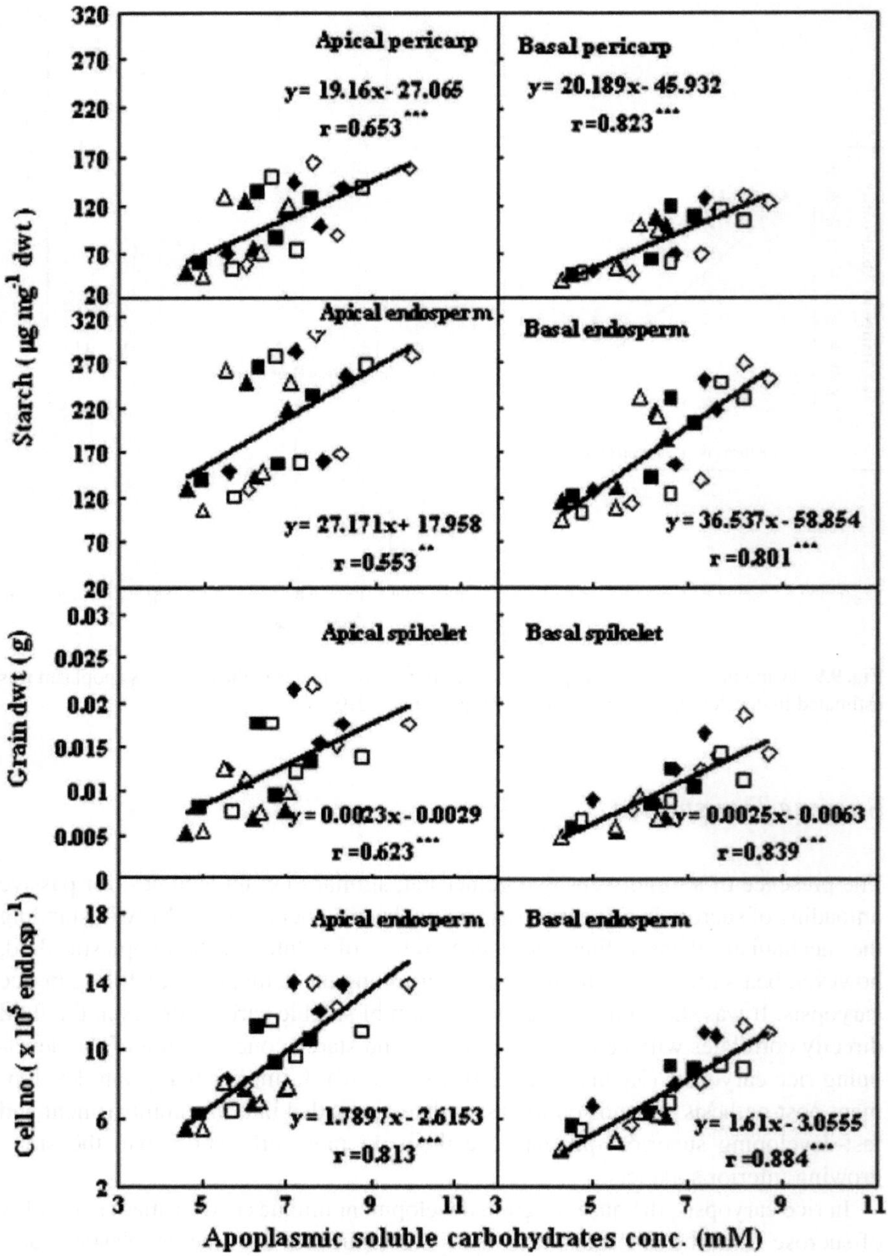

Fig. 9.10 Relationship of concentration of embryonic apoplasmic soluble carbohydrates with starch, weight, and cell number of developing caryopsis in different cultivars of rice during the first half of grain filling period in the dry season of 2006 and wet season of 2007.*, **, *** = significant at 0.05, 0.01 ad 0.001 levels of probability respectively (df = 23). Symbols: *closed* dry season, *open* wet season, *Rhombus*- grain size big (cultivars- Jogesh, Ramachandi), *square*- grain size medium (cultivars-Udayagiri, Mahanadi), *triangle*- grain size small (cultivars- Sidhanta, Ganjeikali), Reference – Kuanar et al. 2010

and *OsSUT5* expression are limited to the first week after flowering. *In situ* cellular localization analyses revealed the presence of transcripts and protein products of *OsSUT1* in nucellar projections of the pericarp, nucellar epidermis, and outer aleurone layer of endosperm, but not in the starchy cells of the endosperm. The role of the transporter is crucial in the release of sucrose from the nucellus and its apoplastic uptake in the aleurone tissue. It is a symporter energetically linked to the H^+-ATPase transporter. Proton motive force generated by hydrolysis of ATP by H^+-ATPase is used for driving the SUT1 symporter for ferrying one proton along with a sucrose molecule across the plasma membrane. The importance of OsSUT1 transporter has been confirmed because anti-sense suppression of *OsSUT1* gene in rice leads to impaired grain filling and seed germination (Scofield et al. 2002). *OsSUT1* is located on chromosome number 3 and its allele *SUT1–2* is linked to improvement in grain filling of extra-heavy weight panicle rice cultivars; the frequency of the allele is 74.0% in *indica* rice compared to only 6.3% in japonica rice (Kato and Horibata 2011).

Role of a Turgor Homeostat in Rice Caryopsis

In rice kernel, the early part of seed development is a pre-storage phase comprising pericarp growth and endosperm cellularization that is followed by maturation or storage phase determined by brisk starch accumulation. The rise in filial demand for assimilates in the endosperm is integrated with an increase of phloem import. Although shreds of evidence are wanting, assimilates consumed in endosperm growth alters the solute concentration of apoplasmic fluid, which is perceived by a turgor homeostat in the apoplastic fluid that drives more intake of solutes. Therefore, turgor motivated mass flow of solutes to the endosperm is linked with the coordinate action of the sucrose transporters, which impacts sink strength of developing kernels. The poor activity of OsSUT1 can reduce assimilate unloading and thereby decrease the mass flow of solutes to the pericarp, ultimately lowering apoplasmic solute concentration. Assimilate concentration in the embryonic apoplast fluid of slow- growing inferior caryopsis of rice panicle is much lower than that of the fast-developing superior spikelets (Kuanar et al. 2010). It was opined that the functional disadvantages of the pericarp could be a factor responsible for poor unloading of assimilates and accumulation of solutes in the pericarp, disrupting supply to the apoplasm. Nevertheless, the concentration of solutes in the apoplasm could be an indicator of the assimilate-filling capacity of rice seed.

References

Adriani DE, Dingkuhn M, Dardou A, Adam H, Luquet D, Lafarge T (2016) Rice panicle plasticity in near-isogenic lines carrying a QTL for larger panicle is genotype and environment-dependent. Rice 9:28. https://doi.org/10.1186/s.12284-016-0101-x

Afza R, Shen M, Zapata-Arias FJ, Xie J, Fundi HK, Lee K-S, Bobadilla-Mucino E, Kodym A (2000) Effect of spikelet position on rice anther culture efficiency. Plant Sci 153:155–159

Aoki N, Hirose T, Scofield GN, Whitfeld PR, Furbank RT (2003) The sucrose transporter gene family in rice. Plant Cell Physiol 44:223–232

Bradford KJ (1994) Water stress and the water relations of seed development: a critical review. Crop Sci 34:1–11

Chen L, Deng Y, Zhu H, Hu Y, Jiang Z, Tang S, Wang S, Ding Y (2019) The initiation of inferior grain filling is affected by sugar translocation efficiency in large panicle rice. Rice 12(1). https://doi.org/10.1186/s12284-019-0333-7

Furbank RT, Scofield GN, Hirose T, Wang X-D, Patrick JW, Offler CE (2001) Cellular localization and function of a sucrose transporter OsSUT1 in developing rice grains. Funct Plant Biol 28:1187–1196

Ishimaru T, Matsuda T, Ohsugi R, Yamagishi T (2003) Morphological development of rice caryopses located at the different positions in a panicle from early to middle stage of grain filling. Func Plant Biol 30:1139–1149

Ishimaru T, Hirose T, Matsuda T, Goto A, Takahashi K, Sasaki H, Terao T, Ishii R, Ohsugi R, Yamagishi T (2005) Expression patterns of genes encoding carbohydrate-metabolizing enzymes and their relationship to grain filling in rice (Oryza sativa L.): comparison of caryopses located at different positions in a panicle. Plant Cell Physiol 46:620–628

Jeng TL, Wang CS, Chen CL, Sing JM (2003) Effects of grain position on the panicle on starch biosynthetic enzyme activity in developing grains of rice cultivar Tainung 67 and its NaN3-induced mutant. J Agric Sci 141:303–311

Kato T (2004) Effect of spikelet removal on the grain filling of Akenohoshi, a rice cultivar with numerous spikelets in a panicle. J Agric Sci 142:177–181

Kato T, Horibata A (2011) Non-random distribution of the alleles for good grain filling at *OsAGPS2* and *OsSUT1* among a wide range of rice (Oryza sativa L.) cultivars. Breed Sci 61:217–220

Kuanar SR, Panigrahi R, Kariali E, Mohapatra PK (2010) Apoplasmic assimilates and grain growth of contrasting rice cultivars differing in grain dry mass and size. Plant Growth Regul 61:135–151

Li G, Pan J, Cui K, Yuan M, Hu Q, Wang W, Mohapatra PK, Nie L, Huang J, Peng S (2017) Limitation of unloading in the developing grains is a possible cause responsible for low stem non-structural carbohydrate translocation and poor grain yield formation in rice through verification of recombinant inbred lines. Front Plant Sci 8:1369

Lim JD, Cho JI, Park YI, Hahn TR, Choi SB, Jeon JS (2006) Sucrose transport from source to sink seeds in rice. Physiol Plant 126:572–584

Mohapatra R, Mohapatra PK (2006) Ethylene control of seed coat development in low and high sterile semidwarf *indica* rice cultivars. Plant Growth Regul 50:47–55

Mohapatra P, Patel R, Sahu S (1993) Time of flowering affects grain quality and spikelet partitioning within the rice panicle. Funct Plant Biol 20:231–241

Mohapatra P, Turner N, Siddique K (2003) Assimilate partitioning in chickpea (Cicer arietinum L.) in a drought-prone environment. In: Management of agricultural drought: agronomy and genetic options. Science Publishers Inc., Enfield, pp 173–188

Mohapatra PK, Masamoto Y, Morita S, Takanashi J, Kato T, Itani T, Adu-Gyamfi JJ, Shunmugasundaram M, Nguyen NT, Fujita K (2004) Partitioning of 13C-labelled photosynthate varies with growth stage and panicle size in high-yielding rice. Funct Plant Biol 31:131–139

Mohapatra PK, Panigrahi R, Turner NC (2011) Chapter 5: Physiology of spikelet development on the rice panicle: is manipulation of apical dominance crucial for grain yield improvement? Adv Agron 110:333–359

Oparka K, Gates P (1981) Transport of assimilates in the developing caryopsis of rice (*Oryza sativa* L.): ultra-structure of the pericarp vascular bundle and its connection with aleurone layer. Planta 151:561–573

Panda B, Badoghar A, Das K, Panigrahi R, Kariali E, Das SR, Dash SK, Shaw BP, Mohapatra PK (2015) Compact panicle architecture is detrimental for growth as well as sucrose synthase activity of developing rice kernels. Funct Plant Biol 42:875–887

Patel R, Mohapatra PK (1996) Assimilate partitioning within floret components of contrasting rice spikelets producing qualitatively different types of grains. Funct Plant Biol 23:85–92

Scofield GN, Hirose T, Gaudron JA, Upadhyaya NM, Ohsugi R, Furbank RT (2002) Antisense suppression of the rice transporter gene, *OsSUT1*, leads to impaired grain filling and germination but does not affect photosynthesis. Funct Plant Biol 29:815–826

Takai T, Fukuta Y, Shiraiwa T, Horie T (2005) Time-related mapping of quantitative trait loci controlling grain-filling in rice (Oryza sativa L.). J Exp Bot 56:2107–2118

Thorne JH (1985) Phloem unloading of C and N assimilates in developing seeds. Annu Rev Plant Physiol 36:317–343

Umemoto T, Nakamura Y, Ishikura N (1994) Effect of grain location on the panicle on activities involved in starch synthesis in rice endosperm. Phytochemistry 36:843–847

Vergara BS (1988) Raising the yield potential of rice. Philippines Technol J 13:3–19

Wang N, Fisher DB (1994) Monitoring phloem unloading and post-phloem transport by microperfusion of attached wheat grains. Plant Physiol 104:7–16

Wang N, Fisher DB (1995) Sucrose release into the endosperm cavity of wheat grains apparently occurs by facilitated diffusion across nucellar membranes. Plant Physiol 109:579–585

Wang F, Cheng F, Zhang G (2008) Impact of cultivar variation in grain density of rice panicle on grain weight and quality. J Sci Food Agric 88:897–903

Westgate ME, Grant DLT (1989) Water deficits and reproduction in maize: response of the reproductive tissue to water deficits at anthesis and mid-grain fill. Plant Physiol 91:862–867

Wu X, Liu J, Li D, Liu CM (2016) Rice caryopsis development I: dynamic changes in different cell layers. J Integr Plant Biol 58:772–785

Xu X, Vergara BS (1986) Morphological changes in rice panicle development: a review of the literature, IRRI research paper series no 117. International Rice Research Institute, Philippines

Xu F, Guo X, Zhang L, Xiong H, Zhu YC, Liu M, Zhou XB (2013) Effects of sink-source structures on filling of superior and inferior spikelets of mid-season hybrid rice. J Agric Sci Technol (Beijing) 15:96–101

Yang J, Zhang J (2010) Grain-filling problem in 'super' rice. J Exp Bot 61:1–5

Yang J, Peng S, Visperas RM, Sanico AL, Zhu Q, Gu S (2000) Grain filling pattern and cytokinin content in the grains and roots of rice plants. Plant Growth Regul 30:261–270

Yang J, Peng S, Zhang Z, Wang Z, Visperas RM, Zhu Q (2002a) Grain and dry matter yields and partitioning of assimilates in japonica/indica hybrid rice. Crop Sci 42:766–772

Yang J, Zhang J, Huang Z, Wang Z, Zhu Q, Liu L (2002b) Correlation of cytokinin lvels in the endosperms and roots with cell number and cell division activity during endosperm development in rice. Ann Bot 90:369–377.

Yang J, Zhang J, Wang Z, Liu K, Wang P (2006) Post-anthesis development of inferior and superior spikelets in rice in relation to abscisic acid and ethylene. J Exp Bot 57:149–160

Yoshida S (1972) Physiological aspects of grain yield. Annu Rev Plant Physiol 23:437–464. https://doi.org/10.1146/annurev.pp.23.060172.002253

You C, Zhu H, Xu B, Huang W, Wang S, Ding Y, Liu Z, Li G, Chen L, Ding C, Tang S (2016) Effect of removing superior spikelets on grain filling of inferior spikelets in rice. Front Plant Sci 7:1161. https://doi.org/10.3389/fpls.2016.01161

Zhang X, Lei J, Zheng D, Liu ZH, Zhang W, Wang S, Ding Y (2016) Amino acid composition of leaf, grain, and bracts of japonica rice (*Oryza sativa* ssp. japonica) and its response to nitrogen fertilization. Plant Growth Regul 82:1–9

Zheng D, Tian Z, Rao Y, Dong G, Yang Y, Huang L, Leng Y, Xu J, Sun C, Zhang G, Li Z, Gao H, Hu X, Guo L, Xiong G, Wang Y, Li J, Qian Q (2017) Rational design of high-yield and superior-quality rice. Nat Plants 3(4):17031

Zhu Q, Cao X, Luo Y (1988) Growth analysis on the process of grain filling in rice. Acta Agron Sin 14:182–112

Chapter 10
Enzymes Controlling Starch Biosynthesis

Abstract Starch synthesis in a developing rice grain has been the major factor for grain filling. The physiological constraints that unduly terminate starch synthesis in the inferior type spikelets, and solutes moving preferentially towards superior spikelets have been subject of intensive investigation. A better understanding of how the sink strength of individual spikelets is governed by time and spatial distribution along the panicle axis holds promises for improving the yield ceiling of rice. Sucrose released into the embryonic apoplasm of developing caryopsis is loaded into endosperm cells by proton motive force generated by H^+ATPase that drives a sucrose-H^+ symporter. In the endosperm cells, sucrose is converted into amylose or amylopectin in the starch biosynthesis pathway through a co-ordinated series of enzyme activities, downregulation of which in inferior type caryopsis undermines starch filling. Altogether, there are 33 enzymes involved in the starch biosynthesis pathway of rice endosperm, out of which sink strength-related effects contributing to variation in grain filling of individual spikelets of panicle have been studied in four key enzymes. These enzymes are sucrose synthase (EC 2.4.1.13 SUS), adenosine diphosphoglucose pyrophosphorylase (EC 2.7.7.27 AGPase), starch synthase (EC 2.4.1.21 SS), and starch branching enzyme (EC 2.4.1.18 BE).

Keywords Starch synthesis · Sucrose · Starch synthesizing enzymes · Sink strength

Sub-Optimal Sink Strength Undermines Grain Filling

Dry unpolished matured rice grain consists of 80–90% starch (Yoshida 1972). Starch largely contributes to grain weight and quality. Therefore starch synthesis in a developing rice grain has been the major factor for grain filling. The physiological constraints that limit starch synthesis in the inferior type spikelets, which ultimately influence grain yield, have been a subject of intensive investigation in the past 2–3 decades. During the grain maturation period phloem assimilates preferentially move towards the apical superior spikelets in the panicle. This selective transport in turn

P. K. Mohapatra, B. B. Sahu, *Panicle Architecture of Rice and its Relationship with Grain Filling*, https://doi.org/10.1007/978-3-030-67897-5_10

limits sink efficiency of basal inferior spikelets for starch synthesis (Mohapatra et al. 1993). Restriction in assimilate partitioning contributes to the limitation of sink efficiency in starch synthesis and assimilates not used in growth accumulate in the failing inferior spikelets (Patel and Mohapatra 1996; Liang et al. 2001). As discussed in the previous chapter, this problem has been more acute in the recently introduced large panicle rice cultivars bearing numerous spikelets in the panicle (Fu et al. 2011, 2013; Panda et al. 2015). The decline in sink strength of these cultivars has been a serious problem in grain filling (Liang et al. 2001; Sun et al. 2015) The inferior spikelets are not able to convert phloem translocated soluble carbohydrates into endosperm starch. Deficiency in sink strength of spikelet disadvantaged in the distribution of time and space on the panicle axis leads to accumulation of unused sugars in the kernel (Mohapatra et al. 1993).

The possession of a high spikelet number per panicle in the newer cultivars has been a trait important for the improvement of panicle yield and yield potential. But the advantage has been undermined because of poor starch synthesis in the later flowered inferior spikelets on the proximal part of the panicle. It is surmised that the spikelet-specific variation in sink efficiency for starch biosynthesis has been responsible for the asynchronous development of spikelets of rice panicle in a hierarchical pattern. A better understanding of how the sink strength of individual spikelets is governed with time and with spatial distribution along the panicle axis can open some avenues for improving the yield ceiling of rice. Enumeration of endosperm cell number multiplication and an assay of the activities of starch synthesizing enzymes of the kernel in the post-anthesis stage have confirmed this proposition (Kuanar et al. 2010; Fu et al. 2013; Panda et al. 2015).

Starch Biosynthesis Begins with the Entry of Phloem Sucrose

During grain development, endosperm growth is sustained primarily by sucrose in the phloem solute to the seed. Sucrose moves by a pressure gradient-driven mass flow in the phloem pericarp before it is actively unloaded into the embryonic apoplasm surrounding the endosperm and embryo. The role of pericarp in phloem unloading is crucial. It was shown that the application of ethylene regulates the concentration of phloem solutes by affecting pericarp activity for assimilating unloading (Mohapatra and Mohapatra 2006; Kuanar et al. 2010; Mohapatra 2012). Sucrose released into embryonic apoplasm is loaded into endosperm cells by proton motive force generated by H^{+}ATPase that drives a sucrose-H^{+} symporter (Furbank et al. 2001). Hence, a co-ordinated activity of the symporter with sink activity of the endosperm for starch synthesis is essential for grain filling; but it does not occur in all spikelets and depending on sink strength, fertile spikelets of rice panicle take up sucrose delivered into the apoplasm. Starch synthesis of individual grains not being identical in the panicle, it is logical to assume unbalanced activities of enzymes

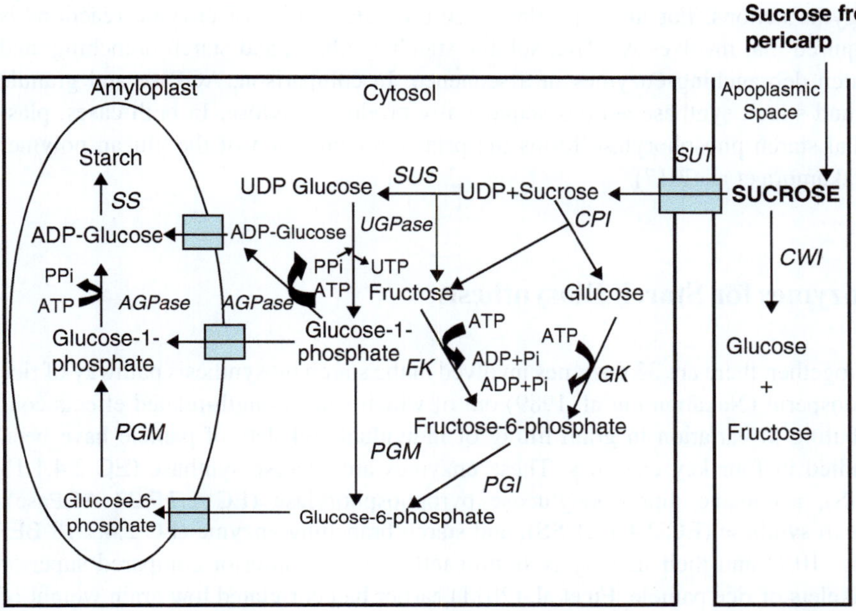

Fig. 10.1 Starch biosynthesis pathway in rice endosperm cell (Panigrahi et al. 2019). The details of enzymes and transporters regulating the pathway are *CWI* Cell wall invertase, *CPI* Cytoplasmic invertase, *SUT* Sucrose transporter, *SUS* Sucrose synthase, *GK* Glucokinase, *FK* Fructose kinase, *PGI* Phosphoglucoisomerase, *PGM* Phosphoglucomutase, *UGPase* UDP-glucose pyrophosphorylase, *AGPase* ADP-glucose pyrophosphorylase, *SS* Starch synthase

responsible for sucrose splitting and the follow-up steps of starch biosynthesis pathway in the endosperm (Fig. 10.1). Up to date, there has been no report on discrimination in the provision of substrate limiting starch biosynthesis of individual spikelets of panicle during grain development.

Synthesis of Grain Starch

In the endosperm cells, sucrose is converted into amylose or amylopectin in the starch biosynthesis pathway. Both amylose and amylopectin are D-glucose homopolymers, where the linear type amylose comprises α-D-glucan units united with one another by 1, 4 glycosidic linkages. In comparison, amylopectins are highly branched glucans, in which glucan branching is formed by α-1, 6 glycosidic bonds. High amylose content ensures better grain quality in rice and vice versa (Matsue et al. 1995), whereas amylopectin contributes structurally to starch granule organization in the endosperm (Jeon et al. 2010). Biosynthesis of both of the glucose homopolymers in the amyloplasts of endosperm cells needs a series of co-ordinated enzyme actions. For amylopectins, a co-ordinated series of enzyme reactions is required that involves AGPase, soluble starch synthase, and starch branching, and starch debranching enzymes in a sequence. In comparison, AGPase and granule bound starch synthase actions sequentially produce amylose. In both cases, plastidial starch phosphorylase forms the primer for initiation of the glucan polymer (Nakamura et al. 2017).

Enzymes for Starch Biosynthesis

Altogether, there are 33 enzymes involved in the starch biosynthesis pathway of rice endosperm (Nakamura et al. 1989) out of which sink strength-related effects contributing to variation in grain filling of individual spikelets of panicle have been studied in four key enzymes. These enzymes are sucrose synthase (EC 2.4.1.13 SUS), adenosine diphosphoglucose pyrophosphorylase (EC 2.7.7.27 AGPase), starch synthase (EC 2.4.1.21 SS), and starch branching enzyme (EC 2.4.1.18 BE) (Fig. 10.1) and their activity is significantly lower in inferior compared superior spikelets of rice panicle. Fu et al. (2011) earlier had correlated low grain weight of inferior spikelet to small sink size (small number of endosperm cells) and poor sink activity (sub-optimal activities of starch synthesizing enzymes) that undermine sink strength of the developing kernel. Intensive research on the function of the enzymes has provided information on how both the fractions of grain starch amylose and amylopectins are synthesized and distributed and how grain quality is determined.

Entry of Sucrose into Starch Biosynthesis Pathway

Accumulation of starch in endosperm cells is primarily sustained by the availability of sucrose in the phloem solute. Upon entering into grain starch biosynthetic pathway sucrose is converted into hexoses by either sucrose synthase (SUS) or cell wall invertase (CWI) (Fig. 10.1). In the embryonic apoplasm sucrose is cleaved irreversibly into glucose and fructose by CWI before they move into the endosperm cells actively with help of hexose transporters (Hirose et al. 2002). The reaction catalyzed by CWI is exothermic and irreversible. Conversely, sucrose remains structurally unchanged while it enters into the endosperm cells directly with help OsSUT transporters and cleaves into monosaccharides by the activity of SUS therein. Unlike CWI, the reaction catalyzed by SUS is freely reversible. Although CWI activity is not the preferred pathway for carbon partitioning of rice caryopsis, its role becomes essential in the early stage of caryopsis development, when endosperm cells multiply rapidly before starch accumulation begins in the amyloplasts (Ishimaru et al. 2005). It is construed that hexoses produced by CWI activity are used in the mitotic cell division of the endosperm. Thus, assimilate partitioning, although very low at this critical stage fixes the kernel size and grain yield (Morey et al. 2018). Also, there are instances of correlation of CWI activity with grain sink strength at the early stage of seed development (Wang et al. 2008). In transgenic maize improvement of CWI activity increased grain starch by 20% (Li et al. 2013), which supports the notion that this enzyme is essential in the formation of cell components when rapid cell division of storage tissue occurs before starch accumulation in the endosperm.

Following the entry of sucrose into the cytoplasm a chain of events begins. Sucrose is converted into UDP-glucose and fructose by SUS activity. A phosphorylase action with the participation of ATP releases the UDP moiety while converting glucose into glucose-1-phosphate. In the sequel, glucose-1-phosphate becomes a target for the AGPase enzyme. AGPase enzyme combines the phosphorylated glucose to ATP for the production of ADP-glucose. Thereafter through the action of starch synthase, the glucose part of ADPG is added to increase the chain length of glucan primers in the amyloplast by the establishment of α1,4 glycosidic linkage to the non-reducing end of the chain. The enzyme starch phosphorylase (Pho EC 2.4.1.1) elongates the glucan polymer chain using glucose-1-phosphate as a substrate (Crofts et al. 2015). Branching of the glucan chains is formed by starch branching enzyme (SBE) by initiating new branches with α1,6 glycosidic linkage. The activities of the enzymes increase temporally to peak level in the first 2 weeks after anthesis before declining steadfastly to a minimum at maturity in the superior spikelets (Fu et al. 2013). In comparison, the rise and fall in activities were much slower with the progress of time for the inferior spikelets (Fig. 10.2). Starch debranching enzymes (DBEs) like pullulanase (PUL EC 3.2.1.41) and isoamylase (ISA EC 3.2.1.68) hydrolyze and remove α1–6 linked branches to give water-insoluble properties of organized amylopectin structure (Jeon et al. 2010).

Sucrose Synthase

Sucrose synthase is the first enzyme mostly responsible for priming the entry of sucrose into the starch biosynthesis pathway. Sucrose synthase activity is followed by hexokinase, UDP-glucose pyrophosphorylase, and phosphoglucomutase preceding activity of starch synthase. Sucrose synthase exhibits high activity in rapidly growing tissues; in general, the activity gives a measure for sink strength. In rice, the activity of sucrose synthase is positively linked to the strength of the developing kernel to accept sucrose in grain filling and convert it into fructose and uridine diphosphate glucose (Kato 1995; Naik and Mohapatra 2000). The poor activity of the enzyme directly limits starch synthesis to the detriment of grain filling in the inferior type spikelets of rice panicle (Mohapatra et al. 2011). In comparison, greater activity of the enzyme ensures the production of a high-density well-filled kernel in

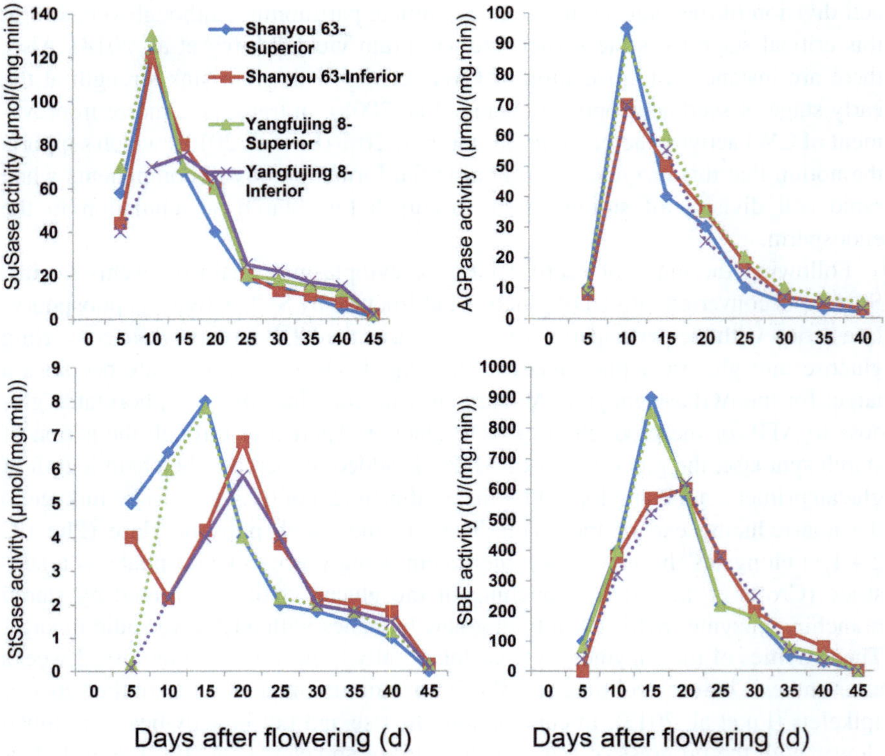

Fig. 10.2 Changes in activities of Sucrose synthase (SuSase), Adenosine diphosphate pyrophosphorylase (AGPase), Starch synthase (StSase, Starch branching enzyme (SBE) in superior and inferior spikelets of super rice cultivars Liangyoupeijiu (*indica*-inclined two-line hybrid) and Yangfujing 8 (*japonica*) during grain maturation (Adopted from Fu et al. 2013)

superior spikelets of rice panicle. Because of the close relationship with grain filling, the activity of sucrose synthase has been accepted as a potential indicator for the development of rice kernels (Counce and Gravois 2006; Tang et al. 2009; Panda et al. 2015). The activity of the enzyme increases rapidly during the initial phase of grain development, when cell multiplication rate is rapid and grain growth is in an exponential stage. The enzyme CWI also splits sucrose into glucose and fructose, but its activity is mostly limited to the first 4 days of grain filling (Hirose et al. 2002), not thereafter. During the later stages, when grain growth achieves greater momentum, the action of sucrose synthase predominates over that of invertase (Naik and Mohapatra 2000).

Equations:

Sucrose synthase

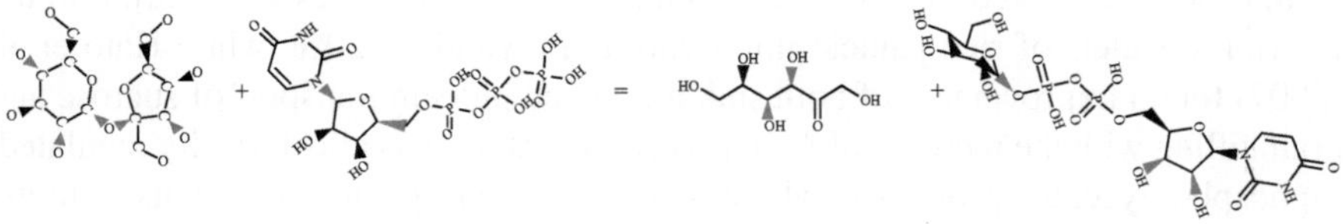

Sucrose + UTP = Fructose + UDP-glucose

Invertase

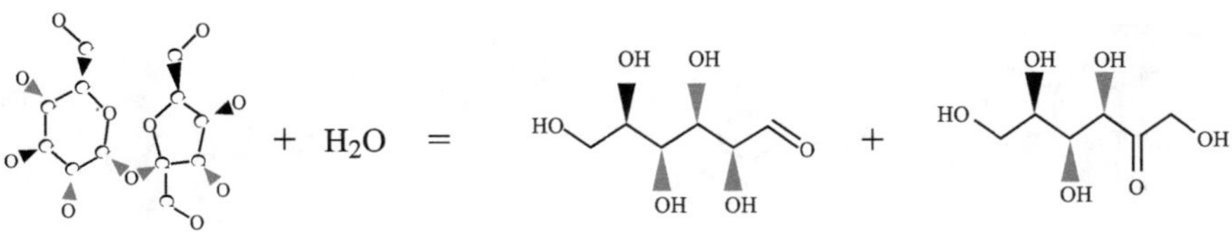

Sucrose + H_2O = Glucose + Fructose

ADP-Glucose Pyrophosphorylase (AGPase)

UDP-glucose is further metabolized into glucose-1-phosphate by phosphorylase action. Glucose and fructose generated by the activities of the two enzymes (SUS and CWI) are later converted into glucose-1-phosphate also and this becomes a substrate for AGPase action. AGPase enzyme controls a rate-limiting step in the starch biosynthesis pathway. It is an allosteric heterotetramer comprising two large and two small identical sub-units each. The enzyme combines glucose-1-phosphate and ATP to produce ADP-glucose and pyrophosphate in the cytoplasm. ADP-glucose enters into amyloplast for the synthesis of long-chain amylose or amylopectins. The formation of ADP-glucose is the first committed step for starch biosynthesis, but there is a difference in glucan chain syntheses of amylose and amylopectin. For amylopectin synthesis, a co-ordinated series of enzymatic reactions involve AGPase, soluble starch synthase, starch branching, and starch debranching enzymes in a sequence. In comparison, AGPase and granule bound starch synthase produce the amylose chain. For both amylose and amylopectin, it is plastidial starch synthase that forms the primer initiation of the glucan chain (Hwang et al. 2016a, b; Nakamura et al. 2017). In both cases, AGPase is the glucosyl donor for the polymerization of glucose to amylose or amylopectin. Plastidial starch phosphorylase is the enzyme that forms the primer for the initiation of the glucan chains in both amylose and amylopectin. 3-phosphoglyceric acid (3PGA) is the positive and inorganic phosphate (Pi) is the negative effectors for the AGPase. Specific reports are linking AGPase activity to the sink strength of rice kernel for starch synthesis. Smidansky et al. (2003) suggested that enhanced AGPase activity promotes seed setting in the inferior spikelets of rice panicle improving grain yield by 20%, while Kato et al. (2007) found improvement of both sink mobilized phloem transport of sucrose and grain filling with the increase of AGPase activity. AGPase is allosterically regulated. 3-phosphoglycerate promotes, whereas inorganic phosphates inhibit its activity. Thus, inorganic phosphate levels equal to or higher than 3-phosphoglycerate downregulate AGPase activity.

Equation for AGPase

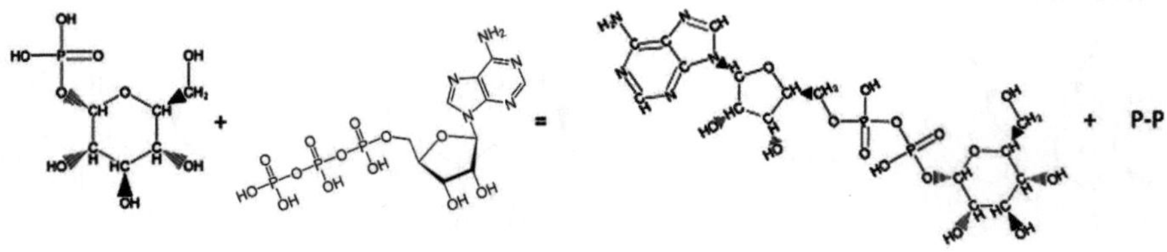

Glucose-1-phosphate + ATP = ADP-glucose + P-P

Starch Synthase

Glucose moiety of ADP-glucose is used to extend chain length of amylose and amylopectin in rice endosperm cell by the formation of linkage to the non-reducing end of α– glucan primer. Two forms of starch synthase, granule bound starch synthase (GBSS) and soluble starch synthases (Soluble SS) are involved in the establishment of the glycosidic linkages; the former is responsible for amylose formation while the latter increases the chain length of amylopectins. Because amylose content and grain quality of superior spikelets are much higher than the inferior spikelets of rice panicle, greater emphasis has been laid on GBSS activity in the regulation of grain development. The activity of the enzyme is necessary for maintaining the integrity of starch granules, which increases up to 21 days post-anthesis i.e. equivalent to grain hardening stage. Compared to superior spikelets, the activity of GBSS is poor in inferior spikelets. Japonica rice cultivars having better grain filling ability exhibit greater activity of the enzyme than the *indica* cultivars. Jeng et al. (2007) revealed that poor expression of GBSS resulted in low grain amylose content that can be improved with chemical mutation.

Starch Branching Enzyme (BE)

As discussed before, both amylose and amylopectins are α–glucan polysaccharide chains where the monomeric glucose moieties are linked with one another. Amylose type glucans have polysaccharide chains mostly linear with a few branches, whereas amylopectin type glucans are highly branched. The latter has a distinct ordered pattern of an arrangement named tandem cluster structure, in which most of the side chains are aligned in parallel and neighboring chains form double helices when linear portions of facing chains reach a length exceeding the degree of polymerization 10. The formation of double helices in the amylopectin contributes immensely to hydrophobicity and crystallinity. During endosperm amylopectin biosynthesis, starch branching enzyme (BE) catalyzes the formation of α -1, 6-glycosidic linkage of glucan that gives rise to a side chain, unlike the α-1, 4-glycosidic linkage responsible for linear chain formation. BE in rice has three isozymes, namely BEI, BEIIa, and BEIIb (Nakamura et al. 2010). BEI type enzyme has a preference for amylose glucans whereas BEII has a higher activity for branched glucans. BEI isozyme transfers long amylose chains for multiple clusters of amylopectins and medium-size chains for amorphous amylopectin lamellae. In comparison, BEIIb transfers short chains that lie at the interface of amorphous and crystalline lamellae of amylopectin. The activities of BE isozymes with a linear glucan decrease with that of the molar size of the amylose. Amylose chains with a degree of polymerization less than 80 have no activity with BEIIa and BEIIb, whereas the preferential substrate for BEI is amylose having a degree of polymerization of nearly 50. Wang et al. (2015) correlated activity and gene expression for BE with grain filling of super rice cultivars where low levels of activity and gene expression restrict grain filling of inferior spikelets.

Expression Profile of the Genes Controlling the Activity of Starch Biosynthetic Enzymes

Sun et al. (2015) using the RNA Seq method analyzed the dynamics of gene expression profiles between superior and inferior grains of large panicle rice. In toto, there is 19,442 gene expression of which are detected during the period of grain development, and between the two types of grains genes involved in sucrose to starch, transformation expressed poorly in the inferior compared to superior grains. Chen et al. (2016) investigated and compared post-anthesis protein expression concerning grain development of inferior and superior spikelets of rice panicle. The study identified differential expression of proteins that function prominently in carbon assimilation processes, such as photosynthesis, starch synthesis, carbohydrate and energy metabolism and amino acid metabolism, closely related to grain filling. The expression of the proteins was mostly down regulated in the inferior spikelets to the detriment of grain filling. Although up regulation in expression of all these proteins can improve grain filling in the failing spikelets, emphasis has been given to the enzymes involved in the starch synthesis of the grains more than the others because activities of the enzymes are directly linked to grain filling. Zhu et al. (2011) found a greater expression profile of starch synthesizing enzyme genes in superior spikelets compared with the inferior spikelets. Thus, metabolic dominance in the assimilate partitioning of superior spikelets over the inferior ones (Xu and Vergara 1986) is incumbent on the expression of genes encoding the starch synthesizing enzymes.

To consolidate the stance further, Zhang et al. (2014) performed comparative proteomics of the two types of contrasting spikelets during grain development and found the low activity of the enzymes at expression level as a major reason causing poor filling of inferior spikelets. Fu et al. (2013) reported that peak values and mean activities of SUS, AGPase, and SS and SBE were lower in inferior spikelets than superior spikelets during grain filling and that grain filling is positively correlated with activities of the first three enzymes. With the help of quantitative real-time RT-PCR (qRT-PCR), Panda et al. (2015 – Fig. 10.3) has explored the expression of genes encoding sucrose synthase, AGPase, and starch synthase to correlate transcript level expression with enzyme activity positively. The expression of the genes encoding the enzymes is poor in the inferior spikelets (Panda et al. 2015) and seed setting rate, grain weight, and activities and expression of the enzymes improve significantly in these spikelets subject to removal of the superior spikelets (You et al. 2016). This stance on the disparity in grain filling is consolidated further with iTRAQ based proteome profile analysis of superior and inferior spikelets at early grain filling stage of rice, where reduced endosperm cell division and sub-optimal activities of key enzymes associated with starch synthesis and nitrogen metabolism are attributed to disadvantages of grain sink strength of inferior spikelets (You et al. 2017). These authors also reported an increased abundance of 14-3-3 protein in inferior spikelets responsible for inhibition in starch synthesis. It is proposed that the protein becomes an important factor in the signaling network for grain development of inferior spikelets, but its effect on endosperm starch synthesis is yet to be understood (Zhang et al. 2015). Similar to the effect of the distribution of spikelets in space on the panicle axis, the expression of genes, as well as activity of the starch synthesizing enzymes, differ considerably between cultivars. A cultivar with lax type panicle architecture, i.e. amenable to the good grain filling, exhibits a greater expression of the genes during caryopsis development compared to a compact panicle cultivar bearing poor grain filling attributes. This concept was tested in an experiment with two by two factorial design comprising four cultivars, such as lax versus compact panicle type and cultivars with different genetic bases versus similar genetic base of recombinant inbred lines (Figs. 10.3, 10.4 and 10.5). The level of expression was always much higher in the spikelets of apical versus basal positions and lax- versus compact-panicle cultivars. Similar results have also been obtained by Wang et al. (2015).

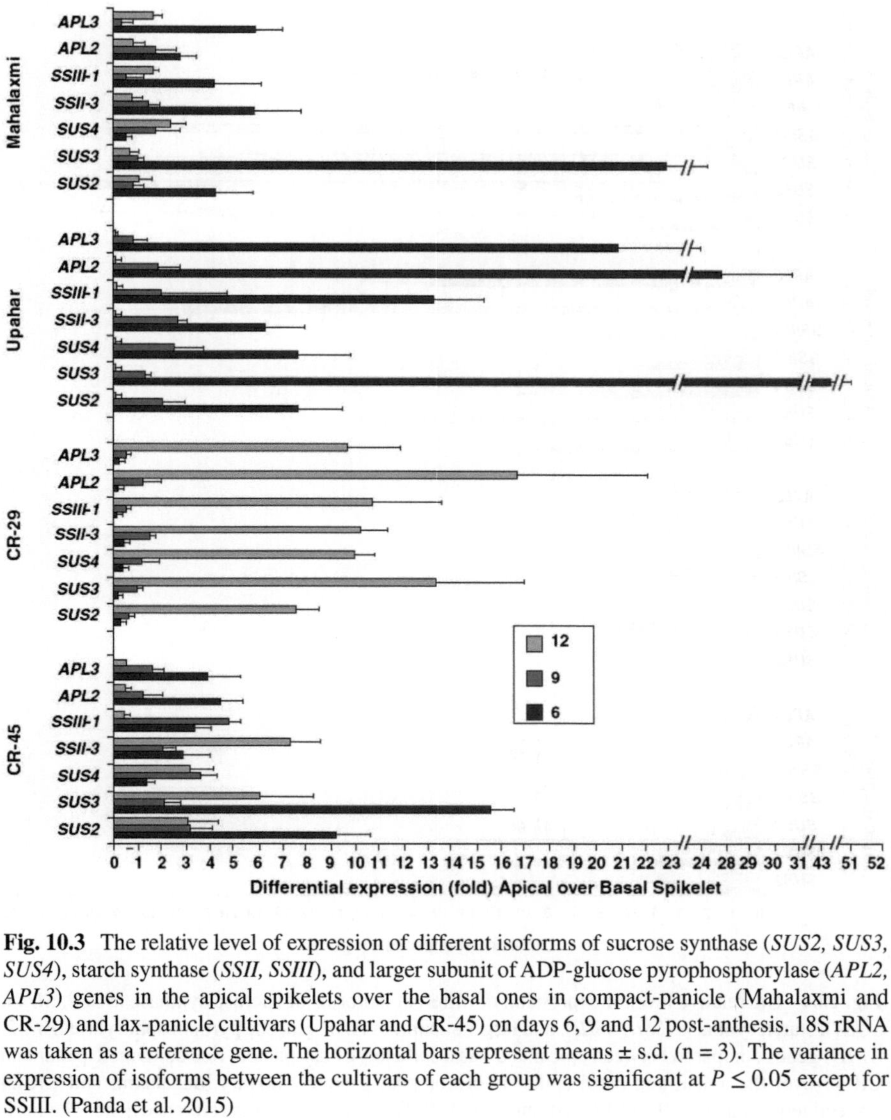

Fig. 10.3 The relative level of expression of different isoforms of sucrose synthase (*SUS2, SUS3, SUS4*), starch synthase (*SSII, SSIII*), and larger subunit of ADP-glucose pyrophosphorylase (*APL2, APL3*) genes in the apical spikelets over the basal ones in compact-panicle (Mahalaxmi and CR-29) and lax-panicle cultivars (Upahar and CR-45) on days 6, 9 and 12 post-anthesis. 18S rRNA was taken as a reference gene. The horizontal bars represent means ± s.d. (n = 3). The variance in expression of isoforms between the cultivars of each group was significant at $P \leq 0.05$ except for SSIII. (Panda et al. 2015)

Sucrose Synthase Gene Expression

Starch biosynthesis in rice endosperm begins 5 days after anthesis with concomitant expression genes encoding enzymes involved in the biosynthetic pathway. For the enzyme sucrose synthase, the SUS gene family in rice comprises six genes each encoding one isozyme of the enzyme (Hirose et al. 2008). The six genes are

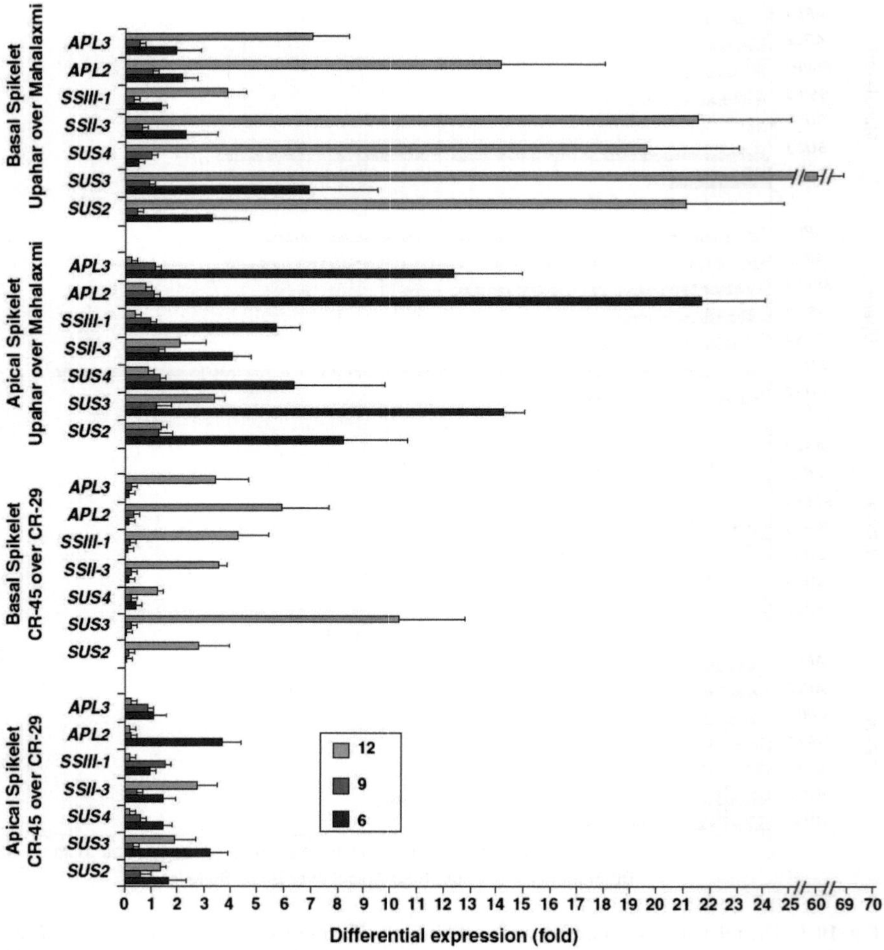

Fig. 10.4 The relative level of expression of different isoforms of sucrose synthase (*SUS2, SUS3, SUS4*), starch synthase (*SSII, SSIII*), and larger subunit of ADP-glucose pyrophosphorylase (*APL2, APL3*) genes compared between apical and basal spikelets of lax-panicle rice cultivars Upahar and CR-45 over the apical and basal spikelets of compact-panicle cultivars Mahalaxmi and CR-29 respectively, on days 6, 9 and 12 post-anthesis. 18S rRNA was taken as a reference gene. The horizontal bars represent mean ± s.d. (n = 3). (Panda et al. 2015)

dispersed throughout the genome. Of the six, the expression of four is tissue and/ or stage-specific. *SUS1* and *SUS2* are involved in the regulation of vegetative growth while *SUS3* and *SUS4* gene expressions control partitioning of assimilates to caryopsis during the grain filling period (Ishimaru et al. 2005). The functions of *SUS5* and *SUS6* are relatively unexplored. Panda et al. (2015) reported a greater abundance of transcripts of the isoforms in superior spikelets compared to inferior

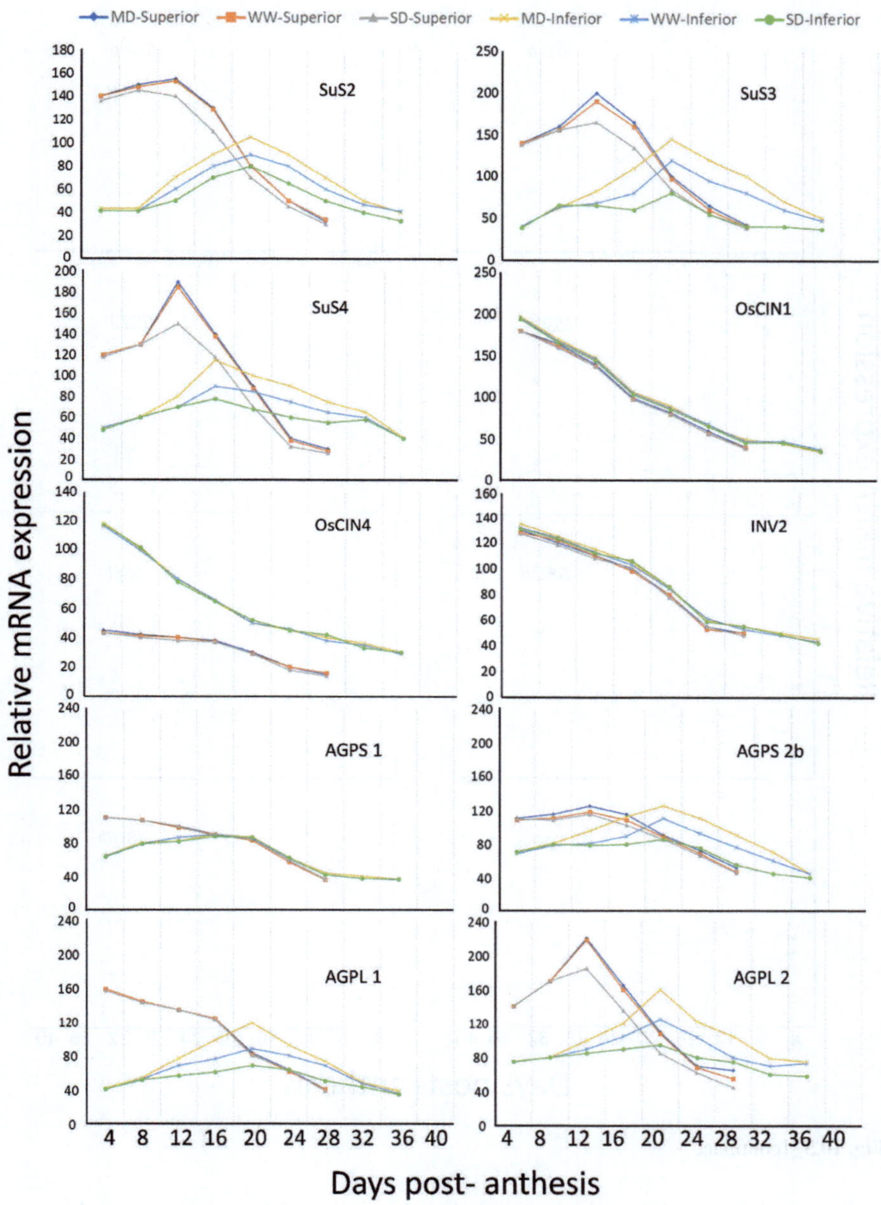

Fig. 10.5 Changes in gene expression levels (% of relative mRNA level) of various isoforms of starch synthesizing enzymes in superior (solid lines) and inferior (dotted lines) spikelets of rice. WW, MD, and SD represent well-watered, moderate soil-drying and severe soil-drying treatments, respectively, during grain filling. (Adopted and modified from (Wang et al. 2015))

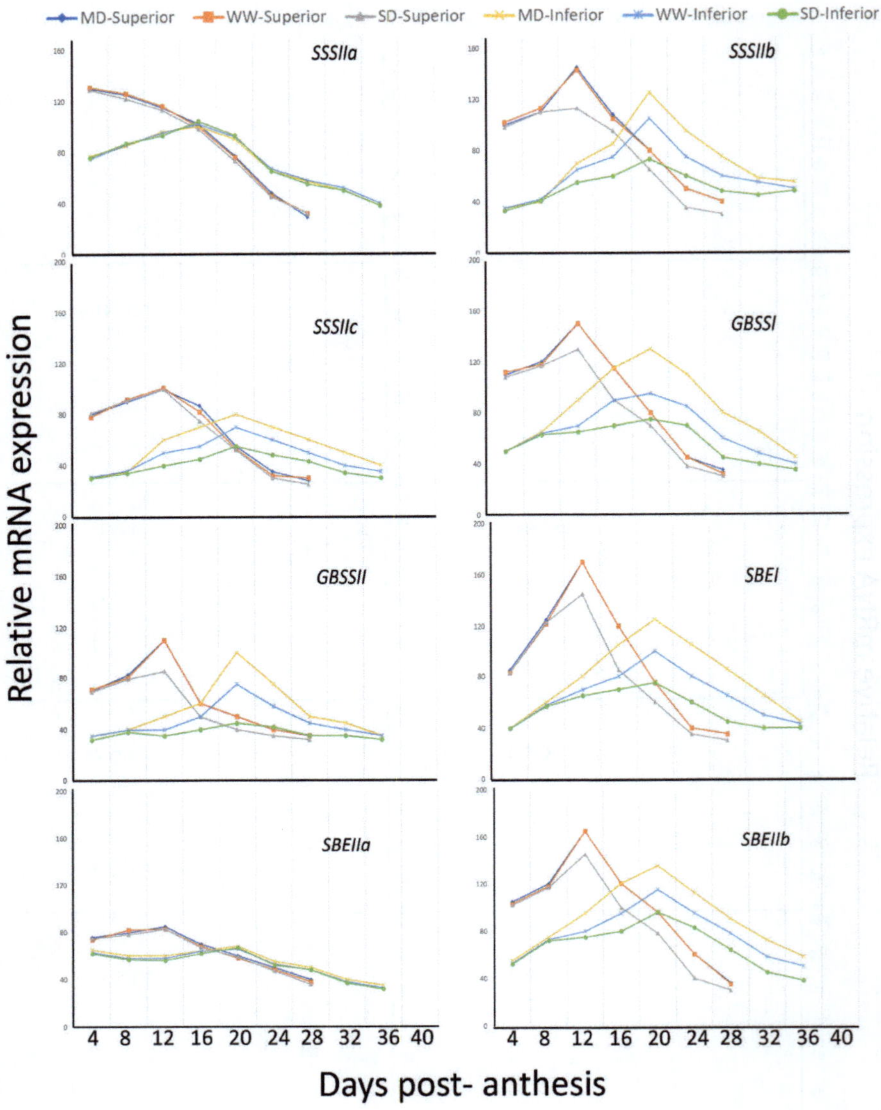

Fig. 10.5 (continued)

spikelets of rice panicle, which revealed sink control of the enzyme in endosperm starch synthesis. The expression of *SUS* genes during grain filling also varies in cultivars bearing different types of panicle architecture. A cultivar with lax type architecture possessing ability for good grain filling exhibits greater expression of the genes than a compact panicle cultivar bearing poor grain filling attributes. Panda et al. (2015) tested this concept by comparing lax versus compact panicle cultivars with different genetic bases (Upahar – lax and Mahalaxmi – compact) and two contrasting panicle types of recombinant inbred lines derived from the same parents. The relative expression levels of the genes encoding various isoforms of the enzymes were compared in qRT-PCR in superior versus inferior spikelets and lax verses compact spikelets; the relative level of expression in the former was several times higher than the latter (Figs. 10.3 and 10.4).

Among the enzyme isoforms, the differential expression of *SUS3* between developing superior and inferior spikelets was much greater than that of any other enzyme isoform. Panda et al. (2015) reported greater differential expression of SUS3 between superior and inferior spikelets than that of any of the other five isoforms studied. Therefore, *SUS3* expression was proposed as an indicator for dry matter partitioning and metabolic dominance of spikelets of rice panicle. The spikelets disadvantaged for grain filling on account of any type of placement or orientation would be constrained for expression of *SUS3* resulting in poor starch synthesis and grain filling. This report was the first of its kind, where hierarchy in spikelet development was linked to a probable factor of genetic expression.

ADP-Glucose Pyrophosphorylase Gene Expression

In addition to the six genes encoding sucrose synthase isoforms, there are 27 genes regulating functions of the six other starch synthesizing enzymes; and they are AGPase, SS, starch branching enzyme (BE), starch debranching enzyme, starch phosphorylase, and starch disproportionate enzyme. The mode of expression of these enzymes is tissue and development-specific (Ohdan et al. 2005). The expression profile of the genes has been divided into four groups. Group 1 genes express early and are responsible for the synthesis of glucan primers necessary for the initiation of starch biosynthesis. Group 2 genes are expressed throughout endosperm development. Group 3 gene expression coincides with the initiation of starch biosynthesis in the endosperm. Group 4 genes exhibit poor expression profile and control starch biosynthesis of the pericarp.

AGPase is heterotetrameric in composition, possessing two large and two small subunits. Six genes are involved in encoding different isoforms of the enzyme subunits, out of which the expression of four is developmental stage-specific, while the other two express constitutively (Akihiro et al. 2005). Two genes *OsAGPS1* and *OsAGPS2* code for the small subunits; while *OsAGPL1*, *OsAGPL2*, *OsAGPL3*, and *OsAGPL4* code for the large subunits. The *OsAGPS2* gene is located on chromosome number 8. *AS2–2* allele at *OsAGPS2* is linked with good grain filling trait in the combination of *SUT1–2* allele of *OsSUT1* (sucrose transporter) gene. As double homozygotes, the combination holds great potential for improvement in grain filling of extra-heavy weight panicle type rice cultivars; its frequency is very low in japonica (3.6%) compared to the indica rice (56.1%) (Kato and Horibata 2011; Kato 2013). The sequence-based variation in *AGPase* gene expression is significantly related to the amylose content and viscosity property of rice, where *AGPS1* correlates with increased amylose and *AGPL4* associated with decreased breakdown viscosity (Lu and Park 2012). According to Ohdan et al. (2005), AGPS2 encodes transcripts for AGPS2a and AGPS2b, where the former possessing 1930 bp carries a transit peptide sequence and the latter with 1794 bp has no transit peptide sequence. The localization of *AGPase* genes in rice has been further characterized by judging the presence of transit peptides and sequence homologies corresponding to other species. These studies identified the presence of OsAGPS1, OsAGPS2a, OsAGPL3, and OsAGPL4 proteins in the plastids and OsAGPL2 and OsAGPS2b in the cytosol. The *AGPase* genes express prominently in high-yielding rice than other cultivars (Hwang et al. 2005). Takaiwa and Qu (2010) have identified the promoter of *AGPase* in rice caryopsis and found its expression high during the seed maturation period in patent no US7700835. However, there is scant information on the variation of AGPase gene expression leading to differences in individual grain weight of rice panicle. In this context, Panda et al. (2015) found expression of *AGPase* gene high in developing grains of lax-panicle rice cultivars compared to the poorly filled grains of compact-panicle cultivars. A relative level of the expression pattern of the enzyme showed sub-optimal expression of *APL2* (*OsAGPL2*) and *APL3* (*OsAGPL3*) in the inferior type rice kernels compared to the superior ones in this work.

Starch Synthase and Starch Branching Enzyme

In rice caryopsis, there are two types of granule bound starch synthase (GBSSI, GBSSII) and four types of soluble starch synthase (SSI-IV). Amylose synthesis in rice amyloplast is dependent on the expression and activity of GBSSI (Vandeputte and Delcour 2004). *GBSSI* gene expression of reaches peak level at 5 days post-anthesis and continues till grain maturity. In comparison, the expression of GBSSII remains low throughout starch biosynthesis. Eight genes are controlling the expression of the four isoforms of soluble starch synthase (Hirose and Terao 2004; Ohdan et al. 2005; Kosar-Hashemi et al. 2007). SSI is the major form of soluble starch synthase, which controls endosperm starch synthesis. SSI is encoded by a single gene unlike the other three having more than two genes for transcription. SSI activity is high during the first few days after anthesis and remains high until the mid-grain filling stage. *OsSSI* gene controls the expression of this isozyme. SSII is coded by three genes (*OsSSIIa*, *OsSSIIb*, *OsSSIIc*). SSIII and SSIV are encoded by two genes each, namely *OsSSIIIa* and *OsSSIIIb*, and *OsSSIVa* and *OsSSIVb* respectively (Hirose and Terao 2004). Significant temporal variation in the expression of the genes is noticed during endosperm development. *SSIIa* and *SSIIIa* expression reach a climax at 5–7 days post-anthesis and remains high at the later stage of development. *SSIIIb* expression is high at the beginning of seed formation but declines later. *SSIVb* transcription increases up to day five post anthesis and declines thereafter. Transcription of *SSI*, *SSIIa* and *SSIIa* is higher than that of *SSIIIb* and *SSIVb* genes. In comparison transcription of the other *SS* genes i.e. *SSIIb*, *SSIIc* and *SSIVa* are very low; the expression is noticeable during the first 5 days of flowering but not later (Ohdan et al. 2005). Panda et al. (2015) reported greater expression of *SSI* and *SSII* in superior spikelets than inferior spikelets during grain development, but no difference could be noticed for *GBSS* expression.

Hwang et al. (2005) found that genes for AGPase, GBSS, SS, and SBE express highly between 7–28 days post-anthesis. Similarly, temporal variation in the expression profiles of the genes encoding the starch branching enzyme, starch debranching enzyme, starch disproportionating enzyme, and starch phosphorylase has been investigated (Ohdan et al. 2005), but their contribution to the spatial or temporal difference in grain development or variation in the starch synthesis of individual spikelets of panicle has rarely been reported. Recently, Sawada et al. (2018) identified the role of three starch branching enzymes BE1, BEIIa, and BEIIb in the starch synthesis of rice endosperm. BEIIb activity controls short-chain amylopectins with 6–13 degrees of polymerization, whereas BE1 activity has a distinct role in the synthesis of intermediate chain length. BEIIa coordinates with BE1 for starch synthesis but not with BEIIb. Wang et al. (2015) reported that activity, as well as expression levels of SBEI and SBEIIb genes, were low in inferior compared to superior spikelets of super rice bearing poor grain filling attribute and this type of variation is similar to that of other enzyme genes like *SUS2, SUS3, SUS4, AGPL1, AGPL2, SSIIb, SSIIc, GBSSI,* and *GBSSII*. Grain weight and expression of the enzyme genes increase with the application of ABA. This work also included evidence for greater expression of *CELL WALL INVERTASE 1* and *4* (*OsCIN1, OsCIN4*) and *VACUOLAR INVERTASE 2* (*INV2*) is superior compared to inferior spikelets. You et al. (2016) also found greater expression of genes encoding sucrose synthase, AGPase, soluble starch synthase, and starch branching enzymes are superior compared with inferior spikelets, although there was not observation for granule bound starch synthase. As of today, the contribution of differential expression of genes encoding these enzymes to variation in grain weight and quality of individual spikelets of rice panicle has remained unresolved. A model has been proposed to show co-ordinated expression of starch synthesizing enzymes genes (Fig. 10.6).

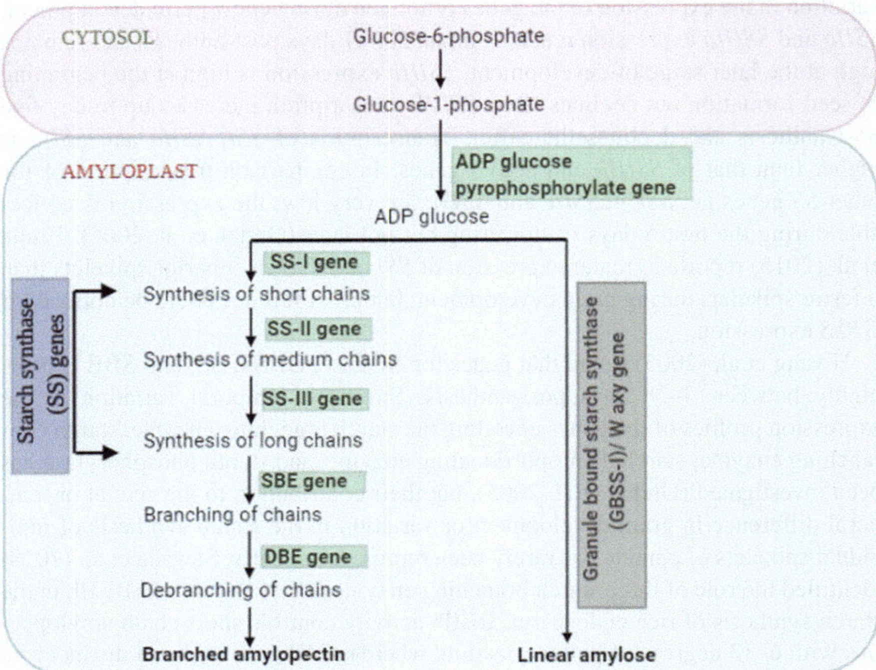

Fig. 10.6 Co-ordination between different starch synthesizing enzyme genes in rice endosperm leading to amylose and amylopectin synthesis

References

Akihiro T, Mizuno K, Fujimura T (2005) Gene expression of ADP-glucose pyrophosphorylase and starch contents in rice cultured cells are cooperatively regulated by sucrose and ABA. Plant Cell Physiol 46:937–946

Chen T, Xu G, Wang Z, Zhang H, Yang J, Zhang J (2016) Expression of proteins in superior and inferior spikelets of rice during grain filling under different irrigation regimes. Proteomics 16:102–121

Counce PA, Gravois KA (2006) Sucrose synthase activity as a potential indicator of high rice grain yield. Crop Sci 46:1501–1507

Crofts N, Abe N, Oitome NF, Matsushima R, Hayashi M, Tetlow IA, Emes MJ, Nakamura Y, Fujita N (2015) Amylopectin biosynthetic enzymes from developing rice seed form enzymatically active protein complexes. J Exp Bot 66:4469–4482

Fu J, Huang Z, Wang Z, Yang J, Zhang J (2011) Pre-anthesis non-structural carbohydrate reserve in the stem enhances the sink strength of inferior spikelets during grain filling of rice. Field Crop Res 123:170–182

Fu J, Xu Y, Chen L, Yuan L, Wang Z, Yang J (2013) Changes in enzyme activities involved in starch synthesis and hormone concentrations in superior and inferior spikelets and their relation with grain filling of super rice. Rice Sci 20(2)

Furbank RT, Scofield GN, Hirose T, Wang XD, Patrick JW, Offler CE (2001) Cellular localization and function of a sucrose transporter OsSUT1 in developing rice grains. Funct Plant Biol 28:1187–1196

Hirose T, Terao T (2004) A comprehensive expression analysis of the starch synthase gene family in rice (*Oryza sativa* L.). Planta 220:9–16

Hirose T, Takano M, Terao T (2002) Cell wall invertase in developing rice caryopsis: molecular cloning of OsCIN1 and analysis of its expression in relation to its role in grain filling. Plant Cell Physiol 43:452–459

Hirose T, Scofield GN, Terao T (2008) An expression analysis profile for the entire sucrose synthase gene family in rice. Plant Sci 174:534–543

Hwang JW, Kim SK, Lee JS, Kim IS (2005) Gene expression of the biosynthetic enzymes and biosynthesis of starch during rice grain development. J Plant Biol 48:448–455

Hwang SK, Koper K, Satoh H, Okita TW (2016a) Rice endosperm starch phosphorylase (Pho1) assembles with disproportionating enzyme (Dpe1) to form a protein complex that enhances the synthesis of malto-oligosaccharides. J Biol Chem 291:19994–20007

Hwang SK, Singh S, Cakir B, Satoh H, Okita TW (2016b) The plastidial starch phosphorylase from rice endosperm: catalytic properties at low temperature. Planta 243:999–1009

Ishimaru T, Hirose T, Matsuda T, Goto A, Takahashi K, Sasaki H, Terao T, Ishii RI, Ohsugi R, Yamagishi T (2005) Expression patterns of genes encoding carbohydrate-metabolizing enzymes and their relationship to grain filling in rice (Oryza sativa L.): comparison of caryopses located at different positions in a panicle. Plant Cell Physiol 46:620–628

Jeng TL, Wang CS, Tseng TH, Sung JM (2007) Expression of granule-bound starch synthase in developing rice grain. J Sci Food Agric 87:2456–2463

Jeon JS, Ryoo N, Hahn TR, Walia H, Nakamura Y (2010) Starch biosynthesis in cereal endosperm. Plant Physiol Biochem 48:383–392

Kato T (1995) Change of sucrose synthase activity in developing endosperm of rice cultivars. Crop Sci 35:827–831

Kato T (2013) A trial to develop novel genotypes of extra-heavy panicle types with good grain filling in rice. J Crop Res 58:15–20

Kato T, Horibata A (2011) Non-random distribution of the alleles for good grain filling at OsAGPS2 and OsSUT1 among a wide range of rice (*Oryza sativa* L.) cultivars. Breed Sci 61:217–220

Kato T, Shinmura D, Taniguchi A (2007) Activities of enzymes for sucrose-starch conversion in developing endosperm of rice and their association with grain filling in extra-heavy panicle types. Plant Prod Sci 10:442–450

Kosar-Hashemi B, Li Z, Larroque O, Regina A, Yamamori M, Morell MK, Rahman S (2007) Multiple effects of the starch synthase II mutation in developing wheat endosperm. Funct Plant Biol 34:431–438

Kuanar SR, Panigrahi R, Kariali E, Mohapatra PK (2010) Apoplasmic assimilates and grain growth of contrasting rice cultivars differing in grain dry mass and size. Plant Growth Regul 61:135–151

Li B, Liu H, Zhang Y, Kang T, Zhang L, Tong J, Xiao L, Zhang H (2013) Constitutive expression of cell wall invertase genes increases grain yield and starch content in maize. Plant Biotechnol J 11:1080–1091

Liang J, Zhang J, Cao X (2001) Grain sink strength may be related to the poor grain filling of indica-japonica rice (*Oryza sativa*) hybrids. Physiol Plant 112:470–477

Lu FH, Park YJ (2012) Sequence variations in OsAGPase are significantly associated with amylose content and viscosity properties in rice (*Oryza sativa* L.). Genet Res 94:179–189

Matsue Y, Odahara K, Hiramatsu M (1995) Differences in amylose content, amylographic characteristics, and storage proteins of grains on primary and secondary rachis branches in rice. Jpn J Crop Sci 64:601–606

Mohapatra PK (2012) Sucrose metabolism and grain development of contrasting spikelets located at different positions of the rice panicle. In: Greco AJ (ed) Progress in food science and technology, vol 1. Nova Science Publishers, Hauppauge, pp 231–241

Mohapatra PK, Patel R, Sahu SK (1993) Time of flowering affects grain quality and spikelet partitioning within the rice panicle. Funct Plant Biol 20:231–241

Mohapatra PK, Panigrahi R, Turner NC (2011) Physiology of spikelet development on the rice panicle: is the manipulation of apical dominance crucial for grain yield improvement? Adv Agron 110:333–359

Mohapatra R, Mohapatra PK (2006) Ethylene control of seed coat development in low and high sterile semidwarf indica rice cultivars. Plant Growth Regul 50:47–55

Morey SR, Hirose T, Hashida Y, Miyao A, Hirochika H, Ohsugi R, Yamagishi J, Aoki N (2018) Genetic evidence for the role of a rice vacuolar invertase as a molecular sink strength determinant. Rice 11:6. https://doi.org/10.1186/s12284-018-0201-x

Naik PK, Mohapatra PK (2000) Ethylene inhibitors enhanced sucrose synthase activity and promoted grain filling of basal rice kernels. Aust J Plant Physiol 27:997–1008

Nakamura Y, Yuki K, Park SY, Ohya T (1989) Carbohydrate metabolism in the developing endosperm of rice grains. Plant Cell Physiol 30:833–839

Nakamura Y, Ono M, Sawada T, Crofts N, Fujita N, Steup M (2017) Characterization of the functional interactions of plastidial starch phosphorylase and starch branching enzymes from rice endosperm during reserve starch biosynthesis. Plant Sci 264:83–95

Nakamura Y, Utsumi Y, Sawada T, Aihara S, Utsumi C, Yoshida M, Kitamura S (2010) Characterization of the reactions of starch branching enzymes from rice endosperm. Plant Cell Physiol 51:776–794

Ohdan T, Francisco PB, Sawada T, Hirose T, Terao T, Satoh H, Nakamura Y (2005) Expression profiling of genes involved in starch synthesis in sink and source organs of rice. J Exp Bot 56:3229–3244

Panda B, Badoghar A, Das K, Panigrahi R, Kariali E, Das SR, Dash SK, Shaw BP, Mohapatra PK (2015) Compact panicle architecture is detrimental for growth as well as sucrose synthase activity of developing rice kernels. Funct Plant Biol 42:875–887

Panigrahi R, Kariali E, Panda BB, Lafarge T, Mohapatra PK (2019) Controlling the trade-off between spikelet number and grain filling: the hierarchy of starch synthesis in spikelets of rice panicle in relation to hormone dynamics. Funct Plant Biol 46:507–523

Patel R, Mohapatra PK (1996) Assimilate partitioning within floret components of contrasting rice spikelets producing qualitatively different types of grains. Funct Plant Biol 23:85–92

Sawada T, Itoh M, Nakamura Y (2018) Contribution of three starch branching isozymes to the fine structure of amylopectin in rice endosperm. Front Plant Sci. https://doi.org/10.3389/fpls.2018.01536

Smidansky ED, Martin JM, Hannah CL, Fischer AM, Giroux MJ (2003) Seed yield and plant biomass increases in rice are conferred by deregulation of endosperm ADP-glucose pyrophosphorylase. Planta 216:656–664

Sun H, Peng T, Zhao Y, Du Y, Zhang J, Li J, Xin Z, Zhao Q (2015) Dynamic analysis of gene expression in rice superior and inferior grains by RNA-Seq. PLoS One 10:e0137168

Takaiwa F, Qu L (2010) AGPase promoter from rice and its uses thereof. United States Patent No.7700835 B2 dated 20 April 2010

Tang T, Xie H, Wang Y, Lü B, Liang J (2009) The effect of sucrose and abscisic acid interaction on sucrose synthase and its relationship to grain filling of rice (*Oryza sativa* L.). J Exp Bot 60:2641–2652

Vandeputte GE, Delcour JA (2004) From sucrose to starch granule to starch physical behavior: a focus on rice starch. Carbohydrate Polymers 58:245–266. https://doi.org/10.1016/j.carbpol.2004.06.003

Wang E, Wang J, Zhu X, Hao W, Wang L, Li Q, Zhang L, He W, Lu B, Lin H, Ma H, Zhang G, He Z (2008) Control of rice grain-filling and yield by a gene with a potential signature of domestication. Nat Genet 40:1370–1374

Wang Z, Xu Y, Chen T, Zhang H, Yang J, Zhang J (2015) Abscisic acid and the key enzymes and genes in sucrose-to-starch conversion in rice spikelets in response to soil drying during grain filling. Planta 241:1091–1107

Xu X, Vergara BS (1986) Morphological changes in rice panicle development: a research development, IRRI research paper series no 117. International Rice Research Institute, e Philippines

Yoshida S (1972) Physiological aspects of grain yield. Annu Rev Plant Physiol 23:437–464. https://doi.org/10.1146/annurev.pp.23.060172.002253

You C, Chen L, He H, Wu L, Wang S, Ding Y, Ma C (2017) iTRAQ-based proteome profile analysis of superior and inferior spikelets at early grain filling stage in japonica rice. BMC Plant Biol 17:100

You C, Zhu H, Xu B, Huang W, Wang S, Ding Y, Liu Z, Li G, Chen L, Ding C, Tang S (2016) Effect of removing superior spikelets on grain filling of inferior spikelets in rice. Front Plant Sci 7:1161. https://doi.org/10.3389/fpls.2016.01161

Zhang Z, Zhao H, Tang J, Li Z, Li Z, Chen D, Lin W (2014) A proteomic study on the molecular mechanism of poor grain-filling of rice (*Oryza sativa* L.) inferior spikelets. PLoS One 9:e89140. https://doi.org/10.1371/journal.pone.0089140

Zhang Z, Tang J, Du T, Zhao H, Li Z, Li Z, Lin W (2015) Mechanism of developmental stagnancy of rice inferior spikelets at the early grain-filling stage as revealed by proteomic analysis. Plant Mol Biol Report 33:1844–1863

Zhu G, Ye N, Yang J, Peng X, Zhang J (2011) Regulation of expression of starch synthesis genes by ethylene and ABA in relation to the development of rice inferior and superior spikelets. J Exp Bot 62:3907–3916

Chapter 11
Hormonal Regulation of Spikelet Development

Abstract Many of the identified panicle development genes of the rice plants are homologous to known hormone genes. The class of plant hormones includes both positive and negative growth regulators. In rice, hormone-related biosynthetic pathways have been targeted for yield improvement since long bereft of any yield advantage hitherto. In the heterogeneous architecture rice panicle spikelet, position-specific grain growth is surmised coincidental to variation in balance between different hormones. The abundance of GA, auxin, cytokinin, ABA, and brassinosteroids improves grain yield whereas enhancement of ethylene synthesis, jasmonic acid production, and IAA perception inhibits grain filling. Hence, it has been obligatory to identify and characterize the expression of individual hormone-regulated panicle development genes to break yield frontiers in rice.

Keywords Spikelet · Hormones · Yield

Hormones as Limiting Factor for the Disparity in Grain Filling of Spikelets

The disparity in grain filling between individual spikelets ordained in a heterogeneous architecture of panicle that exhibits strong apical dominance has been seldom correlated with discrimination in assimilate supply limiting kernel growth of the disadvantaged units (Mohapatra et al. 2011a). The spikelet-specific growth in the panicle is strongly dictated by temporal and spatial distribution for which assimilates are not prejudiced to be the limiting factors. The basally positioned spikelets, although older than the apical spikelets, emerge late from the flag sheath enclosure and delay in anthesis underscores starch synthesis to the detriment of grain filling (Mohapatra and Sahu 1991; Mohapatra et al. 1993). Because assimilates accumulate in the disadvantaged spikelets in the absence of adequate sink-based demand, regulation of spikelet development with hormones or external growth regulators may be an option for prospective improvement of assimilates partitioning and increase yield thereby.

P. K. Mohapatra, B. B. Sahu, *Panicle Architecture of Rice and its Relationship with Grain Filling*, https://doi.org/10.1007/978-3-030-67897-5_11

Role of Hormones in Inter Spikelet Dominance

Plant hormones individually or in consortium play a pivotal role in every aspect of the plant life cycle including grain filling. Ding et al. (2014) suggested that the panicle architecture of rice is influenced by hormones. Many of the identified panicle development genes are homologous to known hormone genes that have been characterized for so long. The hormones cue membrane-bound cellular signals to induce a cascade of developmental responses. The class of plant hormones includes both positive and negative growth regulators and an appropriate balance between the two types dictates the outcome in growth or development. Thus, hormone actions are far from simple and involve several interactions within the group under the control of a master regulator that could be a transcription factor deciding the fate of gene expression. In rice, hormone-related biosynthetic pathways have been targeted for yield improvement since long bereft of any yield advantage hitherto. The concept proposing a growth inhibitor regulating spikelet development in branches of rice panicle gained strength because of the heterogeneous pattern of individual grain growth, where the spikelet on the branch tip is most developed compared to the most laggard and dominated one just below its position and the order of dominance receding in a sequence to the base (Xu and Vergara 1986).

After emergence at the heading stage, the spikelet at the tip reaches anthesis and probably assumes a capacity for inhibitor production first; the inhibitor suppresses the growth of the penultimate spikelet to maximum and the influence declines sequentially down ward with an increase of distance (Mohapatra and Sahu 1991). The inhibitor could be the gaseous hormone ethylene that has been construed as a second messenger for IAA action in apical dominance of spikelets in individual branches of rice panicle (Sekhar et al. 2015b). Strong basipetal transport of IAA synthesized from the apical spikelet inhibits IAA export from the basal spikelet. Diffusion has been forbidden apically, the hormone accumulates in basal spikelet and gives rise to second messenger ethylene for growth inhibition (Fig. 11.1). Opposite to IAA, cytokinin derived from the root moves acropetally and promotes grain development. In addition, the functions of other plant hormones like ABA, gibberellins and brassinosteroids in source sink assimilate partitioning, apical dominance, metabolic activities are also crucial for grain development.

Positive Regulators for Grain Growth

Although it was more plausible to explain inter-spikelet apical dominance based on the production of a growth inhibitor in response to variation in time and space, most of the research conducted also focused attention on positive manipulation of seed growth that involved cytokinin and gibberellins actions. To elucidate the mechanism of grain filling and improve grain yield, intrinsic levels of hormones have been analyzed and hormone-related biosynthetic pathways have been targeted with the use of chemical inhibitors and physical manipulations in rice panicle. The effects of the inhibitors on the expression of hormone associated genes have been studied to conclude that chemicals inhibitory to the biosynthesis of GA, auxin, and brassinosteroids

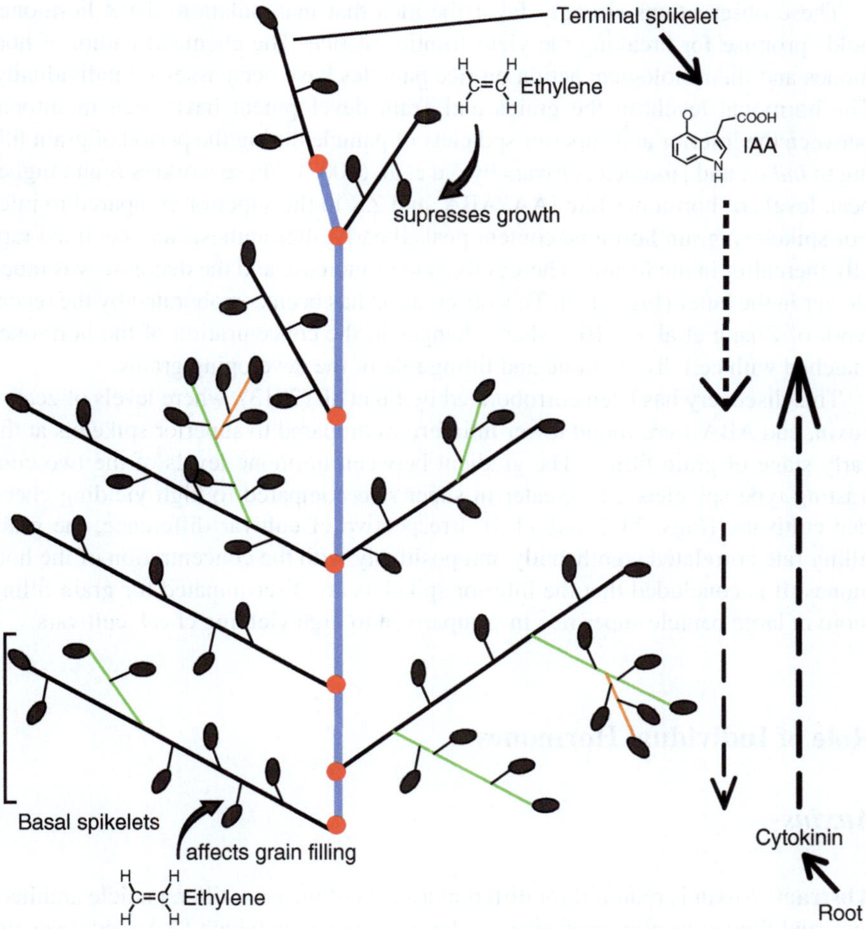

Fig. 11.1 Schematic model for ethylene action regulating inter spikelet dominance for growth in branches of rice panicle. IAA produced in the terminal spikelet of a branch moves basipetally and its signal becomes weaker with an increase of distance from the tip because of acropetal transport of root-derived cytokinin. IAA accumulation in a spikelet evokes ethylene production to the detriment of grain filling.

affect grain yield whereas a similar intervention, which infringes up on ethylene synthesis, jasmonic acid production, and IAA perception promotes grain filling (Tamaki et al. 2015). The hormone-associated genes, functions of which are related to grain yield, mostly strengthen sink activity with the sole exception of *thousand grain weight 6 (TGW6)* reinforced by source activity. Wang et al. (2006) reported that a partial sink removal treatment with excision of some spikelets of panicle branches markedly reduced grain filling and hormone levels of IAA, GA, cytokinin, and ABA and delayed the time taken for the reaching peak level of grain filling rate and hormone content in inferior spikelets compared to superior spikelets. Thus, inferior spikelets discriminated for grain filling because of spatial disadvantage in basal placement or compact-packing are vulnerable for grain filling.

These observations strongly favor the idea that manipulation of the hormones holds promise for breaking the yield frontier of rice. The chemical nature of hormones and their biological action on rice panicles have been assessed individually. The hormonal levels in the grains and grain development have been monitored between the inferior and superior spikelets of panicle during the period of grain filling in *indica* and japonica cultivars by Xu et al. (2007). These workers found higher peak levels of hormones like IAA, ABA, and ZR in the superior compared to inferior spikelets; grain hormone content peaked early after anthesis and declined rapidly thereafter in the former whereas the rate of increase and the decrease was much slower in the latter (Fig. 11.2). This observation has been corroborated by the recent work of Zhang et al. (2016), where changes in the concentration of the hormones matched with cell division rate and filling rate of the developing grains.

This discovery has been corroborated by Fu et al. (2013), where levels of zeatin, auxin, and ABA were found lower in inferior compared to superior spikelets at the early stage of grain filling. The gradient between hormone levels of the two contrasting type spikelets was greater in super rice compared to high yielding check rice cultivars. (Figs. 11.2 and 11.3). Irrespective of cultivar difference, the grain filling rate correlated significantly and positively with the concentration of the hormones. It is concluded that the inferior spikelets are discriminated for grain filling more in large panicle super rice in comparison to high yielding check cultivars.

Role of Individual Hormones

Auxins

Abstract Auxin is required for different aspects of the root, tiller, panicle architecture, and flower development of rice. The mechanism by which IAA modulates the growth of plant organs has been clearly defined by the elucidation of expression of genes and chemical reactions controlling tryptophan-dependent and -independent auxin biosynthesis pathways. Hence disruption of auxin biosynthesis, transport, and signal transduction considerably influence agricultural produce. In rice, several key auxin genes have been discovered that hold immense potential for modulation of auxin biosynthesis pathway for crop improvement. Auxin is a key regulator for the initiation of inflorescence and spikelet primordia including branch meristems in rice panicle. The role of auxin is identically critical for the grain development of rice. In the first two weeks of the post-anthesis period, IAA concentration rises sharply to a peak level and declines briskly to ground level at maturity. The decimation of auxin synthesis, at the early post-anthesis stage, inhibits grain filling.

Keywords Auxin · Biosynthesis · Transport · Signal transduction · Genes · Grain filling

Auxins are a class of plant hormones involved in the elongation of cells in the shoot part of the plant. They co-ordinate various growth processes in a plant's life cycle and are essential for plant body development. The hormone can stimulate

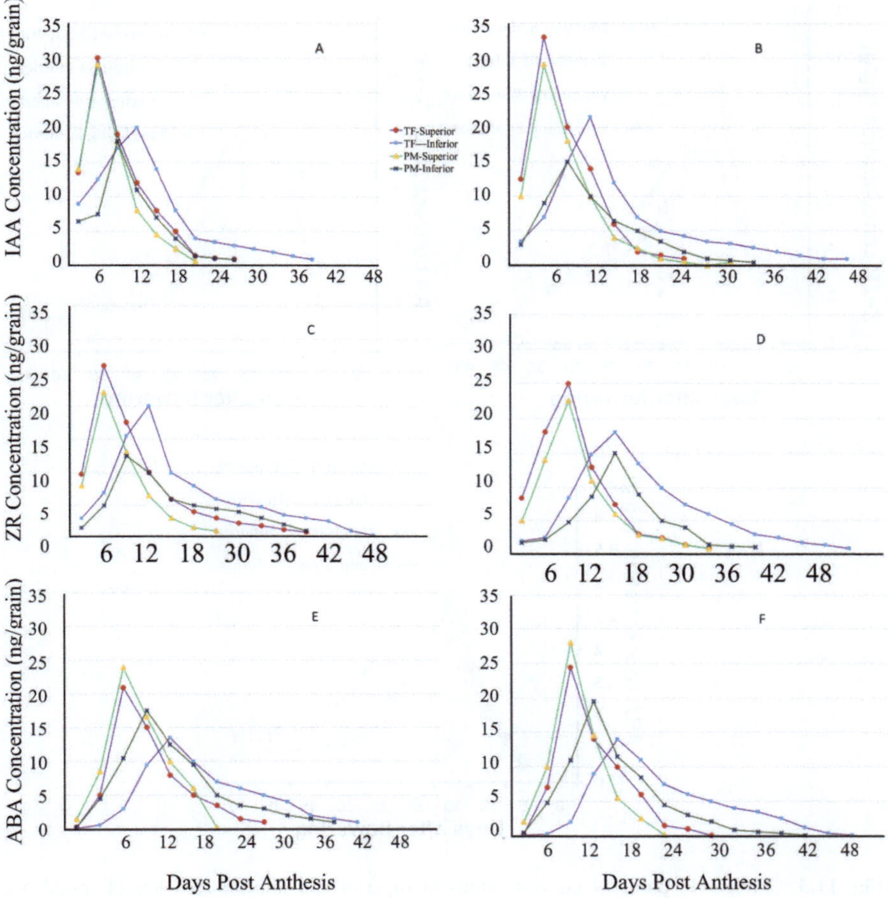

Fig. 11.2 Hormonal changes in the developing grains of rice cultivars. (**A**, **C**, and **E**): Shanyou 63 (*indica* hybrid rice in boxes) and (**B**, **D**, **F**): Wuyujing 3 (japonica cultivar in boxes). The spikelets that flowered in the first two days after heading were taken as superior grains and those that flowered on the 5th and 6th days after heading were sampled as inferior grains. (Adopted from Xu et al. 2007).

differential growth in response to irradiance and gravity. The chemical is present in all parts of the plant, but the concentration is location-specific, which becomes a critical factor for the regulation of cell development. Tight regulation coupled through synthesis, metabolism, and transport determines the cellular concentration of the chemical and guides development of the respective cells.

Auxin is synthesized in the shoot tip. It has one aromatic ring with a carboxylic acid group attached to it (Fig. 11.4). The indole-3-acetic acid is the most abundant and native auxin available in plants. It is the first hormone identified in plants and the most potent auxin that generates the majority of the auxin effects. Four other auxins with minor structural variation with IAA, have been identified in plants, they are 4-chloroindole acetic acid (4-Cl-IAA), 2-phenyl acetic acid (PAA), Indole-3-butyric

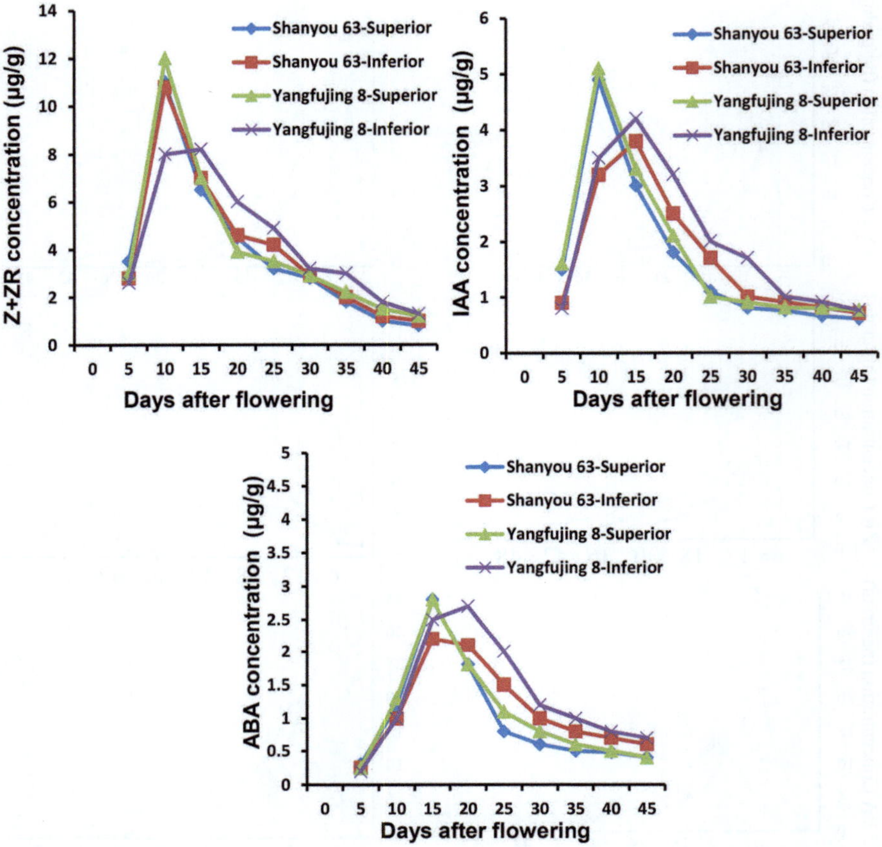

Fig. 11.3 Changes in hormone concentrations in superior and inferior spikelets of tested rice cultivars. Z+ZR= Zeatin+ Zeatin riboside, IAA= 3-indole acetic acid, ABA= Abscisic acid. Shanyou 63 (*indica* hybrid) and Yangfujing 8 (*japonica*) are high-yielding varieties. (Adopted from Fu et al. 2013)

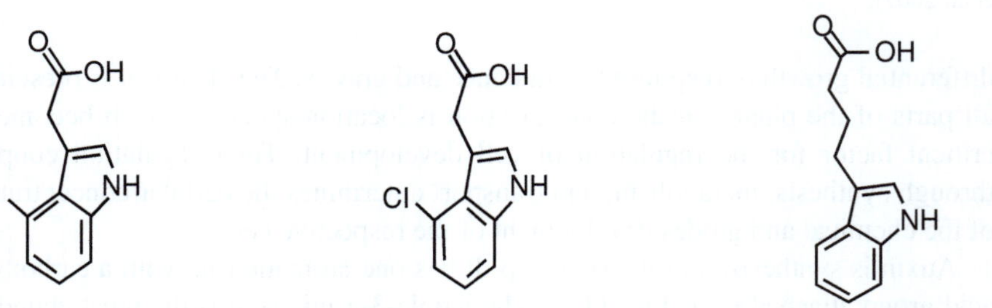

Indole-3-Acetic Acid 4 Chloro Indole-3-Acetic Acid Indole-3- Butyric Acid

Fig. 11.4 Structure of IAA and other related compounds showing auxin activity in plants acid (IBA), and Indole-3- propionic acid (IPA). The commercially available synthetic auxins are naphthalene-1-acetic acid (NAA) and 2,4 dichlorophenxyacetic acid (2,4-D). Compared to natural auxins they are more stable and highly used in horticulture/agriculture. However, these auxins do not share their mechanism of action with native auxins. The natural auxin distribution pattern in plant tissues is dynamic and environment responsive i.e. is achieved through a complex and well- coordinated system of a cell-to-cell active transport of auxin molecules named polar transport of auxins. Typical auxin action in plants is in concert with or in opposition to other hormones, prominently cytokinin and ethylene; the ratio in between being the final determinant for growth of root verses shoot and apical verses branch primordia.

Auxin Biosynthesis

Because auxin is essential for various aspects of plant development, modulation of auxin pathways fosters great potential for crop improvement. The mechanism by which IAA modulates the growth of plant organs has been clearly defined by the elucidation of expression of genes and chemical reactions controlling tryptophan-dependent and - independent auxin biosynthesis pathways (Zhao 2010; Mano and Nemoto 2012). It is understood that local auxin concentration is of prime importance for organ development, and the concentration is determined by combinations of three factors, such as A. biosynthesis, B. conjugation with other small molecules, and C. irreversible chemical degradation (Wang et al. 2018). In the plant kingdom, there are two major pathways proposed for IAA biosynthesis; tryptophan-dependent and tryptophan-independent pathways. Tryptophan-dependent IAA biosynthesis has four pathways, such as i. indole-3-acetamide pathway, ii. indole pyruvic acid pathway, iii. tryptamine pathway and indole-3-acetamide pathway according to Mano and Nemoto (2012). Auxin biosynthesis pathway is given below in Fig. 11.5.

Auxin Transport

Auxin transport in plants is both polar and non-polar. Plant organs normally exhibit apex-base structural polarity and this polarity is based on the polarity of auxin transport. For the entire plant, shoot apex is the primary source of auxin. From the shoot apex, it moves in the xylem unidirectionally by basipetal polar transport towards the root. The longitudinal gradient in auxin distribution from the shoot to the root is the principal factor affecting various developmental processes of plant organs, such as apical dominance, leaf senescence, and stem elongation. The basipetal polar transport proceeds by the cell to cell diffusion, where auxin exits the cell through the plasma membrane and enters the cell below in succession through its plasma membrane. The loss of auxin from the cell is termed auxin efflux and the gain is called auxin influx. Auxin transport through the cells is mediated by auxin-influx and –efflux carriers; the efflux carriers are specifically located on the basal part of the

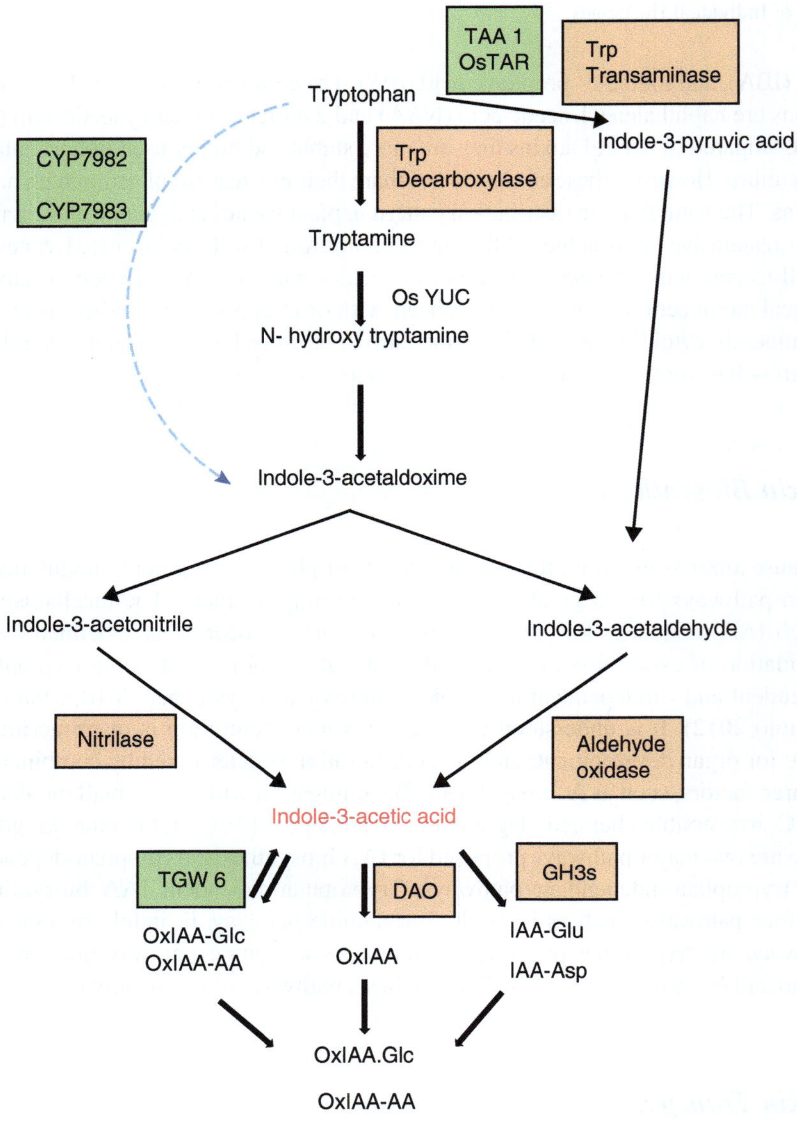

Fig. 11.5 Biosynthesis of Auxin. Auxin is synthesized mainly in tryptophan-dependent pathways in two major chemical steps. In step one tryptophan is converted to Indole-3-pyruvic acid (IPA) by the TAA family of tryptophan transaminases. IPA is converted to IAA by the YUCCA (YUC) flavin-containing monooxygenases in the second step. In rice tryptophan amino transferase OsTAR1 converts tryptophan into IPA (Kakei et al. 2017) and OsYUC controls the reaction in the second step (Yamamoto et al. 2007). Conjugation of auxin to both high and low molecular weight compounds effectively reduces the concentration of the biologically active form of the hormone. The low molecular weight conjugated auxin includes esters of auxins with glucose (IAA-glc) or Myo- inositol and amide conjugates with amino acids (IAA-AA). In comparison, high molecular weight conjugates consist of IAA-glucan and IAA-glyco proteins found in cereal seeds. IAA can become inactive by being oxidized into 2-oxo-indole-3-acetic acid (OxIAA) by dioxygenase for auxin oxidation (DAO) and methylated to MeIAA by indole-3-acetic acid carboxy methyltransferase. TGW6 affects auxin levels by catalyzing the hydrolysis of IAA-glucose (IAA-Glc). *GRETCHEN HAGEN 3*(*GH3*) genes, which encode auxin-amido synthetases, promote the inactivation of IAA. The red shadowbox indicates corresponding enzymes, green shadow box indicates genes involved in the pathway

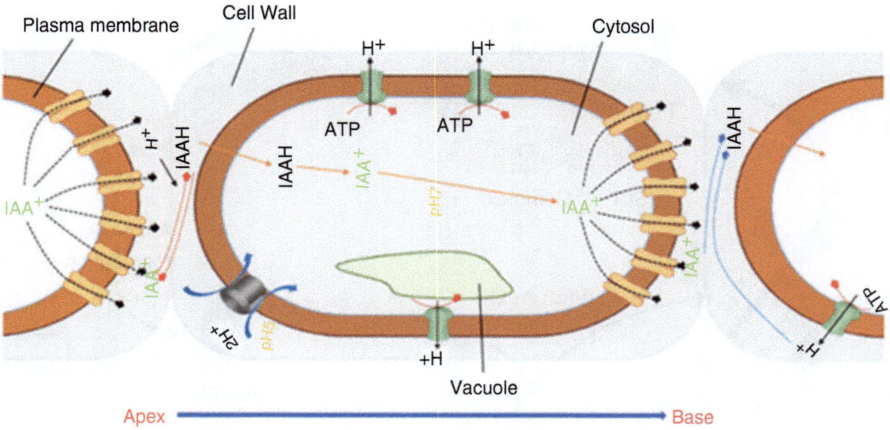

Fig. 11.6 Basipetal Auxin transport in plant cells. Auxin is transported by efflux and influx carriers in the plasma membrane.

cell. In the first step of polar transport, auxin influx occurs by either passive diffusion of protonated IAAH or secondary active transport of auxin using the IAA influx carrier proteins belonging to the AUX/LAX family proteins. IAA entering the cytoplasm dissociate to IAA^{-1}, which is not permeable through the plasma membrane. It needs anion efflux carriers on the basal side for escape. The efflux carriers include PIN and ABC family proteins (Fig. 11.6). The auxin synthesized in leaves is transported non-polarly in the phloem to the rest of the plant organs.

Auxin Signaling

Auxin responses are normally divided into genomic and non-genomic effects as deciphered in activation and synthesis hypotheses respectively. In the non-genomic activation process, auxin response is controlled by auxin binding protein1 (ABP1), located either in the cell surface or in the cytosol. ABP1 combines with IAA and stimulates plasma membrane H^+ATPase for proton pumping. In the genomic pathway, IAA-induced second messengers activate the expression of genes that encode the plasma membrane H^+ATPase. The increase in the number of H^+ATPase transporters increases proton extrusion from the cell. Recently, Wang et al. (2018) have stated the mechanisms of auxin signaling under low and high abundance of the hormone. At low concentration, auxin response factor (ARF) ligands with AUX/IAA repressors and topless (TOP) co-repressors. When concentration is high auxin interacts with TIR/AFB and AUX/IAA promoting ubiquitin-mediated destruction of AUX/IAA. The peptidylprolyl isomerase LTR2 catalyzes the inter-conversion of between the *cis* and *trans* isoforms of peptide bond with the amino acid proline in AUX/IAA proteins (Fig. 11.7). The conformational change generated by LTR2 plays an important role in auxin signaling.

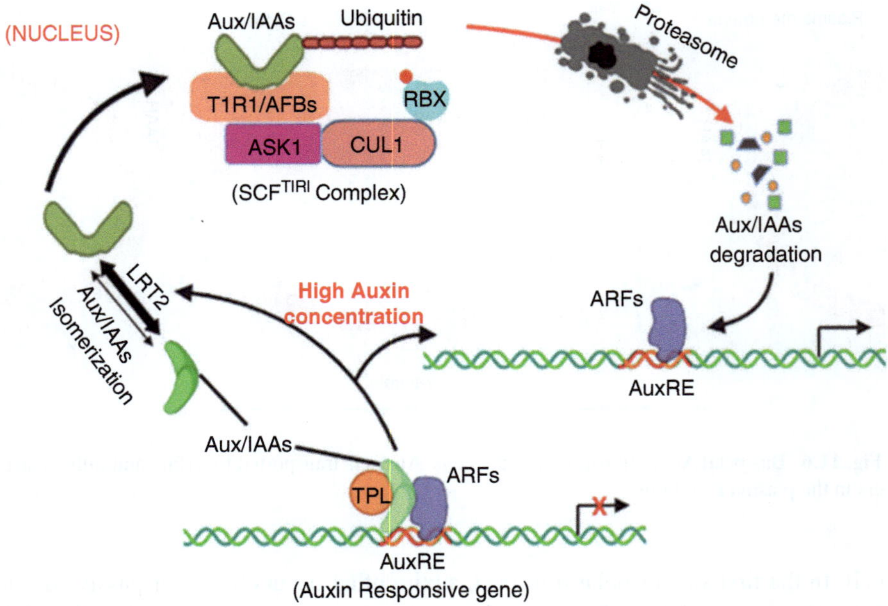

Fig. 11.7 The mechanism of auxin signaling is described in the text. (Adopted from Wang et al. 2018)

Auxin Regulates Rice Growth and Development

Auxin controls many agronomic important traits of rice plants. Auxin is required for different aspects of the root, tiller, panicle architecture, and flower development of rice. Hence disruption of auxin biosynthesis, transport, and signal transduction has a great impact on agricultural produce. In rice, several key auxin genes have been discovered that hold immense potential for modulation of auxin biosynthesis pathway for crop improvement. Although many agronomically important genes discovered are homologous to Arabidopsis, there are some unique to rice and play key roles in auxin homeostasis and signaling (Fig. 11.8). Over-expression of *OsAUX1* (Zhao et al. 2015) and *YUC* (Yamamoto et al. 2007) genes leads to the production of many lateral adventitious roots, which could be beneficial for greater absorption of minerals under stress conditions and higher grain yield per plant. Similarly increase of tiller number and reduction of tiller angles in rice are considered important for crop improvement and these features are regulated by auxin in rice. Gene expression controlling auxin transport and signaling influences tiller production and dimension of tiller angle. It was found that knockdown of auxin transport-related gene *OsPIN1* by RNAi improves tiller number and angle, whereas over expression of *OsPIN2* and *OsPIN3*t also gives a similar response (Chen et al. 2012; Zhang et al. 2012; Xu et al. 2005). Conversely, over-expression of *OsPIN5b* and knockdown of *OsAUX1* genes reduce tiller production (Zhao et al. 2015; Lu et al. 2015; Yu et al. 2015). Tillering is also influenced when auxin signaling is altered; down-regulation of auxin signaling results in the production of more tillers. It was shown that down regulation of auxin receptors OsTIR1 and OsAFB2 via over-expression of rice miRNA393 leads to more tillers, early flowering, and less tolerance to salt and drought (Xia et al. 2012; Li et al. 2017).

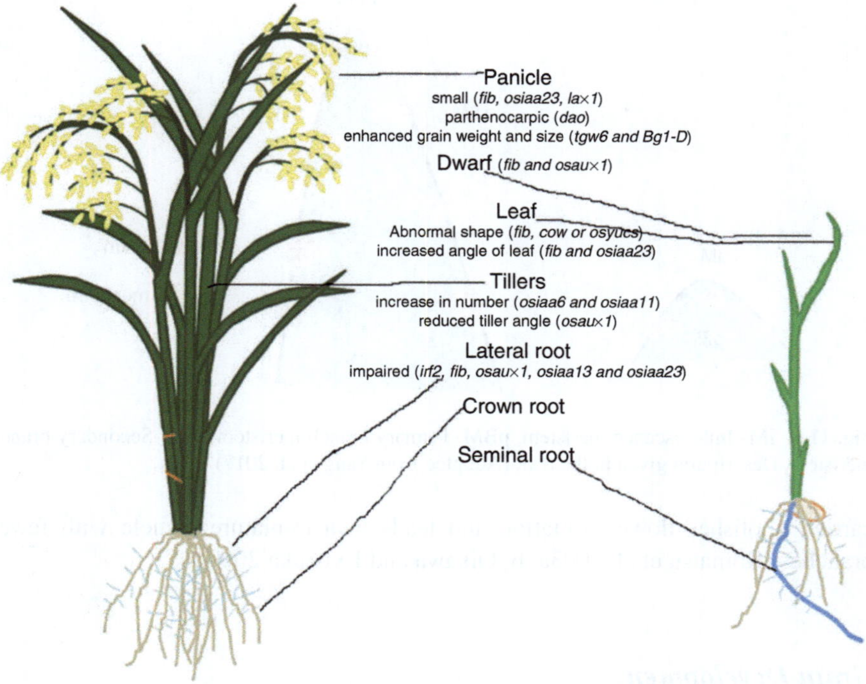

Fig. 11.8 Schematic model for various agriculturally important genes controlling rice plant development. Mutants displaying different phenotypes are recorded on the left side. (Adopted from Wang et al. 2018)

Auxin Response on Panicle Architecture and Grain Development

Panicle Architecture

Although auxin action is essential for all aspects of plant development, seldom experiments have been conducted to identify the role of auxin in rice panicle architecture determination and grain development. These traits are important considerations for rice breeding and crop improvement. Auxin is a key regulator for the initiation of inflorescence and spikelet primordia including branch meristems in rice panicle. Because auxin accumulation is required for reproductive organ initiation; its distribution is intimately associated with the inflorescence branching of rice. Auxin maxima and dynamic relocalization observed at the initiation sites for primary and secondary branches confirm growth regulation of the organs by the hormone (Yang et al. 2017) – Fig. 11.9)

Yoshikawa et al. (2014) reported that loss of function of auxin biosynthesis gene TAA1/ FIB resulted in small panicles possessing fewer branches and spikelets with abnormal organ identities. Similarly, disruption of *LAX1* gene expression, which encodes a basic helix-loop- helix (bHLH) transcription factor for auxin influx

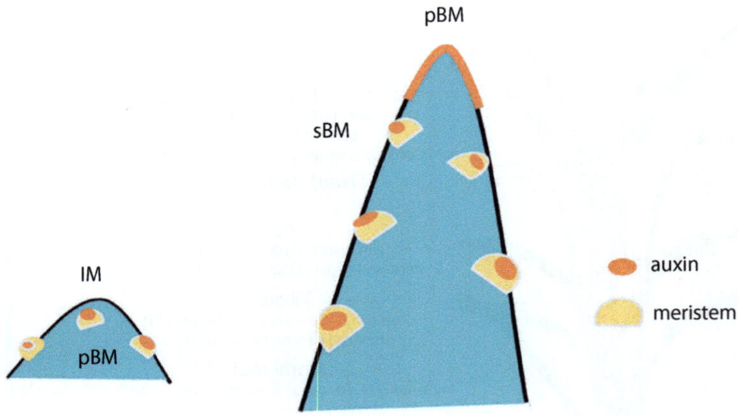

Fig. 11.9 IM- Inflorescence meristem, pBM- Primary branch meristem, sBM-Secondary branch meristem. Description given in the text. (Adopted from Yang et al. 2017)

carrier, abolishes flower initiation and leads to a miniature panicle with fewer branches (Komatsu et al. 2003a, b; Oikawa and Kyozuka 2009).

Grain Development

The role of auxin is very critical for the grain development of rice. In the first two weeks of the post-anthesis period, IAA concentration rises sharply to a peak level and declines briskly to ground level at maturity in the superior spikelets of rice panicle and declines in the same fashion (Zhang et al. 2016) (Fig. 11.10). In contrast, the concentration of the hormone is much lower in the inferior type spikelets and the dynamics of rise and fall in concentration is delayed to the third week of grain development. The slow filling rate of grains in the inferior type spikelets of panicle has been attributed to low IAA content (Zhao et al. 2007; Zhang et al. 2009). Similarly, poor grain filling of inferior spikelets arising from improper cultural conditions is also associated with low content of IAA (Xu et al. 2007). Exogenous application of IAA at the booting stage accelerates growth and development of the superior apical spikelets, but delays them in basal spikelets; as a result of which, the gradient in grain weight between the two contrasting types of spikelets widens (Patel and Mohapatra 1992). IAA content of rice panicle also declines significantly in plants subjected to drought and heat stresses that infringe upon spikelet fertility and grain yield and exogenous IAA application regains the features (Sharma et al. 2018). It was found that endogenous IAA level correlates positively with pollen fertility and yield traits and negatively with lipid peroxidation of spikelets and subjected to stress expression of most of the YUCCA genes and their transcriptional regulators were down- regulated. It is inferred that IAA is not only responsible for apical dominance, but also the survival of spikelets in rice panicle. This inherent apical dominance of spikelets is, however, neutralized and the margin of gradient in growth between

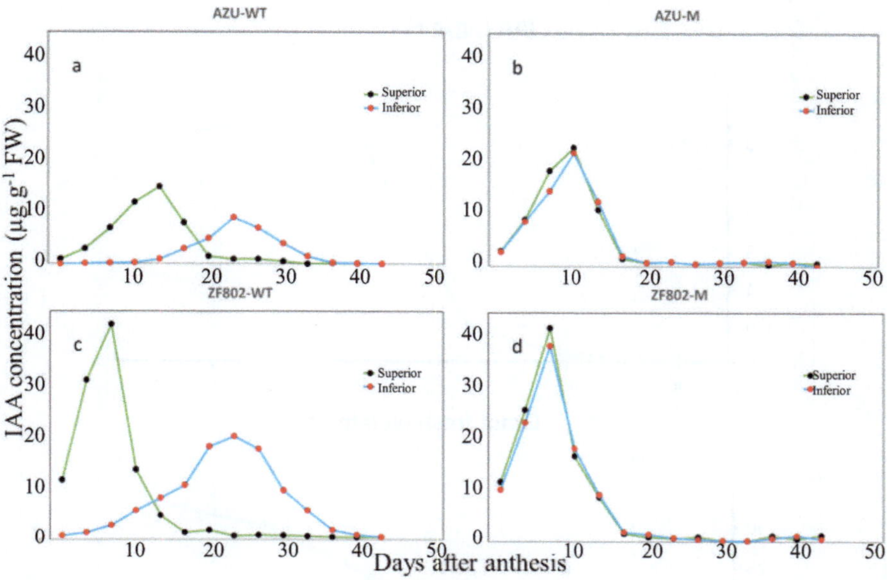

Fig. 11.10 Changes in the concentration of IAA in superior and inferior spikelets of wild-type rice (AZU-WT and ZF802-WT) and their corresponding genetic mutants (AZU-M and ZF802-M) during the grain development period. (Adopted from Zhang et al. 2016)

superior and inferior spikelets is reduced by exogenous cytokinin application that promotes specifically development of the spikelets on basal branches.

The auxin concentration increases significantly from 50 ng g^{-1} to 2.9 μg g^{-1} fresh weigh between days 1 to 14 post-anthesis (Abu-Zaitoon et al. 2012) (Fig. 11.11). The rise in concentration particularly during the first seven days is enormous and concurrent with endosperm cellularisation and early starch deposition. The coincidence of starch deposition with the rise of IAA content suggests a close relationship of the hormone with grain growth. Hormone synthesis begins five days after anthesis. Although some information as surmised above points to a positive role of IAA on grain filling in rice, the evidence obtained often does not support the proposition. The percentage of filled grains in some cultivars does not correlate with the intrinsic IAA level of grains (Yang et al. 2000). Also, Tamaki et al. (2015) showed that application of 2-indol-3-yl-4-oxo-4-phenylbutanoic acid (PEO-IAA; antagonist of auxin receptor) improved grain yield in a concentration-dependent manner.

Auxin Biosynthesis Genes

During grain development, the increase in IAA concentration is supported by over expression of IAA biosynthesis genes *OsYUC9*, *OsYUC11*, and *OsTAR1*. Also, another biosynthetic gene *OsYUC12* expresses early during endosperm development. This gene is co- expressed with *OsIAA29* that encodes an AUX/IAA transcription factor. Auxin

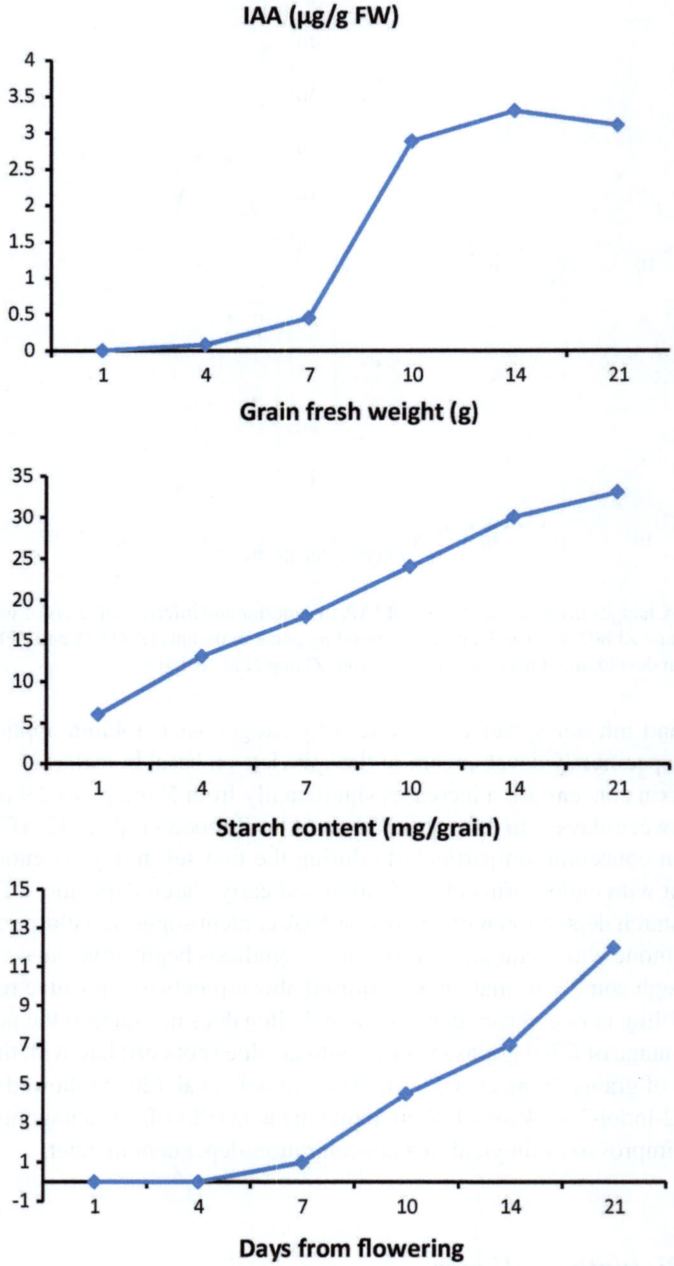

Fig. 11.11 Changes during grain development in (**a**) IAA content of rice grains measured by combined LC tandem MS using 13C IAA as the internal standard (**b**) mean fresh weight and (**c**) starch content. (Adopted from Abu-Zaitoon et al. 2012)

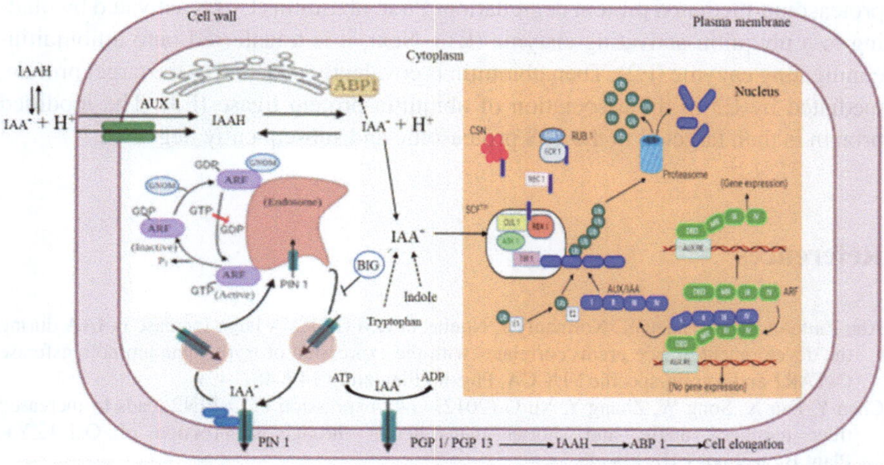

Fig. 11.12 Model of the mechanism of IAA signal transduction.

biosynthesis is regulated by cell wall invertase activity and maximum activity of the enzyme coincides with maximum expression of *OsYUC 12* and *OsIAA 29* genes (French et al. 2014). There are some other auxin-related genes, characterization of which led to the discovery of their role in the grain filling of rice. These auxin genes are *DAO*, *TGW6*, and *BG1* (Wang et al. 2018). The *dao* mutant is male sterile and produces parthenocarpic seeds because of the high accumulation of free IAA in anthers and ovaries (Zhao et al. 2013). Loss of function of IAA- glucose hydrolase gene *TGW6* increases grain weight and yield (Ishimaru et al. 2013). The *BG1* (*BIG GRAIN1*) is a primary auxin response gene that encodes a membrane-bound protein for auxin transport. Activation of *BG1* significantly improves grain size in rice (Liu et al. 2015). Compared to auxins, a possible role of other natural hormones in the characterization of genes involved in rice grain weight and size has been less investigated.

A composite model of auxin siganling mechanism is given in figure above (Fig. 11.12). Asymmetric accumulation of auxin in specific cells is a necessary prerequisite for various developmental processes. Out of many components, tryptophan and indole serve as precursors for auxin synthesis. PIN and AUX 1 are the molecular component for auxin efflux and influx transportation respectively. Besides, PGP1 and PGP19 P- glycoprotein transporter are involved in auxin efflux. Auxin binding protein 1, localized in the endoplasmic reticulum favors auxin binding at high pH. AXR1 (auxin resistant) - is related to the ubiquitylation pathway, which can mark protein for proteasome-mediated degradation. Auxin, after entry into the nucleus, rapidly regulates gene expression by targeting many genes eg. (AUX/IAA), the primary auxin-responsive gene. It contains an auxin response element (AuxRE) with sequence TGTCTC at promoters of many auxin-responsive genes. In the absence of auxin, AUX/IAA family inhibits ARFs (auxin response factors). AUX/IAA and ARFs show many structural similarities. ARFs bind with AuxRE at N- terminal end containing (DNA-binding domain) DBD and facilitates downstream auxin-related gene expression. TIR1 (transport inhibitor response 1) encodes a member of a large family of F-box protein involved in regulated proteasome-mediated protein degradation. First, ubiquitin (Ub) is activated by binding to a ubiquitin-activating enzyme (E1). Next, it is transferred onto a ubiquitin-conjugating enzyme (E2). Then ubiquitin is covalently attached to the target protein, mediated by E3 in the association of ubiquitin-protein ligase (E3). The modified protein is then targeted to the 26S proteasome and subsequently degraded.

References

Abu-Zaitoon YM, Bennett K, Normanly J, Nonhebel HM (2012) A large increase in IAA during the development of rice grains correlates with the expression of tryptophan aminotransferase OsTAR1 and a grain-specific YUCCA. Physiol Plantarum 146: 487–499

Chen Y, Fan X, Song W, Zhang Y, Xu G (2012) Over-expression of OsPIN2 leads to increased tiller numbers, angle, and shorter plant height through suppression of OsLAZY1. Plant Biotechnol J 10: 139–49

Ding C, You J, Chen L, Wang S, Ding Y (2014) Nitrogen fertilizer increases spikelet number per panicle by enhancing cytokinin synthesis in rice. Plant Cell Reports 33: 363–371

French S, Abu-Zaitoon Y, Uddin M, Bennett K, Nonhebel H (2014) Auxin and cell wall invertase-related signaling during rice grain development. Plants 3: 95–112

Fu J, Xu YJ, Chen L, Yuan LM, Wang ZQ, Yang JC (2013) Changes in enzyme activities involved in starch synthesis and hormone concentrations in superior and inferior spikelets and their association with grain filling of super rice. Rice Sci 20: 120–128

Ishimaru K, Hirotsu N, Madoka Y, Murakami, N, Hara N, Onodera H, Kashiwagi T, Ujiie K, Shimizu B, Onishi A, Miyagawa H, Katoh E (2013) Loss of function of the IAA-glucose hydrolase gene TGW6 enhances rice grain weight and increases yield. Nat Genetics 45: 707–711

Kakei Y, Nakamura A, Yamamoto M, Ishida Y, Yamazaki C, Sato A, Narukawa-Nara M, Soeno K, Shimada Y (2017) Biochemical and chemical biology study of rice OsTAR1 revealed that tryptophan aminotransferase is involved in auxin biosynthesis: identification of a potent OsTAR1 inhibitor, pyruvamine2031. Plant Cell Physiol 58: 598–606

Komatsu K, Maekawa M, Ujiie S, Satake Y, Furutani I, Okamoto H, Shimamoto K, Kyozuka J (2003a) LAX and SPA: major regulators of shoot branching in rice. Proc Nat Acad Sci U S A 100: 11765–11770

Komatsu M, Chujo A, Nagato Y, Shimamoto K, Kyozuka J (2003b) FRIZZY PANICLE is required to prevent the formation of axillary meristems and to establish floral meristem identity in rice spikelets. Development 130: 3841–3850

Li J, Sun Y, Du J, Zhao Y, Xia L (2017) Generation of targeted point mutations in rice by a modified CRISPR/Cas9 system. Mol Plant 10: 526–529

Lu G, Coneva V, Casaretto JA, Ying S, Mahmood K, Liu F, Nambara E, Bi YM, Rothstein SJ (2015) OsPIN5b modulates rice (Oryza sativa) plant architecture and yield by changing auxin homeostasis, transport, and distribution. Plant J 83: 913–925

Mano Y, Nemoto K (2012) The pathway of auxin biosynthesis in plants. J Exp Botany 63: 2853–2872

Mohapatra PK, Panigrahi R, Turner NC (2011a) Physiology of spikelet development on the rice panicle: Is the manipulation of apical dominance crucial for grain yield improvement? Adv Agron 110: 333–359

Mohapatra PK, Patel R, Sahu SK (1993) Time of Flowering Affects Grain Quality and Spikelet Partitioning Within the Rice Panicle. Funct Plant Biol 20: 231–241

Mohapatra PK, Sahu SK (1991) Heterogeneity of primary branch development and spikelet survival in rice panicle in relation to assimilates of primary branches. J Exp Botany 42: 871–879

Oikawa T, Kyozuka J (2009) Two-step regulation of LAX PANICLE1 protein accumulation in axillary meristem formation in rice. Plant Cell 21: 1095–1108

Patel R, Mohapatra PK (1992) Regulation of spikelet development in rice by hormones. J Exp Botany 43: 257–262

Sharma L, Dalal M, Verma RK, Kumar SVV, Yadav SK, Pushkar Kushwaha SR, Bhowmik A, Chinnusamy V (2018) Auxin protects spikelet fertility and grain yield under drought and heat stresses in rice. Environ Exp Botany 150: 9–24

Sekhar S, Panda BB, Mohapatra T, Das K, Shaw BP, Kariai E, Mohapatra PK (2015) Spikelet-specific variation in ethylene production and constitutive expression of ethylene receptors and signal transducers during grain filling of compact-and lax-panicle rice (Oryza sativa) cultivars. J Plant Physiol 179: 21–34

Tamaki H, Reguera M, Abdel-Tawab YM, Takebayashi Y, Kasahara H, Blumwald E (2015) Targeting hormone-related pathways to improve grain yield in rice: a chemical approach. PLoS ONE 10: e0131213

Wang F, Cheng F, Zhang G (2006) The relationship between grain filling and hormone content as affected by genotype and source-sink relation. Plant Growth Reg 49: 1–8

Wang Y, Zhang T, Wang R, Zhao Y (2018) Recent advances in auxin research in rice and their implications for crop improvement. J Exp Botany 69: 255–263

Xia K, Wang R, Ou X, Feng Z, Tian C, Duan J, Wang Y, Zhang M (2012) OsTIR1 and OsAFB2 downregulation via OsmiR393 overexpression leads to more tillers, early flowering, and less tolerance to salt and drought in rice. PLoS ONE 7: e30039

Xu M, Zhu L, Shou H, Wu P (2005) A PIN1 family gene, OsPIN1, is involved in auxin-dependent adventitious root emergence and tillering in rice. Plant Cell Physiol 46: 1674–1681

Xu X, Vergara BS (1986) Morphological changes in rice panicle development: A research development. IRRI Research Paper Series No 117. The Philipines

Xu G, Zhang J, Lam HM, Wang Z, Yang J (2007) Hormonal changes are related to the poor grain filling in the inferior spikelets of rice cultivated under non-flooded and mulched conditions. Field Crops Res 101: 53–61

Yamamoto Y, Kamiya N, Morinaka Y, Matsuoka M, Sazuka T (2007) Auxin biosynthesis by the YUCCA genes in rice. Plant Physiol 143: 1362–1371

Yang J, Peng S, Visperas RM, Sanico AL, Zhu Q, Gu S (2000) Grain filling pattern and cytokinin content in the grains and roots of rice plants. Plant Growth Reg 30: 261–270

Yang J, Yuan Z, Meng Q, Huang G, Perin C, Bureau C, Meunier AC, Ingouf M, Bennet MJ, Liang W, Zhang D (2017) Dynamic regulation of auxin response during rice development revealed by newly established hormone biosensor markers. Front Plant Sci 8: 256

Yoshikawa T, Ito M, Sumikura T, Nakayama A, Nishimura T, Kitano H, Yamaguchi I, Koshiba T, Hibara K, Nagato Y, Itoh J (2014) The rice FISH BONE gene encodes a tryptophan aminotransferase, which affects pleiotropic auxin-related processes. Plant J 78: 927–936

Yu C, Sun C, Shen C, Wang S, Liu F, Liu Y, Chen Y, Li C, Qian Q, Aryal B, Geisler M, Jiang DA, Qi Y (2015) The auxin transporter, Os AUX 1, is involved in primary root and root hair elongation and in Cd stress responses in rice (Oryza sativa L.). Plant J 83: 818–30

Zhang W, Cao Z, Zhou Q, Chen J, Xu G, Gu J, Liu L, Wang Z, Yang J, Zhang H (2016) Grain filling characteristics and their relations with endogenous hormones in large-and small-grain mutants of rice. PLoS ONE 11: e0165321

Zhang Q, Li J, Zhang W, Wang R, Zhao J, Li Y, Sun Z, Zhu Z (2012) The putative auxin efflux carrier OsPIN3t is involved in the drought stress response and drought tolerance. Plant J 72: 805–816

Zhao BH, Kai L, Zhang HX, Zhu QS, Yang JC (2007) Causes of poor grain plumpness of two-line hybrids and their relationships to the contents of hormones in the rice grain. Agric Sci China 6: 930–940

Zhao Z, Zhang Y, Liu X, Zhang X, Liu S, Yu X, Ren Y, Zheng X, Zhou K, Jiang L, Guo X, Gai Y, Wu C, Zhai H, Wang H, Wan J (2013) A role for a dioxygenase in auxin metabolism and reproductive development in rice. Dev Cell 27: 113–122

Zhang Z, Xue Y, Wang Z, Yang J, Zhang J (2009) The relationship of grain filling with abscisic acid and ethylene under non-flooded mulching cultivation. J Agric Sci 147: 423–436

Zhao Y (2010) Auxin biosynthesis and its role in plant development. Ann Rev Plant Biol 61: 49–64

Zhao H, Ma T, Wang X, Den Y, Ma H, Zhang R, Zhao J (2015) OsAUX 1 controls lateral root initiation in rice (O Oryza sativa L.). Plant Cell Environ 38: 2208–2222

Gibberellins

Abstract Gibberellins are a large group numbering more than 125 related compounds. The compounds differ more in structure than in biological activities. In rice, gibberellins are mostly associated with stem internode elongation but also required for grain filling and seed germination. The sd1mutant phenotype is synonymous with stem dwarfing in rice; the plant is deprived of bioactive GA synthesis reduces stem height allowing greater partitioning of assimilates in favor of reproductive organs accruing yield benefit. The hormone concentration is high at the early stage of grain filling when endosperm cell number and sink capacity are fixed. The concentration declines rapidly in the developing grain over time up to the maturity stage.

Keywords Bioactive GA · *sd1* · GA20 oxidase · Cell elongation · Grain filling

Properties of Gibberellins

Nearly three decades after auxin discovery in 1927, the second group of plant hormones, the gibberellins was characterized. Unlike auxins, gibberellins are a large group numbering more than 125 related compounds. The compounds differ more in structure than in biological activities. Many of the compounds discovered in plants are biologically inactive. Only a small number of them, such as GA_1, GA_3, GA_4, and GA_7 are biologically active. Non- bioactive gibberellins act as a precursor of the active gibberellins and do not affect cell metabolism. The presence of GA_1 has been widespread in plant species. Gibberellins are mostly associated with stem internode elongation. Additionally, the hormone plays important role in physiological processes of seed germination, flowering, flower and fruit development, and mobilization of stored endosperm reserve. The hormone is seldom transported long distances in the xylem and phloem. The concentration of bioactive GA has been measured by the rates of their biosynthesis and deactivation.

Biosynthesis of Gibberellins

GA metabolism has been studied by identifying the genes encoding metabolizing enzymes through the conventional method of purification of the enzymes from rich sources, functional screening of a cDNA expression library, and molecular approaches using genetically dwarf mutants deficient in GA biosynthesis (Yamaguchi 2008). Expression analyses of genes controlling GA synthesis and signal transduction have revealed that the hormone is synthesized at the site of its action. Chemically gibberellins are isoprenoids compounds. Specifically, they are a large family of tetracyclic diterpene acids, which are synthesized in the terpenoid

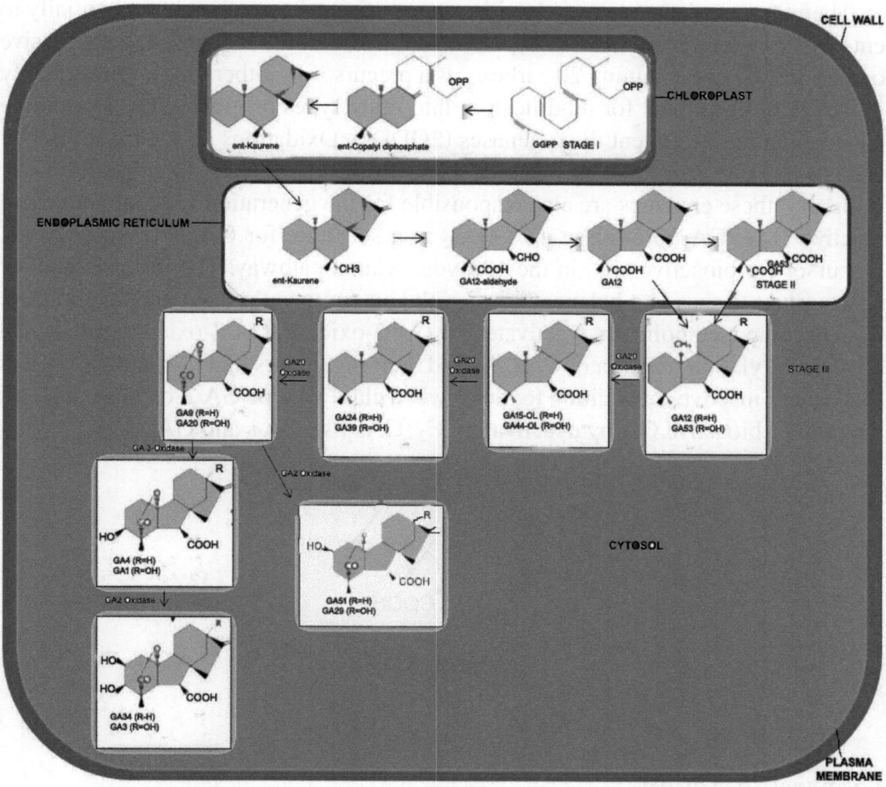

Fig. 11.13 Structure of gibberellins and their biosynthesis (Adopted from Plant Physiology, Taiz and Zeiger). The three stages of gibberellin biosynthesis in plants. Stage 1, Geranyl geranyl diphosphate (GGPP) is converted to ent-kaurene through copalyl diphosphate in plastids. Stages 2, ent-kaurene goes to the endoplasmic reticulum and converted to GA_{12} or GA_{53}. In stage 3, GA_{12} and GA_{53} are converted to other GAs in the cytoplasm. The conversion takes place with a series of oxidations at carbon 20. Hydroxylation by GA20 oxidase enzyme leads to the production of GA_{20}. GA3 oxidase converts GA20 to an active form of GA1. GA2 oxidase action leads to the synthesis of inactive gibberellins.

pathway of plastids and then modified in the endoplasmic reticulum and cytoplasm. They are synthesized from the acetate unit of acetyl coenzyme A by the mevalonic acid pathway leading to the synthesis of geranyl-geranyl pyrophosphate. It is a 20 carbon compound that contributes the carbon atoms for all gibberellins. The trans-geranylgeranyl pyrophosphate by the action of terpene synthase is converted to copalylpyrophosphate with two ring systems and then to ent-kaurene with four ring systems in the plastid. The terpene synthase enzymes are *ent*-copalyl diphosphate synthase and (CPS) and *ent*-kaurene synthase (KS). Ent-kaurene moves into the endoplasmic reticulum, where it is oxidized to produce GA_{12} by two membrane-associated P450 mono- oxygenases. The ent-kaurene oxidase sequentially produces intermediate compounds like kaurenol (alcohol) kaurenal (an aldehyde) and kaurenoic acid, and ent-kaurenoic acid (Fig. 11.13).

Thereafter ent-kaurenoic acid oxidase converts ent-kauronic acid sequentially to ent- hydroxy kaurenoic acid, GA_{12} aldehyde, and finally GA_{12} through successive oxidations. GA_{12} containing 20 carbons, is a precursor for other GAs. It presumably moves into the cytosol for modification into other types of gibberellins by soluble 2-oxoglutarate-dependent dioxygenases (2ODDs). Oxidations on C-20 and C-3 by GA 20-oxidase and GA 3-oxidase respectively convert GA_{12} to bioactive GA_4. Similarly, these enzymes are also responsible for the generation of other active and inactive GAs. GA 13-oxidase uses GA_{12} as a substrate for GA_{53} biosynthesis i.e. precursor for bioactive GA_1 in the 13-hydroxylated pathway. The intrinsic pool of GA in plant tissues is a balance between the amount synthesized and catabolized. GAs become metabolically inactivated by GA 2-oxidase. GA 2-oxidases catalyze a 2β –hydroxylation reaction of C_{19} GAs and their precursors like GA_9 and GA $_{20}$ and make them inactive. In addition to these two, a class III type GA 2-oxidase depletes the pool of bioactive GA by deactivating C_{20} GAs like GA_{12} and GA_{53}.

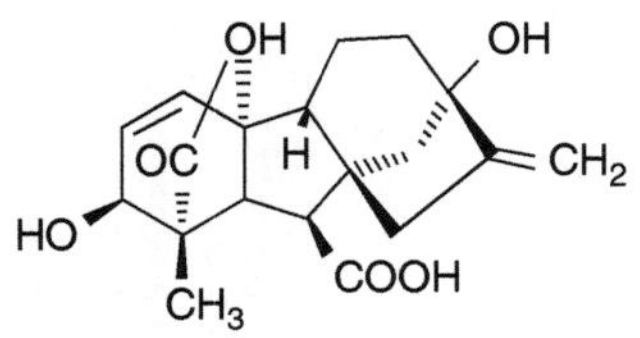

Gibberellin

Gibberellin Action

The possible mechanisms by which gibberellins influence physiological processes in plants are the promotion of cell division in the shoot apex, promotion of cell growth by an increase of starch, fructans, and sucrose hydrolysis into glucose and fructose, and increase of cell plasticity facilitating elongation. In rice, gibberellins influence several aspects of plant development significantly that have relevance to yield and yield-related attributes. Pronounced effects of gibberellins on rapid stem elongation in deepwater rice have been a mechanism of resilience for survival under partial submergence. Poor diffusion of oxygen in water triggers ethylene production in the submerged tissues, which in turn suppresses ABA action making it more responsive to gibberellins for stem elongation. The other major influence gibberellins have on rice growth is in the control of stem growth of semidwarf rice. Spielemeyer et al (2002) observed that the major semi-dwarfing allele *sd1* is extensively used in the new high-yielding semidwarf rice worldwide. The *sd1*mutant phenotype is synonymous with stem dwarfing; the plant is deprived of bioactive GA synthesis and deficiency of gibberellins reduces stem height allowing greater partitioning of assimilates in favor of reproductive organs.

Expression of sd1 Allele

The mechanism of *sd1* expression in controlling stem elongation has been explored further. GA20 oxidase is the enzyme that catalyzes chemical reactions involving successive steps of oxidation in the conversion of GA_{53} to GA_{20} (Fig. 11.13). GA 20 is the precursor for bioactive GA_1 that controls stem growth. *GA20 oxidase* gene is very close and corresponds to *SD1* locus on chromosome 1. Cloning and sequence analyses corroborated that *SD1* encodes and synonymous with the *GA20 oxidase* gene (Ashikari et al. 2002). A loss of function mutation in the *sd1* gene underscores the enzyme synthesis, in the particular synthesis of *GA20 oxidase2*, and blocks the GA biosynthesis pathway resulting in stem dwarfism (Lopez-Cristoffanini et al. 2019). Thus, the rice semidwarf gene *sd1* became known as the Green revolution gene and was used for the development of semidwarf rice IR8 for the success of the Green revolution. Although the precise mechanism of gibberellins action is not known in stem elongation, the hormone promotes cell division and elongation. The elongation process requires loosening of the cell wall and deposition of new materials for the preservation of integrity. Cell elongation occurs due to the reorientation of cellulose microfibrils and cell wall matrix containing polysaccharides and proteins driven by turgor pressure allowing cell extensibility in a particular direction. However, the GA signal transduction pathway contributing to cell wall relaxation is not precisely known.

GA-regulated Gene Promoting Cell Elongation

Microarray analyses of GA-regulated gene expression revealed the existence of three genes that prominently control cell elongation (Jan and Komatsu 2006). *OsXTH8* (*xyloglucan endotransglucosyl/hydrolase 8*) gene is involved in the regulation of cell elongation and cell wall organization. The gene catalyzes endo-cleavage of xyloglucan polymers and subsequent transfer of the reducing ends to new xyloglucan polymers for facilitating cell wall extension. Repression of *OsXTH8* leads to stunted plant growth by severely restricting internode elongation. In comparison, the effects of the *OsPDK1* gene on cell elongation are indirect; the gene controls primary metabolism at the entry point of the TCA cycle regulating the supply of resources for growth. The third gene *OsGAE1* is predominantly expressed in stem apex and juvenile leaves, antisense expression of which leads to shorter plant height. This gene is not involved in GA biosynthesis or signal transduction but is important for the downstream action of the hormone in promoting cell elongation. A model is given below for GA signal transduction (Fig. 11.14).

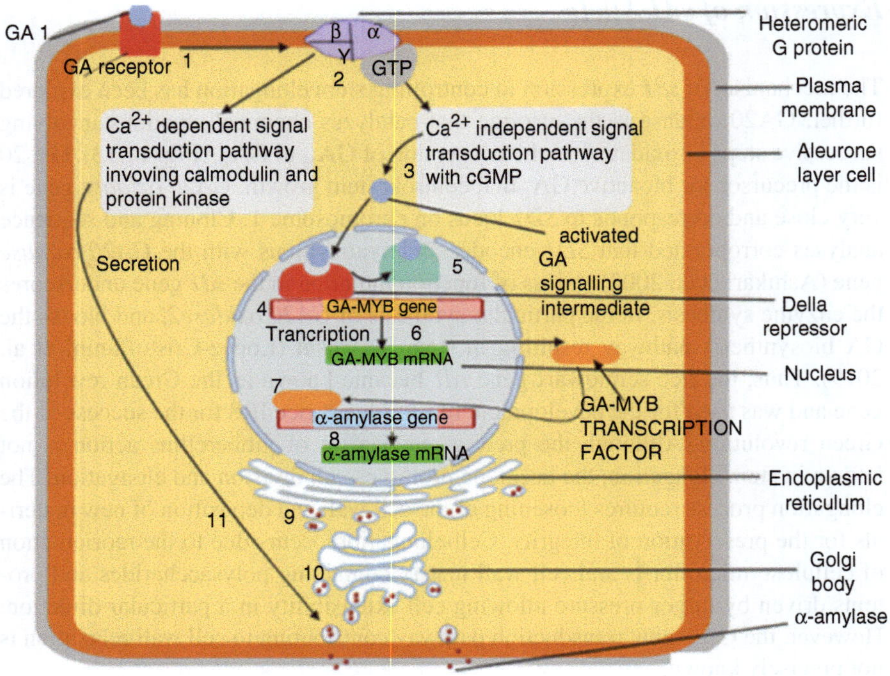

Fig. 11.14 Mechanisms of GA signal transduction in the plant cell. The stimulation of α-amylase gene expression by gibberellins is mediated by a specific transcription factor that binds to the promoter of the α- amylase gene. Treatment of oat aleurone protoplasts with a peptide called Mas7, which stimulates GTP/GDP exchange by G proteins, was found to induce α-amylase gene expression and to stimulate α-amylase secretion, suggesting that such a GTP/GDP exchange on the cell membrane is a reaction en route to the induction of α- amylase biosynthesis by gibberellin. Besides, gibberellin-induced α-amylase gene expression and secretion were inhibited by a guanine nucleotide analog that binds to the α subunit of heterotrimeric G-proteins and inhibits GTP/GDP exchange. Prepared with reference to Ritchie and Gilroy (1998)

Gibberellin Effects on Rice Spikelet Development

There have been fewer studies on the role of gibberellins associated with the processes of grain production, and information available on its effect on the panicle structure of rice is scant. In rice plants the nature of gibberellins varies with the development stage, while GA_1 is absent in the coleoptiles, GA_4 abundance is high during most of the developmental stages (Lopez-Cristoffanini et al. 2019). It was reported that GA application enhances crop response to nitrogen at the heading stage (Mukherjee and Prabhakar 1980). Inferior tillers of rice emerge late and become stunted in growth on account of low endogenous GA and an external application can override the deleterious effect and make them productive (Wang et al.

2017a). The pool of gibberellins in spikelets changes greatly during grain development and is closely associated with grain filling rate (Wang et al. 2006). Spikelets producing good quality grains possess a higher level of GA3 than those developing poor grains. However, Zhang et al. (2016) could not find such a difference in endogenous GA_{1+4} concentrations between the two types of developing spikelets; the concentration remained very often higher in inferior compared to the superior spikelet.

The panicle of a high sterile cultivar like Ratna possesses much lower GA_3 content compared to highly fertile JS52 (Murthy and Murty 1981). The hormone concentration is high at the early stage of grain filling when endosperm cell number and sink capacity are fixed (Yang et al. 2001). The concentration declines rapidly in the developing grain with the passage of time up to the maturity stage (Zhang et al. 2016). The hormone applied externally at the right concentration during the booting stage of the panicle improves grain filling; in particular the inferior spikelets at the basal part of the panicle (Patel and Mohapatra 1992). However, external application of the hormone was found ineffective on spikelet number, grain filling percentage, grain weight, and yield (Pan et al. 2013). In a more recent report, Wang et al. (2019) reported enhancement of seed vigor and yield, when exogenous GA_3 application was given to *indica* hybrid rice at anthesis.

Because of its effects on anther development; gibberellic acid has a special role in spikelet fertility. At anthesis, nearly 80% of the hormone is located in the pollen grains (Pharis and King 1985). In the signaling system, a DELLA protein inhibits GA-associated development, whereas GA abundance induces a protease-mediated degradation of the DELLA protein SLENDER RICE1 (SLR1). The DELLA protein inhibits GA-associated development because of its negative effects on downstream genes in the GA signaling pathway including *GAMYB*, which encodes a transcription factor necessary for positive regulation of GA-responsive genes. A diagrammatic model has been prepared by Kwon and Paek (2016) for GA signaling in spikelet development of rice, where degradation of SLR1 by GA perception activates transcription of *GAMYB*. GAMYB has been a transcription factor in GA signaling, which activates amylase gene expression in the aleurone layer during germination in addition to its role in anther and pollen development of angiosperms. It binds with the promoters of genes encoding exine and Ubisch body formation of pollens and thereby controls anther development. (Fig. 11.15).

Wang et al. (2017b) reported that ectopic expression of *SEPALLATA*-like gene *OsMADS1* in transgenic plants induced abnormal brassinosteroids and gibberellins activities leading to defects in the development of stem and panicle of rice. Overexpression of the gene in transgenic plants caused dwarfism and shorter panicle. The over-expression of brassinosteroid negative regulators in the panicle of transgenic plants might have resulted in brassinosteroid deficiency as well as reduction of GA level due to up-regulation of GA inactivation gene *GA2ox-1*.

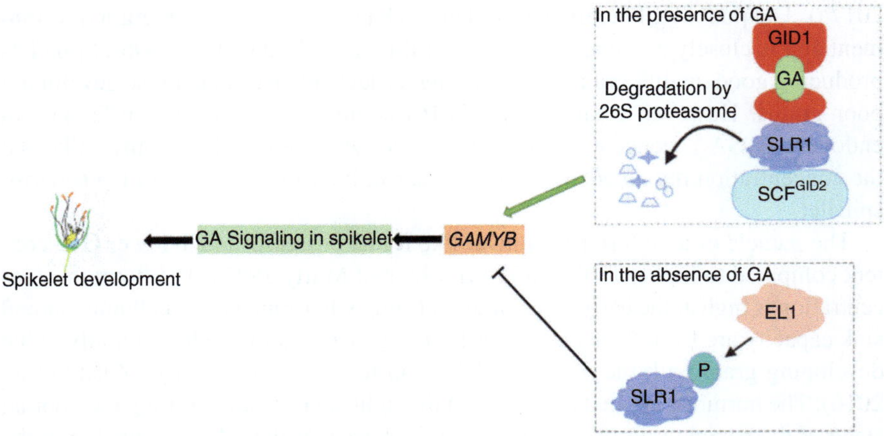

Fig. 11.15 GA signaling in spikelet development of rice panicle. Under GA-deficiency (-GA) casein kinase 1 (EL1) phosphorylates SLENDER RICE1 (SLR1) for an increase of its activity and stability. Next SLR1 represses *GAMYB* expression. Under GA-abundance (+GA), the hormone is perceived by GID1. The resulting conformational change of GID1 permits it to interact with and degrade SLR1 through the SCF^{GID2}-mediated proteasome pathway. In absence of SLR1 activity, *GAMYB* transcription becomes de-repressed; thereafter GAMYB protein positively regulates the transcription of several downstream genes for the formation of the Ubisch body and exine, and cell in the tapetum during spikelet development. (Adopted from Kwon and Paek 2016)

References

Ashikari M, Sasaki A, Ueguchi-Tanaka M, Itoh H, Nishimura A, Datta S, Ishiyama K, Saito T, Kobayashi M, Khush GS, Kitano H, Matsuoka M (2002) Loss-of-function of a rice gibberellin biosynthetic gene, GA20 oxidase (GA20ox-2), led to the rice 'green revolution'. Breeding Sci 52: 143–150

Jan A, Komatsu S (2006) Functional characterization of gibberellin-regulated genes in rice using microarray system. Genomics Proteomics Bioinfo 4: 137–144

Kwon C-T, Paek N-C (2016) Gibberellic acid: a key phytohormone for spikelet fertility in rice grain production. Int J Mol Sci 17: 794

Lopez-Cristoffanini C, Serrat X, Jauregui O, Nogues S, Lopez-Carbonell M (2019) Phytohormone profiling method for rice: effects of GA20ox mutation on the gibberellin content of japonica rice varieties. Front Plant Sci 07 June 2019 https://doi.org/10.3389/fpls.2019.00733

Mukherjee R, Prabhakar B (1980) Effect of gibberellin on rice yield response to nitrogen applied at heading, and quality of seeds. Plant Soil 55: 153–156

Murthy PSS, Murty KS (1981) Variation in concentrations of growth regulators in high and low sterile rice cultures. Indian J Exp Biol 19: 591–592

Pan S, Rasul F, Li W, Tian H, Mo Z, Duan M, Tang X (2013) Roles of plant growth regulators on yield, grain qualities, and antioxidant enzyme activities in super hybrid rice (Oryza sativa L.). Rice 6: 9. https://doi.org/10.1186/1939-8433-6-9

Patel R, Mohapatra PK (1992) Regulation of spikelet development in rice by hormones. J Exp Botany 43: 257–262

Pharis RP, King RW (1985) Gibberellins and reproductive development in seed plants. Ann Rev Plant Physiol 36: 517–568

Ritchie S, Gilroy S (1998) Gibberellins: regulating genes and germination. New Phytol 140: 363–383

Spielmeyer W, Ellis MH, Chandler PM (2002) Semidwarf (sd-1), "Green revolution" rice, contains a defective gibberellin 20-oxidase gene. Proc Nat Acad Sci 99: 9043–9048

Taiz L, Zeiger E (2002) Plant Physiology. 3rd edn. Sinauer Associates, Inc., Sunderland, USA

Wang F, Cheng F, Zhang G (2006) The relationship between grain filling and hormone content as affected by genotype and source-sink relation. Plant Growth Reg 49: 1–8

Wang Y, Ren T, Lu J, Cong R, Hao W, Liu T, Hussain S, Li X (2017a) Exogenously applied gibberellic acid improves the growth and yield performance of inferior rice tillers grown under different nitrogen levels. Acta Physiol Plantarum 39: 5. https://doi.org/10.1007/s11738-016-2307-3

Wang L, Zeng X-Q, Zhuang H, Shen Y, Chen H, Wang Z, Long J, Ling Y, He G, Li Y (2017b) Ectopic expression of OsMADS1 caused dwarfism and spikelet alteration in rice. Plant Growth Reg 81: 433–442

Wang X, Zheng H, Tang Q, Mo W, Ma J (2019) Effects of gibberellic acid application after anthesis on seed vigor of indica hybrid rice (Oryza sativa L.). Agronomy 9: 861. https://doi.org/10.3390/agronomy9120861

Yamaguchi S (2008) Gibberellin metabolism and it's regulation. Ann Rev Plant Biol 59: 225–251

Zhang W, Cao Z, Zhou Q, Chen J, Xu G, Gu J, Liu L, Wang Z, Yang J, Zhang H (2016) Grain filling characteristics and their relations with endogenous hormones in large-and small-grain mutants of rice. PLoS ONE 11: e0165321

Cytokinins

Abstract Cytokinin has been the major factor for organ differentiation in plants. Although cytokinins influence many developmental aspects, the control of cytokinin on cell division is considered most important in plant physiology and is diagnostic for the class of hormones. There are 67 genes involved with cytokinin biosynthesis, degradation, and signaling in the panicle of rice during the early stages of development and they control panicle architecture. The cytokinin pool of rice reproductive meristematic tissues can be maintained by the activity of enzymes involved in its biosyntheses, such as isopentenyl transferase, LONELY GUY, and cytochrome P450 mono-oxygenase. In meristematic tissues, the pool of cytokinin is the major factor that controls cell division and cell proliferation and the cytokinin oxidase enzyme is counterproductive for these activities. Through mutant characterization, it has been shown that shoot apical meristem could be controlled by the fine-tuning of concentrations and the spatial distribution of bioactive cytokinins controlling the expression of *APO*, *DEP*, *WFP*, and *CKX* rice genes. Meristematic tissue apart, cytokinins control growth and development of rice caryopsis, the intrinsic concentration remains high in the early post-anthesis period to account for the rise in cell number and declines with the progression of maturity.

Keywords Zeatin · Cell division · Isopentenyl transferase · Cytokinin oxidase · Meristem · CLV pathway

Properties and Structure of Cytokinins

Cytokinins are a group of organic compounds, usually derivative of adenine, behaving like plant growth substances, which promote cell division and proliferation in shoots and roots and delay the senescence of leaves. It has been the major factor for organ differentiation in plants. The compounds also affect apical dominance, axillary bud growth, floral development, bud dormancy, chloroplast differentiation, and seed germination. Although cytokinins influence many developmental aspects, the control of cytokinin on cell division is considered most important in plant physiology and is diagnostic for the class of hormones. The presence of the compounds in coconut milk and herring sperm DNA stimulating cell division was perceived in the 1940s and 1950s in the USA. Miller et al (1955) identified an adenine derivative 6-furfurylamino purine named kinetin from autoclaved DNA.

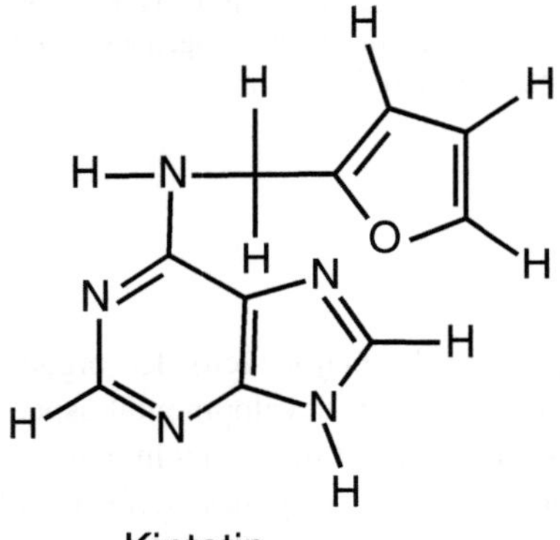

Kintetin

In plants, natural cytokinins are N^6 substituted adenine derivatives where the side chain is attached to the 6 nitrogen of the amino purine. Long after isolation of kinetin, Letham (1973) discovered another adenine or aminopurine derivative named zeatin in the extracts of immature corn endosperm chemically characterized as *trans*-6-(4-hydroxy-3- methylbut-2-enylamino) purine. The structure of zeatin is similar to kinetin. Altogether there are four major cytokinins in the plant kingdom and they are N^6-prenylated adenine derivatives, such as trans-zeatin (tZ), cis-zeatin, dihydrozeatin, and N^6-(Δ^2-isopentenyl) adenine (iP) (Kuroha et al. 2009). The presence of isoprenoid side chains includes all of them into isoprenoid type cytokinin.

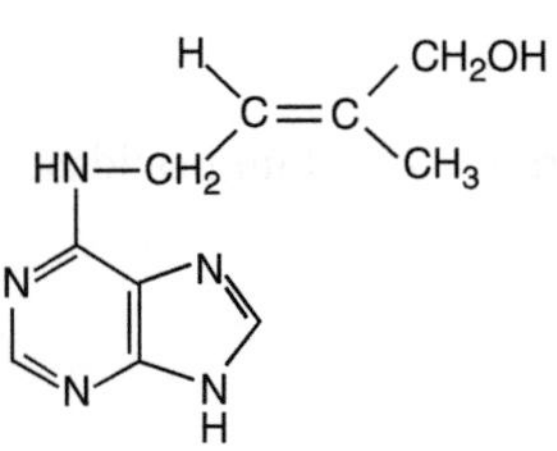

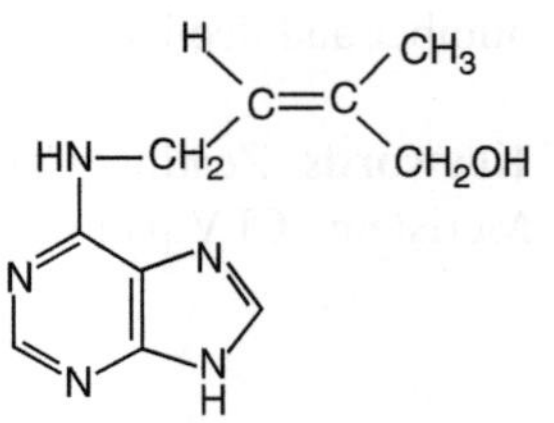

Trans-Zeatin **6-(4-Hydroxy-3-methylbut-2-enylamino) purine** **Cis-Zeatin**

Zeatin stimulates the division of mature cells in the culture medium when added together with auxin. In higher plants, zeatin is the most abundant cytokinin, because of its presence in several plants. Similar to kinetin, zeatin has a side chain attached to the 6 nitrogen of amino purine. The side chain has a double bond that exists in both trans- and cis-forms. The cis form is isomerized to trans by enzyme zeatin cis-/ trans-isomerase. Usually the former is more active in biological assays than the latter. It was reported that in callus cultures trans- zeatin is several folds more active than cis-zeatin (Leonard et al. 1969; Mok et al. 1978). But, according to a recent review cis-zeatin is shown to be as ubiquitous as that of trans-zeatin in the plant kingdom Gajdosova et al. (2011). In different plant parts of rice cis-zeatin abundance ranges from 60.8% to 87.0% of all cytokinin species (Kojima et al. 2009). In addition to cis- and trans-zeatin, two other similar compounds, namely dihydrozeatin and isopentenyl adenine are also found in higher plants. Degradation and conjugation make cytokinin inactive. Degradation occurs by cytokinin oxidase (CKX) activity, whereas conjugation of glucose moiety to 7N or 9 N atom of adenine ring or entry of glucose or xylose moiety to the prenyl side chain of zeatin inactivates cytokinin.

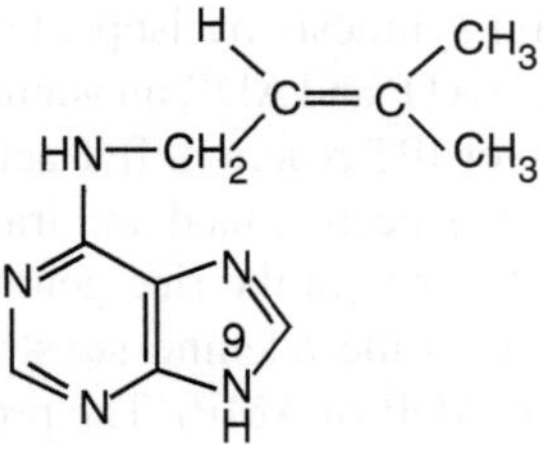

N6 -(Δ^2-isopentenyl)-adenine (IP)

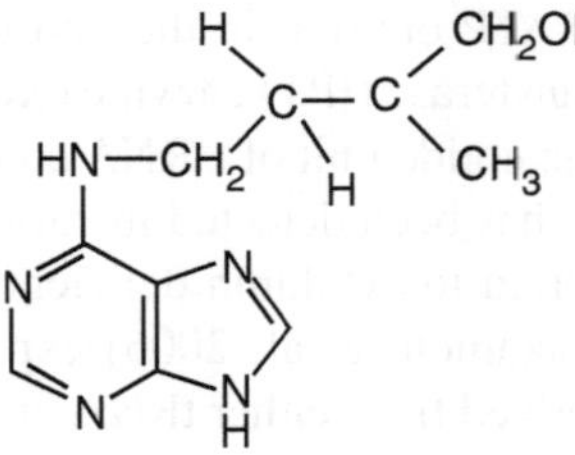

Dihydrozeatin

In contrast, the second type of cytokinin possesses an aromatic chain attached to the N^6 position of adenine and they are 6-benzyladenine, ortho-topolin, and meta-topolin. Isoprenoid cytokinins are mostly common in the plant kingdom whereas detection of the other type is very rare.

Cytokinin Biosynthesis in Rice

Overall, cytokinin levels of rice inflorescence are controlled through its metabolism, biosynthesis, and degradation. Most recently, NanoString nCounter system Yamburenko et al. (2017) studied the expression of 67 genes involved with cytokinin biosynthesis, degradation, and signaling in the panicle of rice during the early stages of development. The expression of several genes involved differs between panicle and vegetative tissue indicating temporal variation in cytokinin action during inflorescence development. The variation is greater in the expression of genes encoding proteins controlling cytokinin metabolism and negative regulators of the pathway than those regulating primary response pathways. Because of its profound effects on inflorescence development and architecture, the expression of these genes could be potential targets for improvement of panicle size and grain yield rice. The cytokinin pool of rice reproductive meristematic tissues can be maintained by the activity of enzymes involved in its biosyntheses, such as isopentenyl transferase, LONELY GUY, and cytochrome P450 mono-oxygenase.

Isopentenyl Transferase (IPT) Gene

Main amino-purine nitrogen base apart, the species of naturally occurring isoprenoid cytokinins have a side chain chemically related to rubber and carotenoid pigments constructed from isoprene units. Compared to many isoprene units of rubber or carotenes, cytokinins contain just one unit. The precursor for the formation of isoprene units is either mevalonic acid or pyruvate plus 3-phosphoglycerate, which is converted to the isoprene unit of dimethylallyl diphosphate (DMAPP). The isopentenyl group of dimethylallyl diphosphate is transferred to the adenosine moiety of ATP or ADP in the first committed step of cytokinin synthesis by isopentenyl transferase (IPT) enzyme (Kakimoto 2003). In addition to ATP and ADP, an adenine nucleotide unit of t-RNA could be an adenine substrate for IPT reaction. IPT activity has been detected in various plant extracts; the gene has been cloned and transported to cytokinin deficient plants. There are eight IPT genes in the rice genome (Sakamoto et al. 2006) expression of which is specific to the adenine substrate derived from either tRNA or adenosine phosphates (ATP, ADP, or AMP). The products of these reactions are iPTP or iPDP are converted to zeatin by a cytochrome P450 hydrolase (CYP735A) enzyme that removes the phosphate group attached to the ribose sugar (Fig. 11.16).

KNOX1 Transcription Factor for IPT *Activity*

A transcription factor KNOTTED1-like (KNOX1) suppresses a gibberellin biosynthesis gene encoding GA20 oxidase, which in turn promotes cytokinin biosynthesis gene expression for isopentyl pyrophosphate transferase. There is evidence that both gibberellins and cytokinins are potential targets for KNOX proteins in rice; synthesis of the former is down-regulated in contrast to up-regulation of the latter by the protein. KNOX protein elicits higher expression of *OsIPT2* and *OsIPT3* genes while downregulating expression GA20 oxidase gene resulting in a favorable balance for cytokinin against gibberellins (Sakamoto et al. 2006b). Recently, Wu et al. (2017) found that incidence of high temperature reduced cytokinin abundance and decreased spikelet number of rice panicle because of its adverse effects on activities of the three biosynthetic enzymes and up- regulation of CKX activity (Fig. 11.17).

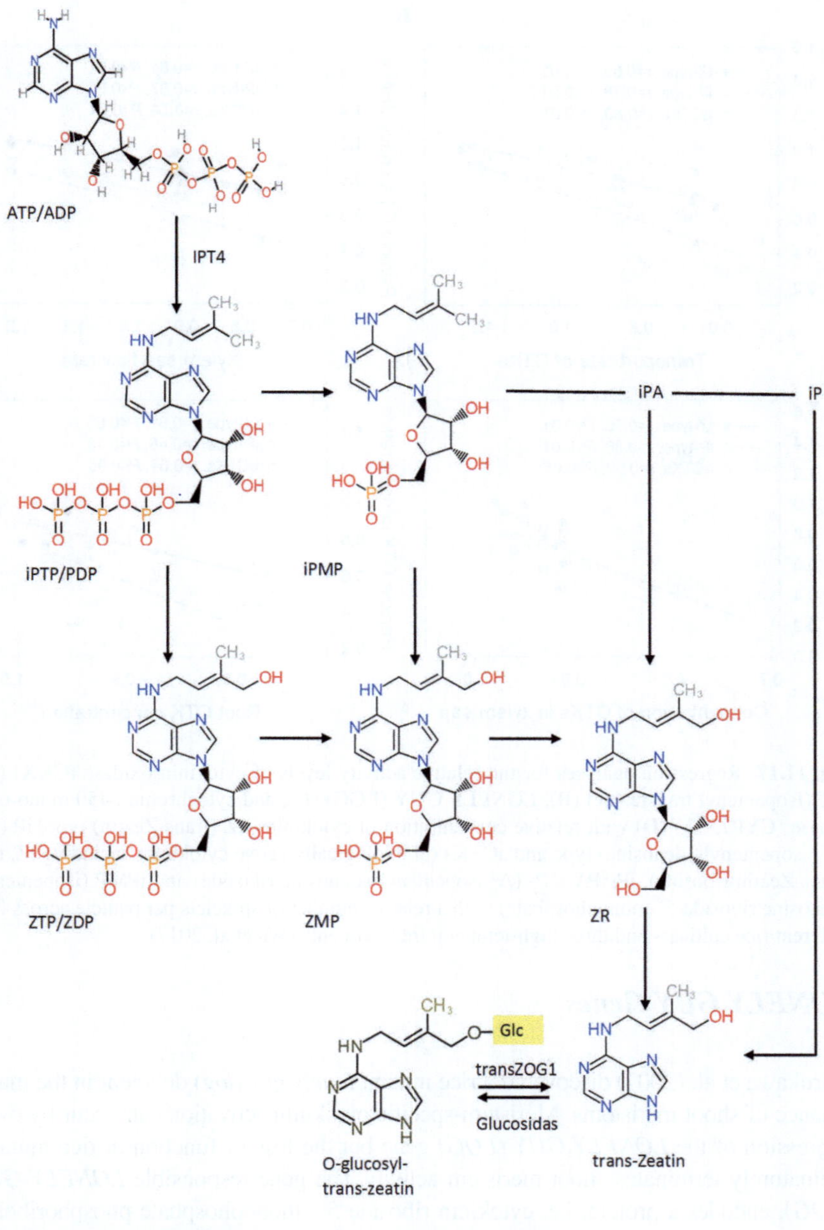

Fig. 11.16 The action of *IPT* gene for production of zeatin

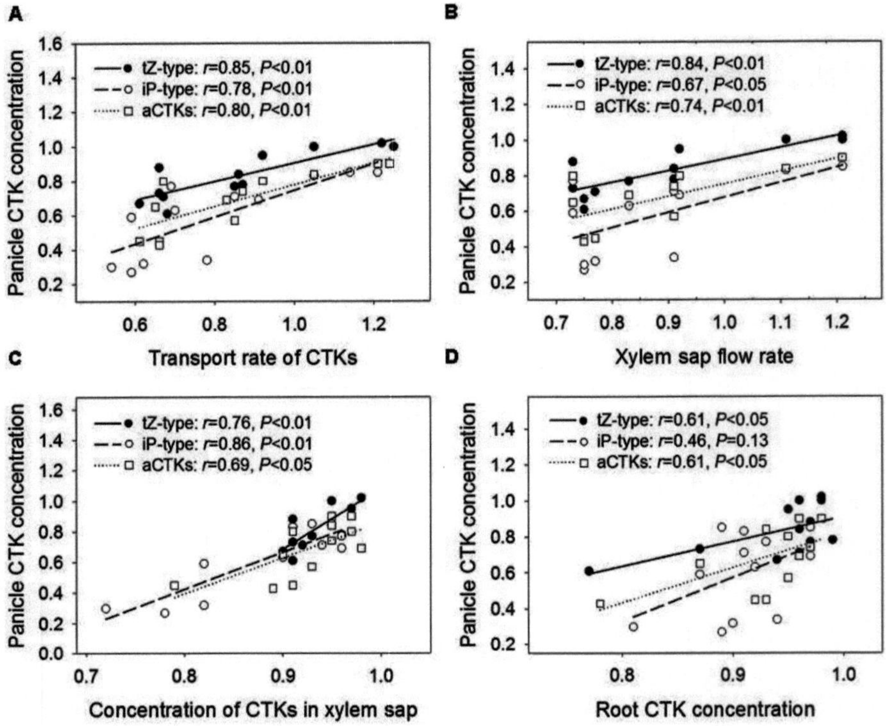

Fig. 11.17 Regression analyses for the relative activity levels of cytokinin oxidase (CKX) (**A**), IPT (Isopentenyl transferase) (**B**), LONELY GUY (LOG) (**C**), and cytochrome P450 mono-oxygenase (CYP735A) (**D**) with relative concentration of cytokinins, tZ (trans-Zeatin) type, iP (N^6-(Δ^2- isopentenyl-adenosine) type and aCTKs (all biologically active cytokinins including tZ, tZR (trans Zeatin riboside), iP, iPA (N^6- (Δ^2-isopentenyl-adenosine riboside) and iPMP (isopentenyl-adenosine riboside-5'-monophosphate) with a relative number of spikelets per panicle across four different rice cultivars and three high-temperature treatments (Wu et al. 2017)

LONELY GUY Genes

Kurukawa et al. (2007) discovered a rice mutant *lonely guy* (*log*) deficient in the maintenance of shoot meristems. Meristem-specific cytokinin activation can occur by overexpression of the *LONELY GUY (LOG)* gene but the loss of function in rice mutants prematurely terminates shoot meristem activity The gene responsible *LONELY GUY (LOG)*, encodes a protein, i.e. cytokinin riboside 5'- monophosphate phosphoribohydrolase enzyme. The enzyme works in the final step of cytokinin synthesis and directly converts inactive cytokinin nucleosides into the biologically active form. Thereby, the gene maintains a buffer stock of cytokinin for the continuous function of the inflorescence meristem in the wild type. In plants, two pathways are producing active cytokinin from nucleotides. The first one is a two-step pathway where cytokinin riboside 5- monophosphates are converted to the corresponding nucleosides and nucleobases by nucleotidase and nucleobases respectively. By contrast, the second one involves a direct pathway for the production of cytokinin nucleobase from the corresponding cytokinin riboside 5-monophosphate. Using phenotypic and metabolic analyses of Arabidopsis mutants increasing or decreasing LOG activities it has been confirmed that the direct pathway via LOG is the major pathway for cytokinin activation and it is responsible for normal growth and development (Kuroha et al. 2009).

The enzyme activates the inactive Zeatin by converting it to the free base form. In the reaction cytokinin nucleobase is released from the cytokinin riboside-5'-monophosphate complex with riboside 5'-monophosphate as a byproduct. Kuroha et al. (2009) identified nine *LOG* genes (*AtLOG1-9*) in Arabidopsis that are homologous to rice. Seven of the AtLOG were localized in cytoplasm and nuclei and perform enzyme activities similar to rice LOG. LOG only hydrolyzes cytokinin riboside-5'-monophosphate, but not the AMP. Zeatin once synthesized can be interconverted from trans to cis forms and vice versa by zeatin cis-trans isomerase enzyme (Bassil et al. 1993). The enzyme requires light FAD or FMN for the reaction. The free base form of trans-zeatin is biologically active unlike its ribotide and riboside derivatives. Cytokinin activity is evident in some other compounds that are readily converted to zeatin, dihydrozeatin, and isopentenyl adenine.

Cytokinin Oxidase (CKX) Gene

In the plant cells, free cytokinin is metabolized to respective nucleoside and nucleotide forms. An enzyme cytokinin oxidase (CKX) cleaves the side chain of zeatin, zeatin riboside, and isopentenyl adenine at the N^6 position of the adenine molecule and makes them hormonally inactive. In meristematic tissues, the pool of cytokinin is the major factor that controls cell division and cell proliferation and CKX activity is counterproductive for these activities. According to Ashikari et al. (2005) rice genome has a QTL, in which *Gn1a* gene encodes cytokinin oxidase/dehydrogenase (OsCKX2); reduced expression of this gene in inflorescence meristem of rice causes cytokinin accumulation for production more number of reproductive organs and better grain yield (Ashikari et al. 2005). The QTL comprising *Gn1* is a recessive allele. It comprises two loci *Gn1a* and *Gn1b*; the former is responsible for encoding cytokinin oxidase/dehydrogenase OsCKX2, reduced expression of which causes cytokinin accumulation in inflorescence for an increase in the number of reproductive organs and higher grain yield (Ashikari et al. 2005). The expression profile of the *OsCKX2* gene was high in leaves, flowers, culm, and inflorescence meristem, but not in roots and embryos. Varietal comparison of rice cultivars contrasting for panicle grain number indicated overexpression of *OsCKX2* in the inflorescence of Koshihikari (low grain number), lower expression in Habataki (higher grain number), and no expression in cultivar 5150 (more than 400 grains) (Murai et al. 2014).

Thus, activities of IPT and CKX are important functions for the determination of the active pool of cytokinins in plant cells. Dihydrozeatin is formed from trans-zeatin by the enzyme zeatin reductase (Mok and Martin 1994) and it is resistant to degradation by cytokinin oxidase (Mok et al. 2000), but its role in panicle development in rice has not been explored.

Cytokinin Signal Transduction

Cytokinins are mobile plant hormones. Cytokinin is synthesized in roots and they travel long distances in the xylem for transmitting signals at the target in shoots. Zhao et al. (2019) recently reported that OsABCG18, a half-size ATP-binding cassette transporter protein of rice is essential for the long-distance transport of root-derived cytokinins. *OsABCG18* encodes a plasma membrane protein and manifests expression in the vascular, stem, and leaf midribs. Its overexpression significantly improves cytokinin levels in shoots and grain yield. Cytokinin signal transduction has been explained for shoot apical meristem development in rice. Similar to other plants rice makes use of the two-component system, originally discovered in bacteria, for signal perception and transduction. Using several complementary approaches Tsai et al. (2012) characterized the two components system in rice. The signaling pathway is similar to the system in Arabidopsis. The initial pathway for cytokinin signal transduction is a multi-step phosphorelay that includes a sensory histidine kinase receptor, a histidine phospho-transfer protein, and a response regulator. The membrane-bound histidine kinase receptors like CRE1/ AHK4, AHK 2, and AHK 3 perceive the signal when cytokinin binds to the extracellular part of the receptors named CHASE domain. They relay the signal from the cytoplasmic membrane to the nucleus. The histidine residue comprising the input domain auto-phosphorylates in response to the environmental signal and thereafter the phosphate group is transferred to an Asp residue on the receiver domain of the response regulator protein. Phosphorylation of the receiver modulates the activity of the output domain of the response regulator, which itself may be a transcription factor with the ability to mediate downstream signaling of the hormone. Instead, the modulated output domain may follow a phosphorelay type signaling with an extra set of phospho-transfer proteins called Hpt. In this multistep phosphorelay system, there are three components; a hybrid sensor histidine kinase, in which both histidine kinase and receiver protein are present in a single protein, a histidine-containing phosphotransfer protein (Hpt), and a separate response regulator. The phosphate group is transferred in sequence from histidine to aspartate to histidine to aspartate in a multiple relay system (Fig. 11.18).

In rice, there are five histidine protein kinases (*OsHK1-4*, *OsHKL1*), fifteen type A (*OsRR1-15*), seven type B (*OsRR16-22*), and five pseudos (*OsPRR1-5*) response regulators, and two phospho-transfer (*OsAHP1-2*) genes. The type B response regulators work as a transcription factor that regulates gene expression. Type A response regulator expression is induced by type B response regulators and they function as negative feedback regulators for cytokinin response. OsAHP phosphorylates types A and B response regulator proteins, which in turn initiates transcription of cytokinin response factor (*CRF*) genes. *CRF* transcription promotes cytokinin signaling with help of the His-Asp phosphorelay system (Azizi et al. 2015).

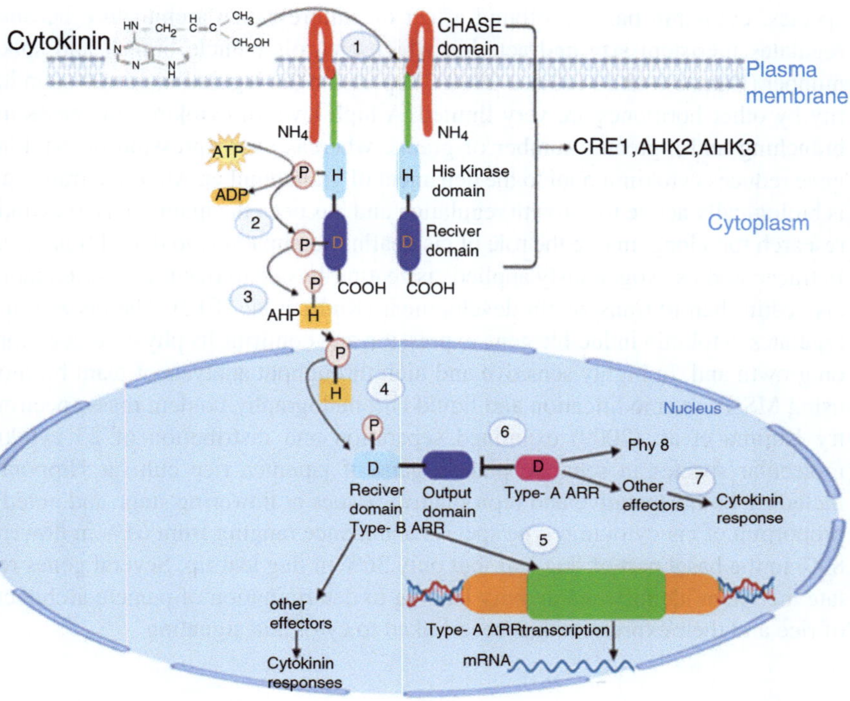

Fig. 11.18 Model for cytokinin action on rice grain filling. 1. Cytokinin binds CRE1, which is likely to occur as a dimer. Cytokinin binds to chase domain 2. Cytokinin binding activates histidine kinase activity of receptor which transfers phosphate (P) to aspartate residue on the fused domain 3.P is then transferred to a histidine present on AHP protein 4.phosphorylated AHP moves to the nucleus and transfer P to an aspartate residue on the receiver domain of type - B ARR 5. Phosphorylated type-B ARR induces transcription of genes encoding type-A ARR 6. Type-A ARR is likely also to be phosphorylated by the AHP proteins 7. The phosphorylated type-A ARR interacts with various effectors to mediate the changes in cell function appropriate to cytokinin.

Cytokinin Response in Rice

As discussed previously, the panicle architecture of rice is determined by spatial and temporal control of phase change from the rachis and branch meristems and subsequent transformation of the latter to floral meristem (spikelet meristem). Delay in spikelet meristem specification increases spikelet number of the panicle. During the transition, multiple flowering signals are transmitted to a group of stem cells in the shoot apex, where the cells are converted and developed into reproductive structures. During the transition, the expression of several genes controlling meristem identity as well activity is critical for rice inflorescence architecture. The role of cytokinin is related to the activity of the meristematic cells because of its primary role in cell division, differentiation, and proliferation, and subsequently the maintenance of primordium development in rice panicle (Ashikari et al. 2005). In grass species, cytokinin has a profound effect on inflorescence architecture because it regulates meristem size and activity, which controls panicle branch and spikelet number (Barazesh and McSteen 2008). Reports on the regulation of meristem identity by other hormones are very limited. A high level of cytokinin increases more branching and a greater number of grains, whereas overexpression of cytokinase gene reduces cytokinin pool to the detriment of grain number. Although trans-zeatin is biologically active for growth regulation and has been the main focus of cytokinin research for a long, in rice the role of cis-zeatin is comparable to that of trans-zeatin. In tracer studies exogenously applied cis zeatin riboside to rice is converted more to cis-zeatin than to trans-zeatin development (Kudo et al. 2012). The cis-zeatin up-regulates cytokinin inducible gene expression and confirms its physiological impact on growth and. In highly sensitive and high throughput analysis of plant hormones using MS-Probe modification and liquid chromatography-tandem mass spectrometry Kojima et al. (2009) examined separation and distribution of 23 cytokinin molecular species in separate plant organs of japonica rice cultivar Nipponbare including both vegetative and reproductive tissues at flowering stage and noted the proportion of cis-cytokinin type species abundance ranging from 61% in flowers to 89% in the basal part of flag leaf, but only 36% in flag leaf tip. Several genes regulate meristem identity and activity leading to determination of panicle architecture of rice and their expression has been linked to cytokinin signaling.

CLV Pathway Genes Controlling Cytokinin Abundance of Rice

Unlike other hormones, cytokinin biosynthesis mostly occurs in the root apical meristem. It moves acropetally from root to shoot in xylem vessels along with water and mineral elements. In shoot and root tips cytokinin has been the major cell division factor. It has a positive growth regulatory response for the shoot, but negative for the root development. During cell division, auxin and cytokinin control the activity of cyclin- dependent protein kinase (CDK), the enzyme that regulates eukaryotic cell cycle operation. In rice, various parts of the inflorescence are generated from the stem cell population of the shoot meristematic tissue which is defined by a regulatory loop between the *CLAVATA(CLV)* and *WUSCHEL (WUS)* genes (Schoof et al. 2000). WUS is a stem cell promoting transcription factor, whereas CLV3 is a differentiation-promoting peptide. The auto- regulatory feedback loop, based on a retrogressive relationship of the expression of the two genes in the CLV3-WUS pathway has evolved as the central regulatory pathway. The pathway coordinates cell proliferation with differentiation. The CLV pathway has been well- characterized in the regulation of meristem development in Arabidopsis and a similar system is functionally conserved not only in rice but also in other plants like maize and tomato; the pathway has been a target of selection for crop domestication and yield enhancement (Somssich et al. 2016).

With the help of mutants, several CLV pathway genes have been identified in rice suggesting manipulation of cytokinin abundance for controlling meristematic cell activities of the inflorescence. *FON4* (*floral organ number 4*) is one of such genes that have been considered orthologous to *CLV3* of Arabidopsis (Chu and Zhang 2007). To regulate meristem size three *CLV* genes act together in the pathway. The genetic mutants of *CLV*, such as *clv1*, *clv2,* and *clv3* cause inflorescence fasciations and enlargement of floral meristems. The dominant *CLV1* encodes a receptor-like protein that consists of three domains, namely a. an extracellular leucine-rich repeat (LRR) domain, b. a small transmembrane domain, and c. a cytoplasmic serine/threonine kinase domain. *CLV2* produces transcripts for the first two parts, but not for the third. The role of *CLV3* is to encode a small peptide ligand (CLV3) to interact with the disulfide-linked CLV1/CLV2 receptor complex, and to restrict the proliferation of the stem cell population in an appropriate size in shoot apical meristem. In contrast, WUS protein is a homeobox transcription factor; it is a positive regulator in maintaining shoot meristem cell activity. It promotes the expression of *CLAVATA3 (CLV3)*, which in turn results in feedback inhibition of the *WUS* domain for transcription. In other words, CLV3 expression is up-regulated by WUS, whereas CLV3 down-regulates WUS expression. Thus, *wus* mutant bears phenotype opposite to *clv* mutant. The negative feedback loop of WUS-CLV is involved in the reproductive maintenance of rice with the identification of *CLV-like* genes *FON1* and *FON2* (Ohmori et al. 2013). Conversely *WUS* domain is over- expressed by AHK4/WOODENLEG receptors for cytokinin sensitivity. Besides, WUS represses the expression of type A *RESPONSE REGULATOR* genes of the cytokinin signal transduction pathway and represses meristematic cell multiplication. The signaling cascade forming a negative feedback loop between *CLV3* and *WUS* genes is described in Figs. 11.19 and 11.20.

Genetic Mutations Regulating Cytokinin Biosynthesis

APO 1 Allele

There are several other instances involving genetic mutants of rice where the activity of shoot apical meristem could be controlled by the fine-tuning of concentrations and the spatial distribution of bioactive cytokinins. The expression of *ABERRANT PANICLE ORGANIZTION1(APO1)*, *APO2,* and *PAP2* genes and their role in inflorescence architecture determination have been discussed before. Compared to the wild type, fewer spikelets are produced in the *APO1* loss of function mutants; whereas elevated transcription of *APO1* increases inflorescence meristem activity and spikelet number in rice panicle (Terao et al. 2010, Ikeda-Kawakatsu et al. 2009). Although not explicit, *APO1* and *APO2* genes delay spikelet meristem identity and promote branch meristem identity and ultimately increase spikelet number through an elevated pool of cytokinin in the tissue. In contrast, the *PANICLE PHYTOMER 2* (*PAP2*) gene is a positive regulator of spikelet meristem identity.

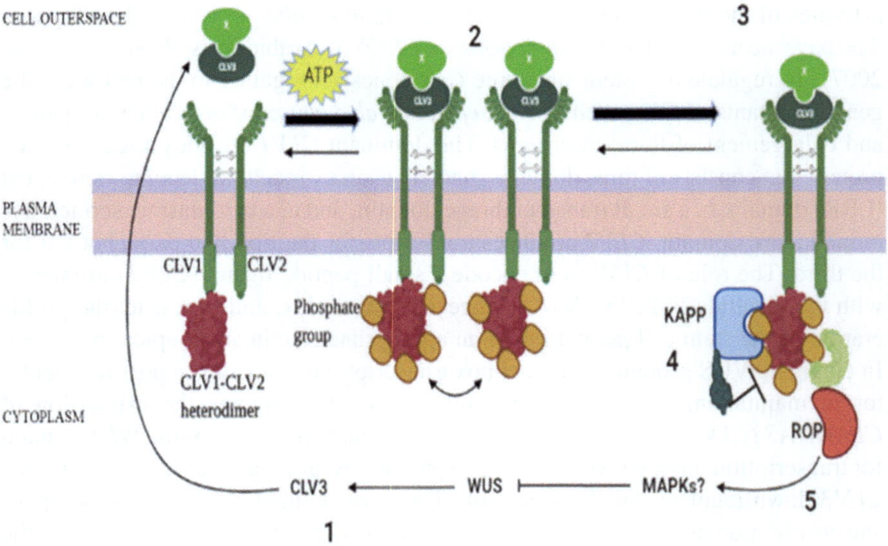

Fig. 11.19 Model of the CLAVAT1/CLAVATA2 (CLV1/CLV2) receptor kinase signaling cascade forming a negative feedback loop with the *WUS* gene. (Adopted from Plant Physiology- Taiz and Zeiger 2002). 1. *WUS* gene expression promotes the expression of the *CLV3* gene. 2. Multimer CLV3, binds in the extracellular domain of CLV1/CLC2 heterodimer and induce auto-phosphorylation of the cytoplasmic domain of CLV1. 3. CLV1 in its phosphorylated state binds to downstream effector molecules KAPP (kinase-associated protein phosphate) and ROP (rho-GTPase). 4. KAPP negatively regulates CLV1. 5. ROP act through MAPK (mitogen-activated protein kinase) cascade to repress *WUS* gene expression, forming a negative feedback loop.

DEP 1 Allele

The poor expression of *OsCKX2* leading to high grain number formation in rice panicle is comparable to that of *dep1* mutant. *DEP1* (*DENSE AND ERECT PANICLE 1*) encodes a phosphatidylethanolamine-binding protein-like domain protein (Huang et al. 2009). A gain of function mutation at this allele causing truncation of the protein enhances meristematic activity resulting in reduced length of internodes of inflorescence and increase of grain number per panicle with potential for 40% increase of yield. It was suggested that the dominant *DEP1* allele acts upstream of *OsCKX2* to reduce its expression level in the panicle meristem and thereby decreased spikelet number. Thus, the *DEP1* gene in rice is a dominant- negative regulator for the expression of the *OsCKX2* gene preserving the cytokinin level of the tissue for greater cell proliferation and larger spikelet number (Huang et al. 2009, Sreenivasulu and Schnurbusch 2012).

WFP Allele

Similarly, a disruption in the miRNA156-directed regulation in the binding site of the gene in a semi-dominant allele of *OsSPL14* gene in the QTL *WFP (WEALTHY FARMER'S PANICLE)* leads to the production of a greater number of panicle branches and higher grain number (Jiao et al. 2010, Miura et al. 2010). Li et al.

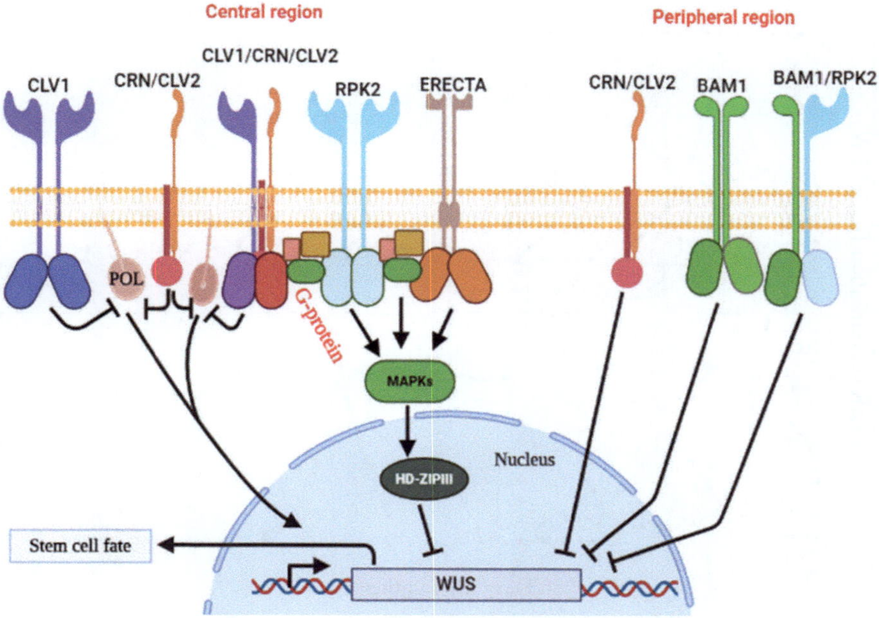

Fig. 11.20 Molecular components of the CLV-WUS pathway in *Arabidopsis*. Different plasma leucine-rich repeat membrane-localized receptor-like kinase (LRR-RLKs) and receptor-like proteins (RLPs) for *CLV* interaction in central and peripheral zones of meristem are shown with their corresponding interacting proteins and down-stream effectors. The arrows indicate the mode of signaling via the receptors that eventually leads to repression of *WUS* and promotion of stem cell fate. CRN= RLK CORYNE, ERECTA = LRR-RKL type, BAM = BARELY ANY MERISTEM, LRR-RKL type, RPK = RECEPTOR LIKE PROTEIN KINASE, POL= POLTERGHEIST protein phosphates. (Adopted from Somssich et al. 2016)

(2011) characterized two allelic *large panicle* (*lp*) mutants possessing significantly increased panicle size. They produced more primary branches and more grains. With map-based cloning, it was found that *LARGER PANICLE (LP)* encodes an endoplasmic reticulum localized Kelch repeat F-box protein. The protein interacts with SKP-1 like protein in an F-box domain-dependent manner. *OsCKX2* expression is down-regulated in the *lp* -mutants suggesting an adverse influence of *LP* on cytokinin pool of inflorescence meristem. It is known that grain number is fixed in rice panicle at the early reproductive stage that coincides with the cessation of meristematic activity. Loss of function of *OsCKX2* expression can be attributed to gain in cytokinin pool in the stem cells of inflorescence meristem and consequent enhancement of reproductive structure formation.

Cytokinin Effects on Grain Filling of Rice

Because of its effects on meristem size and activity, cytokinins have a profound influence on panicle development and architecture, and grain filling of rice. Nitrogen fertilizers applied to rice plant increases spikelet/grain number per panicle primarily

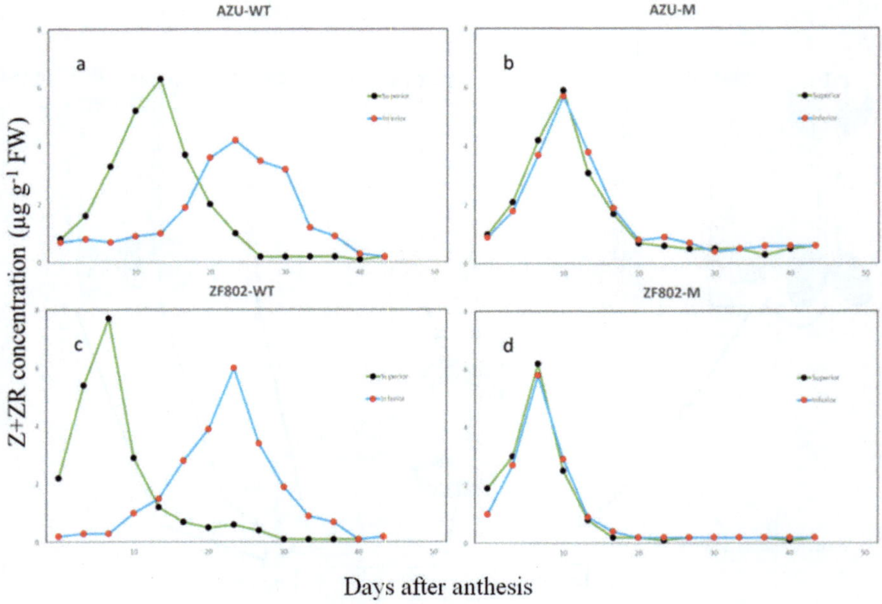

Fig. 11.21 Changes in the concentration of Z+ZR in superior and inferior spikelets of wild-type rice (AZU-WT and ZF802-WT) and their corresponding genetic mutants (AZU-M and ZF802-M) during grain development period. (Adopted from Zhang et al. 2016)

because it improves cytokinin synthesis (Ding et al. 2014). In addition to grain number and panicle size, they help to regulate the individual weight of grains. Rice varieties with small panicle sizes have correspondingly low cytokinin abundance. Similarly, panicle size decreases, and fewer spikelets are produced on account of poor cytokinin abundance in heat-stressed rice plants (Wu et al. 2017). Cytokinin concentration of the spikelet correlates significantly with kernel development; in particular, the hormone level in endosperm regulates cell number and cell division activity in the early phase of grain development (Yang et al. 2002, Zhang et al. 2009).

In super rice, the concentration of the hormone improves and grain filling of the inferior spikelets increases by alternate wetting and drying irrigation of the field through elevation of cytokinin levels (Zhang et al. 2010). The dynamics of cytokinin abundance during grain filling have been elaborated by Zhang et al. (2016). In the early stage of grain filling, zeatin and zeatin riboside concentration were high in the superior spikelets of rice panicle producing well-filled high-density grains compared to the inferior spikelets, where grain filling is very poor. In the superior kernels, cytokinin concentration rises to a peak level within the first ten days post-anthesis and declines sharply thereafter. In contrast, the concentration rises slowly to peak level at three weeks post-anthesis in the inferior type spikelets and declines thereafter to maturity (Fig. 11.21) The temporal changes in concentration of cytokinins matched closely with the rates of cell division and grain filling, which were faster and occurred earlier in the superior to the inferior spikelet. It is implied that the superior spikelets could generate more cells with help of the endogenous cytokinin pool at the early stage of grain development and build sink capacity greater than the inferior spikelet. A rise in cytokinin pool at later stages of grain development was not conducive for an increase of grain sink for assimilates storage of inferior spikelets.

References

Ashikari M, Sakakibara H, Lin S, Yamamoto T, Takashi T, Nishimura A, Angeles ER, Qian Q, Kitano H, Matsuoka M (2005) Cytokinin Oxidase Regulates Rice Grain Production. Science 309: 741–745

Azizi P, Rafii M, Maziah M, Abdullah SNA, Hanafi MM, Latif MA, Rashid AA, Sahebi M (2015) Understanding the shoot apical meristem regulation: A study of the phytohormones, auxin, and cytokinin, in rice. Mechanisms Dev 135: 1–15

Barazesh S, Mcsteen P (2008) Hormonal control of grass inflorescence development. Trends Plant Sci 13: 656–62

Bassil NV, Mok DW, Mok MC (1993) Partial purification of a cis-trans-isomerase of zeatin from the immature seed of Phaseolus vulgaris L. Plant Physiol 102: 867–72

Chu H, Zhang D (2007) The shoot apical meristem size is regulated by FON4 in rice. Plant Signaling Behavior 2: 115–116

Gajdošová S, Spíchal L, Kamínek M, Hoyerova K, Novak O, Dobrev PI, Galuszka P, Klima P, Gaudinova A, Zizkova E, Hanus J, Dancak M, Travnicek B, Pesek B, Krupicka M, Vankova R, Strnad M, Motyoka V (2011) Distribution, biological activities, metabolism, and the conceivable function of cis-zeatin-type cytokinins in plants. J Exp Botany 62: 2827–2840

Ding C, You J, Chen L, Wang S, Ding Y (2014) Nitrogen fertilizer increases spikelet number per panicle by enhancing cytokinin synthesis in rice. Plant Cell Rep 33: 363–371

Huang X, Qian Q, Liu Z, Sun Z, He S, Luo G, Xia, Chu C, Fu X (2009) Natural variation at the DEP1 locus enhances grain yield in rice. Nat Genet 41: 494–497

Ikeda-Kawakatsu K, Yasuno N , OikawaT,Iida S, Nagato Y, Maekawa M, Kyozuka J (2009) Expression level of ABERRANT PANICLE ORGANIZATION1 determines rice inflorescence form through control of cell proliferation in the meristem. Plant Physiol 150: 736–747

Jiao Y, Wang Y, Xue D,Wang J, Yan M, Liu G, Dong G, Zeng D, Lu Z, Zhu X, Qian Q, Li J (2010) Regulation of OsSPL14 by OsmiR156 defines ideal plant architecture in rice. Nat Genet 42: 541–544

Kojima M, Kamada-Nobusada T, Komatsu H, Takei K, Kuroha T, Mizutani M, Ashikari M, Ueguchi-Tanaka M, Matsuoka M, Suzuki K, Sakakibara H (2009) Highly sensitive and high-throughput analysis of plant hormones using MS-probe modification and liquid chromatography-tandem mass spectrometry: an application for hormone profiling in Oryza sativa. Plant Cell Physiol 50: 1201–1214

Kudo T, Makita N, Kojima M, Tokunaga H, Sakakibara H (2012) Cytokinin activity of cis- zeatin and phenotypic alterations induced by overexpression of putative cis-zeatin-O- glucosyltransferase in rice. Plant Physiol 160: 319–331

Kuroha T, Tokunaga H, Kojima M, Ueda N, Ishida H, Nagawa S, Fukuda H, Sugimoto K, Sakakibara H (2009) Functional analyses of LONELY GUY cytokinin-activating enzymes reveal the importance of the direct activation pathway in Arabidopsis. Plant Cell 21: 3152–3169

Kurakawa T, Ueda N, Maekawa M, Kobayashi K , Kojima M, Nagato Y, Sakakibara H, Kyozuka J (2007) Direct control of shoot meristem activity by a cytokinin-activating enzyme. Nature 445; 652 https://doi.org/10.1038/nature05504

Leonard NJ, Hecht SM, Skoog F, Schmitz RY (1969) Cytokinins: synthesis, mass spectra, and biological activity of compounds related to zeatin. Proc Nat Acad Sci 63: 175–182

Letham DS (1973) Cytokinins from Zea mays. Phytochemistry 12: 2445–2455
Kakimoto T 2003. Biosynthesis of cytokinins. J Plant Res 116: 233–239
Li M, Tang D, Wang K, Wu X, Lu L, Yu H, Gu M, Yan C (2011) Mutations in the F-box gene LARGER PANICLE improve the panicle architecture and enhance the grain yield in rice. Plant Biotechnol J 9: 1002–1013
Mok DW, Martin RC (1994) Cytokinin metabolic enzymes. Cytokinins. Chemistry, Activity, and Function. CRC Press, Boca Raton, FL, pp 129–137
Murai N (2014) Review: Plant growth hormone cytokinins control the crop seed yield. Am J Plant Sci 5: 2178–2187
Miura K, Ikeda M, Matsubara A, Song X-J, Ito M, Asano K, Matsuoka M, Ktano H, Ashikari M (2010) OsSPL14 promotes panicle branching and higher grain productivity in rice. Nat Genet 42: 545–549. https://doi.org/10.1038/ng.592
Miller CO, Skoog F, Von Saltza MH, Strong F (1955) Kinetin, a cell division factor from deoxyribonucleic acid1. J Am Chem Soc 77: 1392–1392
Mok MC, Mok DW, Armstrong DJ (1978) Differential cytokinin structure-activity relationships in Phaseolus. Plant Physiol 61: 72–75
Mok DWS, Martin RC (1994) Cytokinin metabolic enzymes. In: Cytokinin: Chemistry, Activity and Function, Mok DWS, Mok MC Eds., CRC Press, Boca Raton, pp 129–138
Mok MC, Martin RC, Mok DWS (2000) Cytokinin biosynthesis, metabolism, and perception. In Vitro Cell Dev Biol—Plant 36: 102–107
Ohmuri Y, Tanaka W, Kojima M, Sakakibara H, Hirano HY (2013) WUSCHEL-RELATED HOMEOBOX4 is involved in meristem maintenance and is negatively regulated by the CLE gene FCP1 in rice. Plant Cell 25: 229–241
Sakamoto T, Sakakibara H, Kojima M, Yamamoto Y, Nagasaki H, Inukai Y, Sato Y, Matsuoka M (2006) Ectopic expression of KNOTTED1-like homeobox protein induces expression of cytokinin biosynthesis genes in rice. Plant Physiol 142: 54–62
Schoof H, Lenhard M, Haecker A, Mayer KFX, Jürgens G, Laux T (2000) The stem cell population of Arabidopsis shoot meristems is maintained by a regulatory loop between the CLAVATA and WUSCHEL genes. Cell 100: 635–644
Sreenivasulu N, Schnurbusch T (2012) A genetic playground for enhancing grain number in cereals. Trends Plant Sci 17: 91–101
Somssich M, Je BI, Simon R, Jackson D (2016) CLAVATA-WUSCHEL signaling in the shoot meristem. Development 143: 3238–3248
Taiz L, Zeiger E (2002) Plant Physiology. 3rd edn. Sinauer Associates, Inc., Sunderland, USA.
Tsai Y-C, Weir NR, Hill K, Zhang W, Kim HJ, Shiu SH, Schaller GE, Kieber JJ (2012) Characterization of genes involved in cytokinin signaling and metabolism from rice. Plant Physiol 158: 1666–1684
Terao T, Nagata K, Morino K, Hirose T (2010) A gene controlling the number of primary rachis branches also controls the vascular bundle formation and hence is responsible to increase the harvest index and grain yield in rice. Theoretical Appl Genetics 120: 875–893
Wu C, Cui K, Wang W, Li Q, Fahad S, Hu Q, Huang J, Nie L, Mohapatra PK, Peng S (2017) Heat-induced cytokinin transportation and degradation are associated with reduced panicle cytokinin expression and fewer spikelets per panicle in rice. Front Plant Sci 8: 371. https://doi.org/10.3389/fpls.2017.00371
Yamburenko MV, Kieber JJ, Schaller GE (2017) Dynamic patterns of expression for genes regulating cytokinin metabolism and signaling during rice inflorescence development. PLoS ONE 12: e0176060
Yang J, Zhang J, Huang Z, Wang Z, Zhu Q, Liu L (2002) Correlation of cytokinin levels in the endosperms and roots with cell number and cell division activity during endosperm development in rice. Ann Botany 90: 369–77
Zhang H, Tan G, Yang L, Yang J, Zhang J, Zhao B (2009) Hormones in the grains and roots in relation to post-anthesis development of inferior and superior spikelets in japonica/indica hybrid rice. Plant Physiol Biochem 47: 195–204
Zhang H, Chen T, Wang Z, Yang J, Zhang J (2010) Involvement of cytokinins in the grain filling of rice under alternate wetting and drying irrigation. J Exp Botany 61: 3719–33

Zhang W, Cao Z, Zhou Q, Chen J, Xu G, Gu J, Liu L, Wang Z, Yang J, Zhang H (2016) Grain filling characteristics and their relations with endogenous hormones in large-and small- grain mutants of rice. PLoS ONE 11: e0165321
Zhao J, Yu N, Ju M, Fan B, Zhang Y, Zhu E, Zhang M, Zhang K (2019) ABC transporter OsABCG18 controls the shoot ward transport of cytokinins and grain yield in rice. J Exp Botany 70: 6277–6291

Abscisic Acid (ABA)

Abstract Abscisic acid is a sesquiterpenoid negative plant hormone, which co-ordinates with the positive regulators to determine the time and extent of plant growth and development. In addition to leaf abscission, it controls bud and seed dormancy, stomatal closure, and adaptive features that delay growth in presence of abiotic and biotic stresses. ABA is synthesized in chloroplast and amyloplast. ABA is biologically active in *cis* isomeric form but not in *trans* form. There has been little information available on the role of ABA in the determination of rice panicle architecture, but it has a significant effect on the grain filling of rice spikelets in the post-anthesis period. ABA level peaks up to the mid-grain filling stage before a gradual decline in the late maturation phase. ABA has a positive influence on sieve tissue unloading and uptake rate in sink cells of developing rice caryopsis. Bereft of this advantage dry matter accumulation is slowed down in the poorly-filled inferior spikelets located on secondary branches of the panicle. Moderate soil drying in the field increases ABA concentration in developing caryopsis. It increases the activities of endosperm starch synthesizing enzymes for the improvement of grain filling.

Keywords Abscisic acid · Negative growth regulator · *cis*-ABA · Grain filling · Assimilates

Introduction

Abscisic acid is a sesquiterpenoid plant hormone primarily believed to be involved with the abscission of leaves and fruits. Unlike the positive growth regulators IAA, cytokinin, and gibberellins, it is a negative regulator. This negative regulator co-ordinates with the positive regulators and the balance between them determines the time and extent of plant growth and development. In addition to leaf abscission, many of the growth inhibitory phenomena controlled by ABA include bud and seed dormancy, stomatal closure, and adaptive features that delay growth in presence of abiotic and biotic stresses. Survival at stake, the plant synthesizes ABA for resilience under stress. The hormone is ubiquitous in vascular plants, but not as much in plants of the lower non-vascular groups. In the higher plant ABA is synthesized in all organs including roots. All plant cells having chloroplast or amyloplast synthesize ABA, in particular when subjected to challenging environments.

Chemically ABA is a fifteen carbon compound that resembles the terminal aromatic ring structure of carotenoids. ABA is present as a cis- or trans-isomer depending on the orientation of the carboxyl group at carbon atom 2; the former is biologically active, but not the latter. The cis- and trans-isomers are inter-convertible in plant tissues. ABA has one more asymmetric carbon in the 1st position of the aromatic ring structure. It results in S and R enantiomers (+ and – respectively) and inter conversion is not possible between them. S- enantiomer is active in quick ABA response process like stomatal closure whereas both S and R forms are active in long-term response during seed maturation.

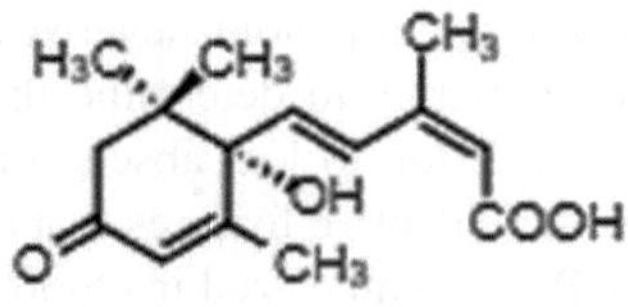

Abscisic Acid

Chemical Structure and Biosynthesis of Abscisic Acid

In the plant cells, ABA is synthesized in the chloroplast and plastids. ABA synthesis is divided into three stages (Milborrow 2001). The synthesis begins with the formation of the isoprene unit isopentenyl phosphate (IPP), which is derived from the pyruvate-triose (Methyl- Erythritol Phosphate- MEP) pathway (Milborrow 2001), where glyceraldehydes 3 phosphate combines with pyruvate to form methylerythritol phosphate in the first stage of ABA synthesis (Fig. 11.22). IPP can be derived from acetyl-CoA by the mevalonate pathway as well (Fig. 11.23). In stage two, the precursor IPP units assemble for the formation of uncyclicized C_{40} carotenoid phytoene through farnesyl diphosphate. Zeaxanthin is the first committed step in ABA biosynthesis. Zeaxanthin undergoes a series of enzyme-catalyzed epoxidations and isomerizations via violaxanthin and ends with cleavage of 9'cis-neoxanthin to form xanthoxal (also called xanthoxin), the 15 carbon skeleton of ABA in stage three. The cleavage is conducted by 9-cis-epoxy-carotenoid-dioxygenase (NCED) enzyme. Xanthoxal is oxidized by xanthoxal dehydrogenase into abscisic aldehyde, which undergoes oxidation further by abscisic aldehyde oxygenase into abscisic acid. ABA is inactivated by being oxidized to form phaseic acid and dihydrophaseic acid. Inactivation also occurs by conjugation with monosaccharides in producing ABA-β –D-glucose ester (Taiz and Zeiger 2002). See Fig. 11.24 for different steps of ABA biosynthesis from IPP.

ABA Signal Transduction in Plant Cell

Although ABA influence on plant development is ubiquitous, ABA-dependent gene transcription cueing changes in ion transport is not elucidated properly. ABA-induced ion transport is regulated by a variety of protein kinases and phosphatases.

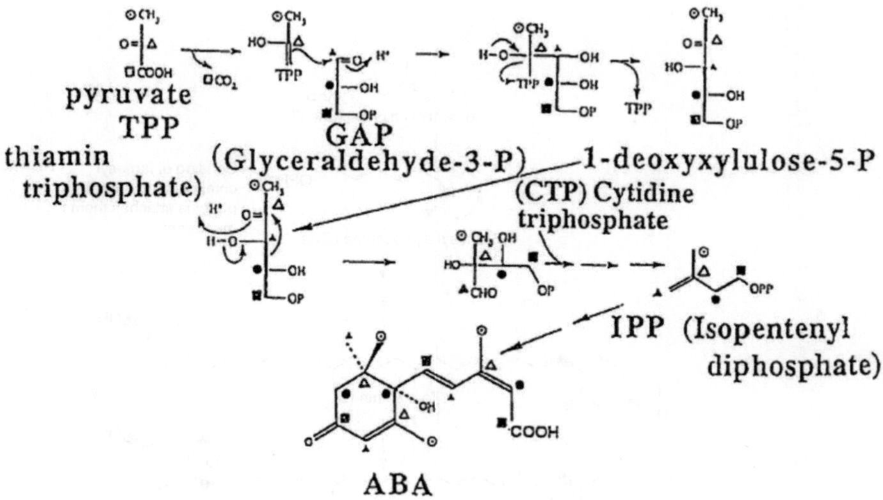

Fig. 11.22 The MEP scheme for the biosynthesis of isopentenyl diphosphate (IPP) from pyruvate and glyceraldehyde 3-phosphate in chloroplasts. (Adopted from Milborrow 2001)

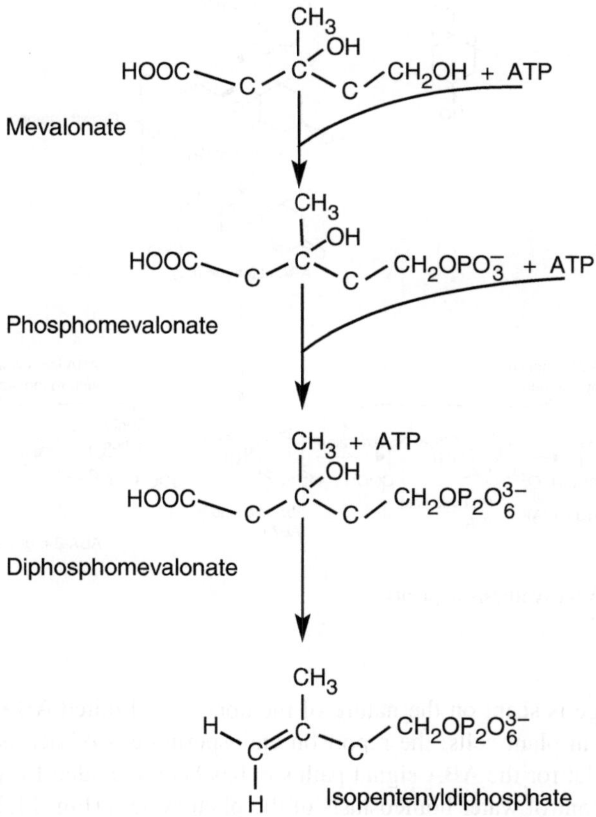

Fig. 11.23 The biosynthesis pathway for the formation of isopentenyl diphosphate (IPP) from mevalonate in the cytosol. (Adopted from Milborrow 2001)

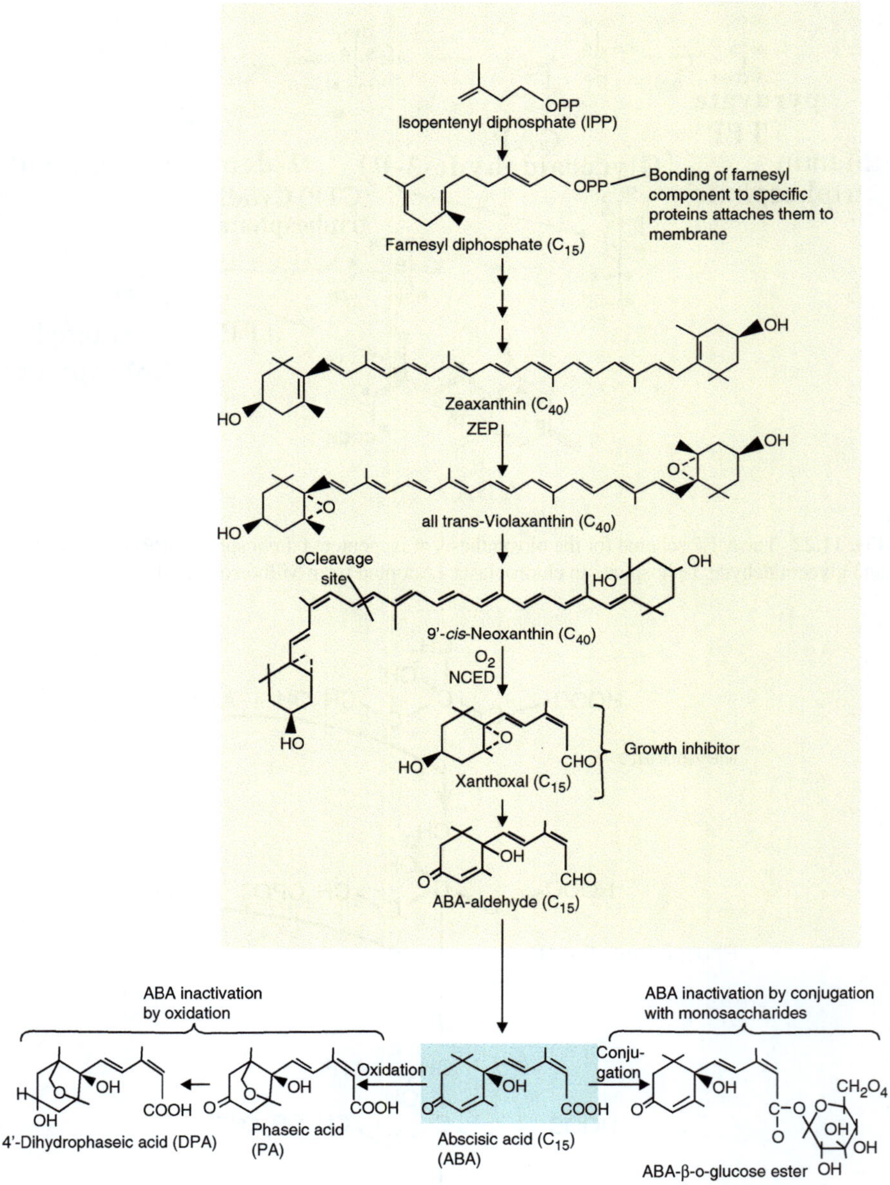

Fig. 11.24 ABA biosynthesis in plants

The knowledge is scant on the nature of membrane de-limited ABA signal transduction steps in plant cells; the report on endosperm cells of rice is still meager. Herein, a model for the ABA signal pathway has been included for guard cells of stomata that control water homeostasis of the plant system (Fig. 11.25). There are two pathways.

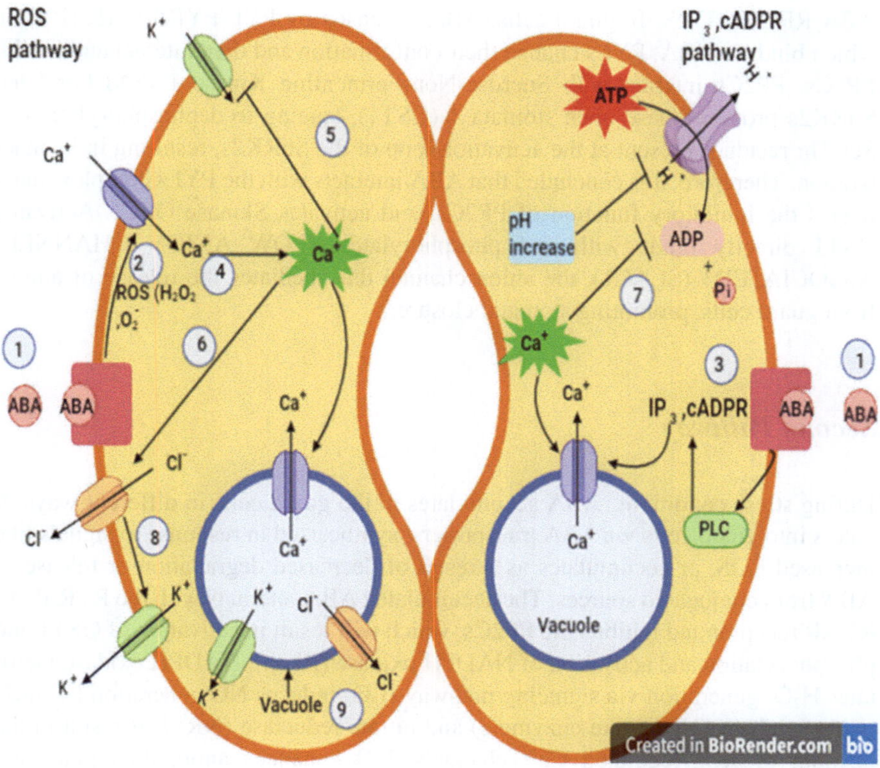

Fig. 11.25 ABA signaling in guard cells of stomata 1. ABA binds to its receptors. 2. ABA binding activates plasma membrane Ca^{2+} channels. 3. ABA increases the levels of cyclic ADP and IP_3 which activates additional calcium channels on the tonoplast. 4. The influx of calcium initiates intracellular calcium oscillations and promotes the further release of calcium from vacuoles. 5. The rise in intracellular calcium blocks potassium inside channels. 6. Calcium rise leads to the opening of chloride channels on plasma membrane (PM) by membrane depolarization. 7. The PM proton pump is inhibited by the ABA-induced increase in cytoplasmic calcium and rise in intracellular pH, further depolarizing the membrane. 8. Membrane depolarization activates K^+ out channels. 9. K^+ and anions to be released across the plasma membrane are first released from vacuoles to the cytoplasm.

First Pathway

Stomata are broadly known for mediating photosynthetic CO_2 exchange and for the efficient use of water for generating the transpiration pull for the ascent of sap. Stomatal movement mediated by ROS has unique and distinct receptors and early signaling elements but they also have common components, for instance, plasma membrane anion channels and K^+ channels through which solute fluxes drive water influx/efflux during actual stomatal movement. H_2O_2 plays a key role in ABA-induced stomatal closure. ABA in cells is sensed by the ABA receptors. The regulatory network of ABA involves three major components of the ABA receptor; the PYRABACTIN (PYR1)/PYR1-LIKE (PYL)/REGULATORY COMPONENTS OF

ABA RECEPTORS. In guard cells, ABA is sensed by PYL/PYR/RCAR (PYLs), which binds to ABA. PYLs change their conformation and then interact and inhibit PP2Cs. PP2Cs interact with Sucrose-Non-Fermenting Kinase 1 (SNF1)-related SnRK2s protein kinase open stomata 1 (OST1), leading to dephosphorylation of Ser/Thr residues present at the activation loop of the SnRK2s, resulting in its inactivation. Therefore, it is concluded that ABA interacts with the PYLs complex, inactivates the inhibitory function of PP2Cs, and activates Skinase OST1. Activated OST1 directly binds with and phosphorylates SLOW ANION CHANNEL-ASSOCIATED1 (SLAC1), the anion channel that mediates the release of anions from guard cells, promoting stomatal closure.

Second Pathway

During stress conditions, ABA accumulates in the guard cells in different ways. It enters into guard cells via ABA transporters, synthesized in response to signals like increased ROS, or accumulates as a result of decreased degradation or release of ABA from conjugated sources. The accumulated ABA interacts with the PYR/PYL/RCAR receptor and inhibits the PP2Cs, which will result in activation of OST1 and phosphorylation and activation of NADPH oxidase (Rboh). NADPH oxidase facilitates H_2O_2 generation via signaling pathway·H_2O_2 induces NO generation by nitric oxide synthase (NOS)-like enzyme(s) and nitrate reductase (NR) that result in the opening of ROS-regulated Ca^{2+} channels. NO enhances antioxidant gene and enzyme activity via MAPK signaling pathways. H_2O_2 directly induces ROS- regulated Ca^{2+} channels (NO-independent signaling) thereby increasing Ca^{2+} (Cyt). Elevated Ca^{2+}(Cyt) induces the expression of abscisic acid-responsive element-binding (AREB) protein that binds to the ABA-responsive element (ABRE) motif in the promoter region of ABA-inducible genes. The expression of ABA-responsive genes requires a combination of an ABRE and a coupling element (CE) for a functional promoter-Ip3, DAG. Calcium release from intracellular stores can be induced by a variety of second messengers, including inositol 1,4,5-trisphosphate (IP3), cyclic ADP-ribose (cADPR), and self-amplifying (calcium- induced) Ca^{2+} release. ABA stimulates nitric oxide (NO) synthesis in guard cells, which induces stomatal closure in a cADPR-dependent manner, indicating that NO is a secondary messenger in this response pathway. An increase in cytosolic calcium, partly derived from intracellular stores, is responsible for ABA-induced stomatal closure.

ABA Regulates the Growth and Development of Rice Panicle

Although ABA controls various aspects of plant life, there has been little information available of the role of ABA in the determination of rice panicle architecture, especially activities of branch and rachis meristems leading to fixation of branch

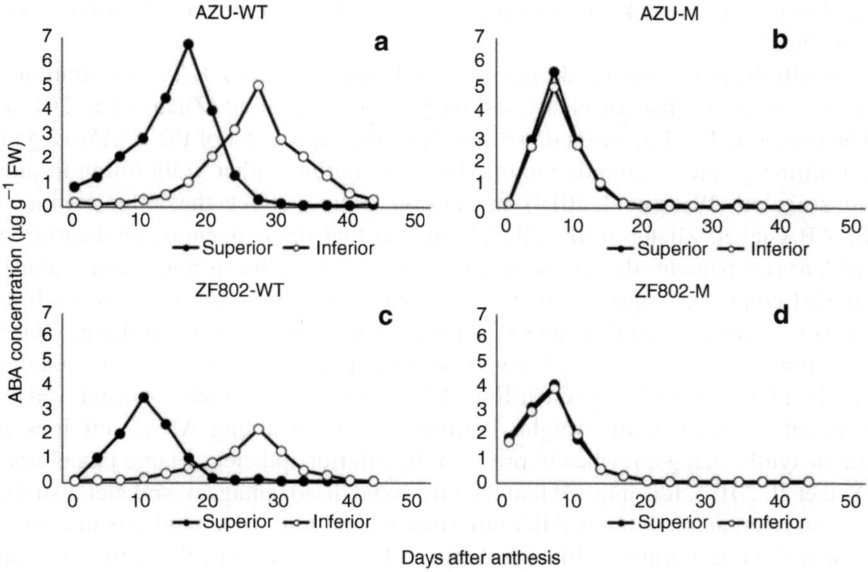

Fig. 11.26 Changes in the ABA concentration of superior and inferior spikelets of wild- type rice (AZU-WT and ZF802-WT) and their corresponding genetic mutants (AZU-M and ZF802-M) during the grain development period. (Adopted from Zhang et al. 2016)

and spikelet number. However, ABA has a significant effect on the grain filling of rice spikelets in the post- anthesis period. Kato et al. (1993) measured the endogenous free abscisic acid content of developing grains of two rice cultivars with different grain size in the post-anthesis period; the ABA level peaked up to mid grain filling stage before a gradual decline in the late maturation phase and the concentration was higher in the cultivar with larger size grain. It was concluded that the hormone promoted assimilate partitioning through its influence on assimilate accumulation to cause the genetic difference in grain growth. Thus, ABA could have a positive influence on sieve tissue unloading and uptake rate in sink cells of developing grains. Bereft of this advantage dry matter accumulation is slowed down in the poorly-filled inferior spikelets located on secondary branches of panicle (Tsukaguchi et al. 1999). Factually, the inferior spikelets have low ABA content of endosperm and inhibition of grain filling in them (Fig. 11.26 - Zhang et al. 2016) and it can be alleviated by exogenous application (Zhang et al. 2009). Similarly, spikelet thinning, in particular removal of superior spikelets, increases ABA, cytokinin, and auxin contents in inferior spikelets that drive higher grain filling (You et al. 2016). In a recent study, Fei et al. (2019) reported that vascular bundle in the panicle neck is a crucial trait in rice production that differs between the *indica* and japonica rice sub-species and identified a locus *qLVB9* in QTL analysis, which appeared synonymous with *DENSE AND ERECT PANICLE 1*. The truncated *dep1* allele down regulated expression of *AP2-like* gene family. They showed that double mutant for the *DEP1* and *OsAP2-39* decreased endogenous ABA level and insensitivity to exogenous ABA treatment. It was concluded that *qLVB9/DEP1* affects large vascular bundles of panicle neck through influence on ABA signaling via regulation of *AP2* gene family.

Similarly, moderate soil drying in the field increases the ABA concentration of inferior spikelets that correlates positively with grain filling (Zhang et al. 2012a). The moderate level of stress down-regulates the expression of the *ABA8ox2* gene permitting greater accumulation of ABA potentiating higher grain filling in inferior spikelets (Wang et al. 2019). In addition to stress effects that stimulate intrinsic ABA level, Zhang et al. (2012a) showed that the exogenous application of ABA to rice panicles during the first two weeks post-anthesis under continuously flooded condition improves activities of starch synthesizing enzymes, such as, sucrose synthase, ADP-glucose pyrophosphorylase, starch synthase, starch branching enzyme, and acid invertase significantly in the inferior spikelets (Table 11.1) for higher grain filling. When superior spikelets are removed by physical excision, grain weight, hormone levels including ABA, activities of starch synthesizing enzymes improve in the inferior spikelet of large panicle rice (You et al. 2016). It is implied that the positional disadvantage of spikelet distribution on rice panicle hinders ABA biosynthesis and low ABA synthesis in consortium with other hormones like cytokinin and auxin determine the nature of grain filling. ABA regulates the growth of the failing spikelets through some proteins and phospho-proteins participating in carbon, nitrogen, and energy metabolism (Zhang et al. 2012b).

In this experiment, plants were grown in cement tanks that were filled with soil and were continuously flooded. The panicles were treated with $20x10^{-6}$ M ABA (T1), $20x10^{-6}$ M fluridone (T2), or $20x10^{-6}$ M ABA + $20x10^{-6}$ M fluridone (T3) daily for 5 d after full heading. Control plants (CK) were treated with de-ionized water. ABA concentration (nmol g^{-1} DW) and activities of sucrose synthase (SuSase), adenosine diphosphate-glucose pyrophosphorylase (AGPase), starch synthase (StSase), starch branching enzyme (SBE), soluble acid invertase (AI), and insoluble AI were determined 5 d and 10 d after treatments, respectively. The activities of the enzymes were expressed as U mg^{-1} protein min^{-1} for SBE and nmol mg^{-1} protein min^{-1} for others. The treatments improved the concentration of ABA and activities of the enzymes more in inferior spikelets compared to superior spikelets. The data are presented as an average between the two *indica* hybrid cultivars YLY-6 and LYP-9 because they showed the same tendencies (Table 11.1-Adopted from Zhang et al. 2012a).

ABA Improves the Activities of Endosperm Starch Synthesizing Enzymes

Later it was observed that under moderate soil drying and alternate wetting, not only ABA improves activities of starch synthesizing enzymes, but also over-expresses genes encoding the starch synthesizing enzymes particularly in the inferior spikelets (Wang et al. 2015). Because ABA level increased under moderate soil drying compared to well-watered soil, it promoted grain filling rate, grain weight, and sink

Table 11.1 The effects of applied abscisic acid (ABA) and fluridone on ABA concentration and activities of starch synthesizing enzymes in superior and inferior spikelets of rice. Adopted from Zhang et al. (2012a)

ABA concentrati on/enzyme activity		5 d after treatment				10 d after treatment		
	CK	T1	T2	T3	CK	T1	T2	T3
Inferior spikelets								
ABA hull	0.54±0.04	0.69±0.05	0.35±0.02	0.55±0.04	0.67±0.06	0.87±0.07	0.51 ±0.03	0.71±0.05
ABA kernel	0.72±0.04	0.97±0.06	0.33±0.02	0.70±0.06	0.94±0.06	1.20±0.07	0.66±0.04	0.91±0.06
SuSase	41.5±3.21	49.6±3.18	32.2±2.63	40.7±2.15	33.5±2.24	40.6±2.44	21.8 ±2.01	32.9±2.15
AGpase	22.8±1.57	30.5±2.45	14.6±1.36	23.5±2.47	31.5±2.75	42.6±3.19	22.9±1.86	30.5±2.27
StSAse	11.5±0.52	14.8±0.74	8.52±0.31	10.8±0.72	9.56±0.59	12.4±0.85	7.22±0.42	9.61±0.59
SBE	312±17.3	389±22.5	244±18.6	296±15.9	428±27.6	527±33.6	316±24.4	418±25.4
Soluble Al	11.8±1.23	12.2±1.31	10.8±1.24	10.9±0.76	10.2±0.74	10.5±0.82	9.84±0.72	9.86±0.73
Insoluble Al	10.4±0.87	10.6±0.75	9.86±0.73	9.98±0.86	9.25±0.75	9.46±0.72	9.25±0.061	9.06±0.55
Superior spikelets								
ABA hull	0.55±0.06	0.76±0.06	0.33±0.02	0.53±0.06	0.64±0.00	0.85±0.07	0.46±0.06	0.66±0.05
ABA kernel	1.24±0.11	1.28±0.10	0.96±0.08	1.21±0.09	0.87±0.06	0.91 ± 0.08	0.68±0.06	0.85±0.05
SuSase	53.4± 4.12	56.6±4.54	42.5±3.25	52.3±3.66	30.8±2.51	31.7 ± 2.16	24.5±2.14	30.2±2.23
AGPase	42.5± 2.56	44.3±3.17	36.6±2.41	41.5±2.53	23.6±1.24	24.5±1.55	22.6±2.04	23.2±1.84
StSase	14.6±0.57	15.2±0.94	12.4±0.31	14.4±0.36	9.43±0.45	10.2±0.64	6.59±0.33	9.27±0.53
SBE	610 ±45.5	623±50.3	522±32.5	598±22.6	253±12.8	259±16.4	207±14.7	248±15.5
Soluble Al	11.2±1.0	11.6±0.94	10.7±0.84	10.9±0.76	10.4±0.83	10.7±0.65	10.1±0.75	10.2±0.78
Insoluble Al	9.23±0.75	9.52±0.63	9.03±0.62	9.11±0.54	8.46±0.66	8.59±1.34	8.22±0.56	8.65±0.57

activity, in terms of the activities of sucrose synthase, ADP glucose pyrophosphorylase, starch synthase, and starch branching enzyme in inferior spikelets by higher expression levels of *SuS2, SuS3, SuS4, AGPS2b, AGPL1, AGPL2, SSSIIb, SSSIIc, GBSSI, GBSSII, SBEI*, and *SBEIIb*. Conversely under severe soil drying these parameters were markedly decreased in both superior and inferior spikelets. Water stress, no doubt increases ABA level for gain in grain filling but too high an elevation under severe drought is counterproductive for sink activity. Although mechanism underlying moisture deficiency treatments are elusive, an increase in the level of ABA up to a limit benefits grain filling at the cost of depression of ethylene production (Yang and Zhang 2006) that led to overexpression of the genes for starch synthesizing enzymes and conversion efficiency from sucrose to starch. Elevation of ABA beyond the normal level in developing grains can depress cytokinin abundance and adversely influence starch synthesis. Tang et al. (2009) corroborated that ABA-mediated grain filling in rice is consequential to an increase of sucrose synthase activity and greater expression of sucrose synthase protein.

Endogenous ABA Biosynthesis High in Superior Spikelets

The higher expression of genes encoding starch biosynthesis also corresponded with similar over-expression of genes encoding ABA biosynthesis, such as *9-CIS-EPOXYCAROTENOID DIOXYGENASE 1 (NCED1) and 9-CISEPOXYCAROTENOID DIOXYGENASE 5 (NCED5*). Decades before the molecular biology work, it was shown that exogenous ABA application promoted endosperm growth by improving partitioning of assimilates to the grains (Kato 1993, Kato and Takeda 1993). Endogenous ABA concentration is as well relative mRNA levels of *NCED1* and *5* rise rapidly to a peak level in the developing grains of superior spikelets of rice panicle in the first 12 days post-anthesis and declines sharply thereafter to null level at maturity. In comparison, ABA concentration and relative mRNA levels of *NCED1* and *5* are lower and the dynamics of rise and fall in the levels is much delayed in the inferior type spikelet, where grain filling is poor (Wang et al. 2015). It was also observed that the dynamics of fluctuation in hormone levels of the spikelets remain similar despite the variation in the soil moisture level of the growth medium and genotype difference (Fig. 11.27).

ABA Sucrose Synergy Regulates Endosperm Starch Synthesis

It is understood that ABA improves grain filling rice by enhancement of activities of sucrose synthase and other starch synthesizing enzymes and expression of genes encoding the enzymes (Tang et al. 2009; Zhang et al. 2012a). It envisages a close coupling of ABA biosynthesis with grain starch synthesis embracing sucrose as

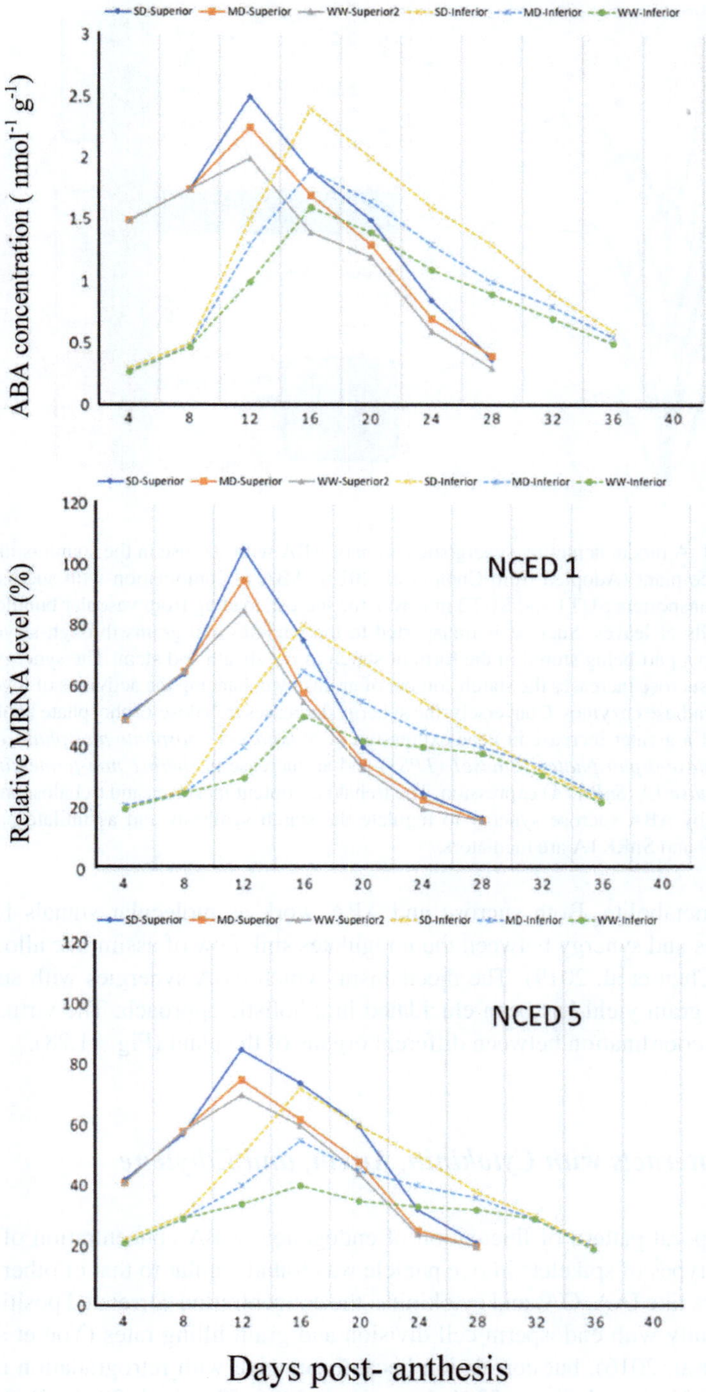

Fig. 11.27 Changes in abscisic acid (ABA) contents (**a**) and expression of *NCED1* (NCED = 9-cis epoxycarotenoids Dioxygenase) (**b**) and *NCED5* (**c**) in superior (solid lines) and inferior (dotted lines) spikelets of rice. WW, MD, and SD represent well-watered, moderate soil drying, and severe soil drying treatments, respectively, during grain filling. (Adopted from Wang et al. 2015)

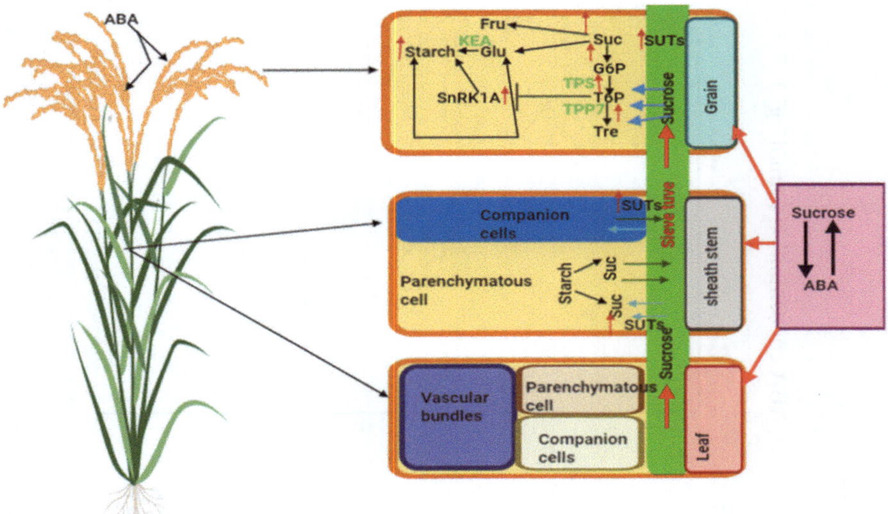

Fig. 11.28 A model depicting synergistic action of ABA with sucrose in the source-sink relationship of rice plant (Adopted from Chen et al. 2019). ABA in combination with sucrose induces sucrose transporters SUT1 and SUT2 in leaves for sucrose loading from vascular bundles to companion cells of leaves. Sucrose is transported to the panicles and grains through sieve tubes in addition to a part being stored in the form of starch in the sheath and stem. The synergy between ABA and sucrose increases the starch content of grains by enhancing the activities of invertase and sucrose synthase enzymes. Conversely, the synergy decreases trehalose-6-phosphate (T6P) content because of a greater increase in gene expression of *trehalose-6-phosphate phosphatase7* (*TPP7*) than *trehalose-6-phosphate synthase1* (*TPS1*) and an increase in *sucrose non-fermenting related protein kinase 1A* (*SnRK1A*) expression. The trehalose content increases and trehalose metabolism improves by ABA sucrose synergy to regulate the starch synthesis and assimilate partitioning where T6P and SnRK1A are mediators.

primer metabolite. Both sucrose and ABA work as molecular signals for stress responses and synergy between them regulates sink flow of assimilate allocation to grains (Chen et al. 2019). The mechanism, which ABA synergies with sucrose to enhance grain yield has been elucidated in a holistic approach. The virtual model involves coordination between different organs of the plant (Fig. 11.28).

ABA Interacts with Cytokinin, Auxin, and Ethylene

The temporal pattern of fluctuation of endogenous ABA concentration of the two extreme types of spikelets in rice panicle was found similar to that of other positive regulators like IAA, GA, and cytokinins, the concentration correlated positively and significantly with endosperm cell division and grain filling rates (You et al. 2016; Zhang et al. 2016), but correlation became negative with retrogradation regulator ethylene (Yang and Zhang 2006: Yang et al. 2006: Zhu et al. 2011). Between the superior and inferior spikelets, the ABA content of endosperm in the former is higher and rises temporally to a peak level in the first two weeks after anthesis before declining sharply thereafter; in contrast ABA content is low and there is a delay of one week in the rise and fall of the material in the latter (Wang et al. 2015). Because of similarities in temporal fluctuation in rise and fall, it was proposed that plant hormones cytokinins, IAA, and ABA work together closely during grain development in regulating sink size mediating through endosperm cell division and enlargement (You et al. 2016), and elevating the level of hormones at an early stage of development can increase grain filling. Shreds of evidence documented by Yang et al. (2006) have significantly and positively correlated endosperm cell division and grain filling rates with ABA contents. However, this relationship between the hormone and post-anthesis grain growth was contradicted later (Zhu et al. 2011). As of today, there has been no explanation satisfactory for any complementary or supplementary action of ABA with cytokinin, although the pattern of fluctuation in concentration of the hormones is synchronous during the grain development period.

References

Chen T, Li G, Islam MR, Fu W, Feng B,Tao L, Fu G (2019) Abscisic acid synergizes with sucrose to enhance grain yield and quality of rice by improving the source-sink relationship BMC Plant Biol 19:525 https://doi.org/10.1186/s12870-019-2126-y

Fei C, Geng X, Xu Z, Xu Q (2019) Multiple areas investigation reveals the genes related to vascular bundles in rice. Rice 12: 17

Kato T (1993) Variation in grain-filling process among grain positions within a panicle of rice (Oryza sativa L.). SABRAO J 25: 1–10

Kato T, Sakurai N, Kuraishi S (1993) The changes of endogenous abscisic acid in developing grain of two rice cultivars with different grain sizes. Japanese J Crop Sci 62: 456–461

Kato T, Takeda K (1993) Endogenous abscisic acid in developing grains on primary and secondary branches of rice (Oryza sativa L.). Japanese J Crop Sci 62: 636–637

Milborrow B (2001) The pathway of biosynthesis of abscisic acid in vascular plants: a review of the present state of knowledge of ABA biosynthesis. J Exp Botany 52: 1145–1164

Tang T, Xie H, Wang Y, Lü B, Liang J (2009) The effect of sucrose and abscisic acid interaction on sucrose synthase and its relationship to grain filling of rice (Oryza sativa L.). J Exp Botany 60: 2641–52

Taiz L, Zeiger E (2002) Plant Physiology. 3rd edn. Sinauer Associates, Inc., Sunderland, USA.

Tsukaguchi T, Horie T, Koshioka M (1999) Dynamics of abscisic acid levels during grain- filling in rice: comparisons between superior and inferior spikelets. Plant Prod Sci 2: 223–226

Wang Z, Xu Y, Chen T, Zhang H, Yang J, Zhang J (2015) Abscisic acid and the key enzymes and genes in sucrose-to-starch conversion in rice spikelets in response to soil drying during grain filling. Planta 241: 1091–1107

Wang GQ, Li HX, Feng L, Chen MX, Meng S, Ye NH, Zhang J (2019) Transcriptomic analysis of grain filling in rice inferior grains under moderate soil drying. J Exp Botany 70: 1597–1611

Yang J, Zhang J (2006) Grain filling of cereals under soil drying. New Phytol 169: 223–236

Yang J, Zhang J, Wang Z, Liu K, Wang P (2006) Post-anthesis development of inferior and superior spikelets in rice in relation to abscisic acid and ethylene. J Exp Botany 57: 149–160

You C, Zhu H, Xu B, Huang W, Wang S, Ding Y, Liu Z, Li G, Chen L, Ding C, Tang S (2016) Effect of removing superior spikelets on grain filling of inferior spikelets in rice. Front Plant Sci 7:1161. https://doi.org/10.3389/fpls.2016.01161

Zhang H, Tan G, Yang L, Yang J, Zhang J, Zhao B (2009) Hormones in the grains and roots in relation to post-anthesis development of inferior and superior spikelets in japonica/indica hybrid rice. Plant Physiol Biochem 47: 195–204

Zhang H, Li H, Yuan L, Wang Z, Yang J, Zhang J (2012a) Post-anthesis alternate wetting and moderate soil drying enhance activities of key enzymes in sucrose-to-starch conversion in inferior spikelets of rice. J Exp Botany 63: 215–227

Zhang Z, Chen J, Lin S, Li Z, Cheng R, Fang C, Chen H, Lin W (2012b) Proteomic and phosphoproteomic determination of ABA's effects on grain-filling of Oryza sativa L. inferior spikelets. Plant Sci 185: 259–273

Zhang W, Cao Z, Zhou Q, Chen J, Xu G, Gu J, Liu L, Wang Z, Yang J, Zhang H (2016) Grain filling characteristics and their relations with endogenous hormones in large- and small-grain mutants of rice. PLoS ONE 11(10): e0165321. https://doi.org/10.1371/journal.pone.0165321.

Zhu G, Ye N, Yang J, Peng X, Zhang J (2011) Regulation of expression of starch synthesis genes by ethylene and ABA in relation to the development of rice inferior and superior spikelets. J Exp Botany 62: 3907–3916

Ethylene

Abstract Ethylene is a gaseous plant hormone that acts at trace levels throughout the life cycle of plants by stimulating or regulating several developmental features like the ripening of fruits, opening of flowers, leaf abscission, and senescence of leaves and seed germination. The low molecular weight of the hormone makes it lighter than air under any physiological condition permitting quick diffusion within and outside plant tissue. Ethylene is synthesized from methionine in the Yang cycle for which, the rate-limiting enzyme in ACC synthase. The protein factors responsible for ethylene perception and signal transduction have been identified in *Arabidopsis*; the system is nearly homologous in rice. Although meristematic tissues are active in ethylene production, studies are scant for temporal changes in ethylene levels of young rice panicle before emergence from the boot of flag leaf sheath. In comparison, there has been a wealth of information available on ethylene-responsive genes, expression of which inhibits rice caryopsis growth. Intrinsic ethylene concentration of rice spikelets culminates at anthesis and declines thereafter with the progression of grain growth. High ethylene production in rice panicle has been a major obstacle for breaking of yield barrier and improvement of grain yield. Ethylene production negatively correlates with spikelet grain filling. Spikelets disadvantaged for spatial freedom produce more ethylene. Under such conditions, the hormone keeps the nascent caryopsis morbid for some time after fertilization and underscores the activity of endosperm starch synthesizing enzymes later to the detriment of grain filling.

Keywords Ethylene · Yang cycle · ACC synthase · Ethylene receptors and transducers · Grain filling

Introduction

Ethylene was identified and confirmed as a natural plant hormone after gas chromatography was introduced into ethylene research in 1959 (Burg and Thimann 1959). It is a gaseous plant hormone that acts at trace levels throughout the life cycle of plants by stimulating or regulating several developmental features like the ripening of fruits, opening of flowers, leaf abscission, and senescence of leaves and seed germination (Abeles et al. 1992). All parts of higher plants like leaves, stems, fruit, flowers, roots, seeds, and tubers can produce ethylene, although the level of production depends on the type of tissue and stage of development. Development specific events inducing ethylene production are germination of seeds, abscission of leaves, ripening of fruits, and maintenance of seedling apical hook, leaf abscission, and senescence of flowers. The meristematic regions of plants are the most active sites for ethylene production. Additionally, ethylene production is stimulated in plants by environmental stresses and physical and chemical injuries. Abiotic stresses like drought, flooding, salinity, and chilling and biotic factors like pathogen attack cue ethylene production.

Chemical Structure

Ethylene is the simplest olefin-type hydrocarbon with four hydrogen atoms bound to a pair of carbon atoms that are connected by a double bond. Ethylene molecular weight is 28. The low molecular weight makes it lighter than air under any physiological condition permitting diffusion as a gaseous hormone. In the plant tissues, it is released easily and diffuses in the gas phase of the intercellular space and outside the tissue. In the ethylene chemical structure, all six atoms (2 carbons and 4 hydrogens) are coplanar. The strong binding of the two carbons by double bond confers rigidity and stability of the structure. Ethylene can be oxidized to ethylene oxide, further oxidation of which leads to the formation of oxalic acid and CO_2 thereafter.

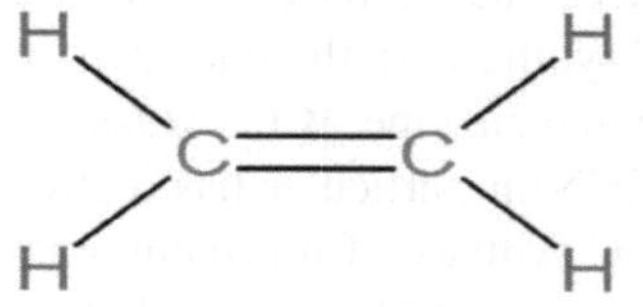

Ethylene

Ethylene Synthesis and Breakdown

The ethylene biosynthesis pathway is named Yang cycle in recognition of the scientist S F Yang who elucidated the pathway. The amino acid methionine is the precursor for ethylene. It is converted to S-adenosyl-L-methionine (SAM, also called

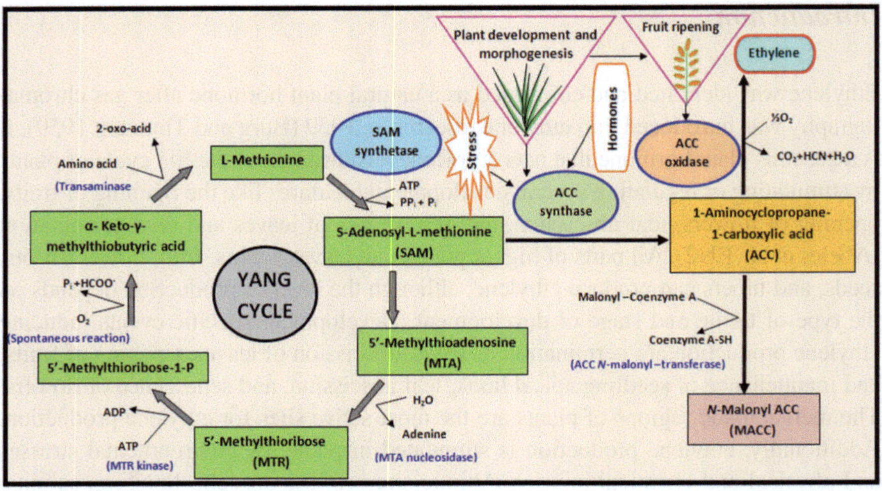

Fig. 11.29 Ethylene biosynthetic pathway and the Yang cycle. The amino acid methionine is the precursor of ethylene. The rate-limiting step in the pathway is the conversion of AdoMet to ACC, which is catalyzed by the enzyme ACC synthase. The last step in the pathway, the conversion of ACC to ethylene, requires oxygen and is catalyzed by the enzyme ACC oxidase. The CH3—S group of methionine is recycled via the Yang cycle and thus conserved for continued synthesis. Besides being converted to ethylene, ACC can be conjugated to N-malonyl ACC

Ado-met) by the enzyme Met-Adenosyl transferase with the participation of ATP, which de-phosphorylates to donate the adenosine part. In the next step, Ado-Met is converted to 1-aminocyclopropane-1- carboxylic acid (ACC) and S-methylthioadenosine by the enzyme ACC synthase. This is a rate-limiting step for ethylene biosynthesis in plants, where the activity of the enzyme determines the rate of ethylene production. There are several regulators for the activity of ACC synthase. Developmental events like fruit ripening and flower senescence and stress factors like chilling injury, drought, and wounding promote and application of chemicals like aminooxy acetic acid (AOA) and aminoethoxy-vinyl glycine (AVG) inhibit the activity of ACC synthase. In the final step, oxygen is required for oxidation of ACC to ethylene by the enzyme ACC oxidase. The by-products of the reaction are CO_2 and HCN. HCN in particular has a strong potential to damage cell metabolism. Similar to ACC synthase, fruit ripening promotes ACC oxidase activity. The activity of this enzyme is inhibited by cobalt, silver, high temperature, and anaerobiosis condition. CH_3-S group attached to adenosine in S- methylthioadenosine is recycled in Yang cycle for methionine formation and maintenance of the cycle operation (Fig. 11.29).

Ethylene Signaling

There are some protein receptors on the membrane for the perception of ethylene. The pathway of ethylene signaling in plants has been investigated mostly in Arabidopsis and some other dicots through intensive studies of genetic mutants and the biochemical processes. In comparison, there have been no systematic studies in monocots including rice. There have been several instances of ethylene sensitivity of rice. In the signaling system of rice, most of the components homologous to Arabidopsis have been found, but many of the processes are unique some of the ethylene receptors are functionally more divergent (Ma et al. 2010). Ethylene's role is crucial for the adaptation of semi-aquatic rice to hypoxia. Ma et al. (2013) identified and characterized some ethylene response mutants based on specific ethylene response phenotypes where ethylene inhibits root growth and promotes coleoptiles growth in dark-grown rice seedling named as a double response (Ma et al. 2013). This response in rice is different from the normal triple response of Arabidopsis seedlings. Ethylene sensitivity of rice at the grain development stage has been extensively elaborated, particularly in dense panicle cultivars (Sekhar et al. 2015b, Panda et al. 2016). Yang et al. (2015) have compared the ethylene response phenotypes of dark-grown seedlings of Arabidopsis, rice, and some other monocots like wheat, maize, and sorghum and summarized homologous genes of the ethylene signaling components. Despite extensive similarities, rice has evolved some new features not seen in Arabidopsis.

In the plant cells, a linear ethylene signaling pathway begins with the perception of the hormone by a family of receptors in the endoplasmic reticulum and concludes with transcriptional regulation inside the nucleus. The receptors are negative regulators of the signaling pathway and they were identified in genetic mutants of Arabidopsis. The first ethylene insensitive mutant identified was recessive *etr1* that did not show a triple response to ethylene. The wild type with dominant *ETR1* allele permitting ethylene perception is similar to bacterial two-component histidine kinase receptor. *Arabidopsis* genome encodes four other proteins similar to ETR1 and they are ETR2, ERS1, ERS2, and ERS4. The signal conceived by the receptor is conveyed to the nucleus for transcriptional regulation. Downstream the receptor there is a Raf-like MAPKKK (mitogen-activated protein kinase, kinase, kinase) CTR1 (Constitutive Triple Response 1), in which kinase activity is monitored by the regulators. CTR1 acts downstream of the membrane-bound ethylene receptors.

In normal conditions, when ethylene abundance is low in the tissue, the active receptors couples with the N-terminal region of EIN2 (ETHYLENE INSENSITIVE 2) to form a complex. It stimulates the kinase activity of CTR1 that phosphorylates the C-terminal domain of EIN2 and thereby its cleavage. EIN2 is a transmembrane protein of the endoplasmic reticulum located downstream of CTR1. In reality, it is the essential transducer of the ethylene signal. Its activation leads to degradation of EIN2 by F-box protein ETP1/2 (EIN2 TARGETING PROTEIN 1 and 2) and depression of ethylene response pathway. Under ethylene abundance, the hormone stabilizes EIN2 at the protein level from degradation by the F-box proteins. It binds with the ETR1

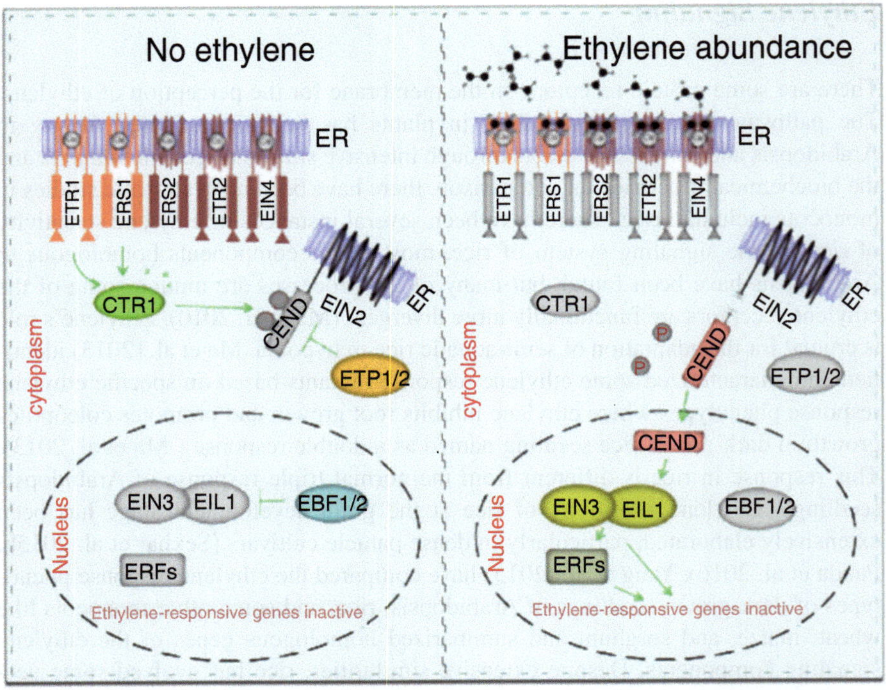

Fig. 11.30 Ethylene signaling pathway. (Adopted from Yang et al. 2015)

dimer with the participation of RAN1 copper transporter protein resulting in its inactivation as well as the inactivation of CTR1. Inactive receptors and CTR1 result in de-phosphorylation of EIN2 leading to its C-terminus cleavage. The split EIN2 C-terminal fragment (CEND) enters the nucleus and participates in the stabilization and accumulation of transcription factors EIN3 and EIL1 (Yang et al. 2015). They function as master transcription factors to activate the expression of targets like ERF1 (Ethylene response factor 1) which in turn regulates the expression of ethylene-responsive genes. The signal cascade leads to transcription of *ERF1* and other ethylene-sensitive genes. The transcriptional cascade also represses gene expressions resulting in ethylene-responsive cell functions. In the absence of ethylene, EIN3 and EIL1 are degraded by ubiquitin-mediated F-box protein EBF1/2 (EIN3 BINDING F-BOX PROTEIN 1 and 2). The signal pathway is given in Fig. 11.30.

Dynamics of Ethylene Production During Rice Panicle Development

Although meristematic tissues are active in ethylene production, studies are scant for temporal changes in ethylene levels of young rice panicle before emergence from the boot of flag leaf sheath. The role of ethylene inhibiting grain filling of rice spikelets was first reported by Mohapatra et al. (2000) and corroborated by other

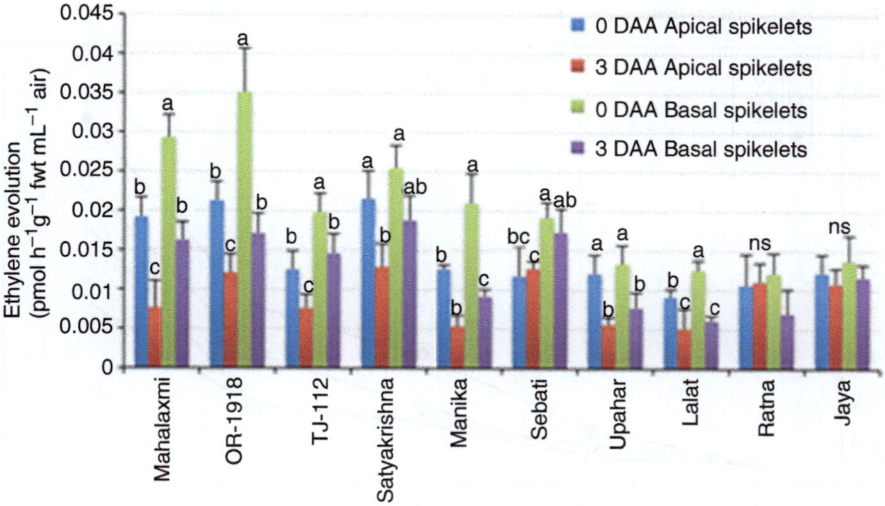

Fig. 11.31 Ethylene evolution by the superior (apical) and inferior (basal) spikelets of the rice cultivars including Mahalaxmi, Or-1918, TJ-112, Satyakrishna, Manika, Sebati, Upahar, Lalat, Ratna, and Jaya. The measurements were done on 0 and 3 days after anthesis (DAA). The superior or inferior spikelets, as the case may be, were incubated in rubber stopper test tubes in dark under moist conditions for two hours. Following this, 1 ml of the head-space gas was drawn with a syringe and injected into the GC column for the quantification of the ethylene released from the spikelets. The individual bars are the mean ± SD of five estimations. In a cultivar, the difference between the two means marked by at least one common alphabet is not statistically significant. ns = The difference between means not statistically significant in a cultivar. (Adopted from Sekhar et al. 2015a)

workers later (Yang et al. 2006, Zhu et al. 2011). It was seen that ethylene concentration in spikelets is low before anthesis, rises rapidly to a peak at the early stage of grain filling before decreasing sharply with the linear period of grain growth (Liu et al. 2008, Panda et al. 2009, Zhang et al. 2009a). Intrinsic ethylene production reaches climax at anthesis probably because the splitting of anther box gives rise to a burst of ethylene. Zhang et al. (2017) found ethylene content of young panicle diminishing with panicle differentiation from spikelet primordium to pollen mother cell differentiation stage and the content increased slowly thereafter. Sekhar et al. (2015a) reported high ethylene evolution at anthesis and substantial fall in concentration thereafter in the following three days in both superior and inferior spikelets of rice panicle of several rice cultivars to corroborate temporal decline in the concentration of the hormone during grain maturation period (Fig. 11.31).

Ethylene Controls Spikelet Development

In the heterogeneous architecture of panicle, spikelet development is staggered in the basipetal direction and spikelet-specific ethylene production follows the same sequential pattern. Spikelets in the apical region reach anthesis early and produce ethylene to slacken the development of later flowering inferior spikelets in the basal

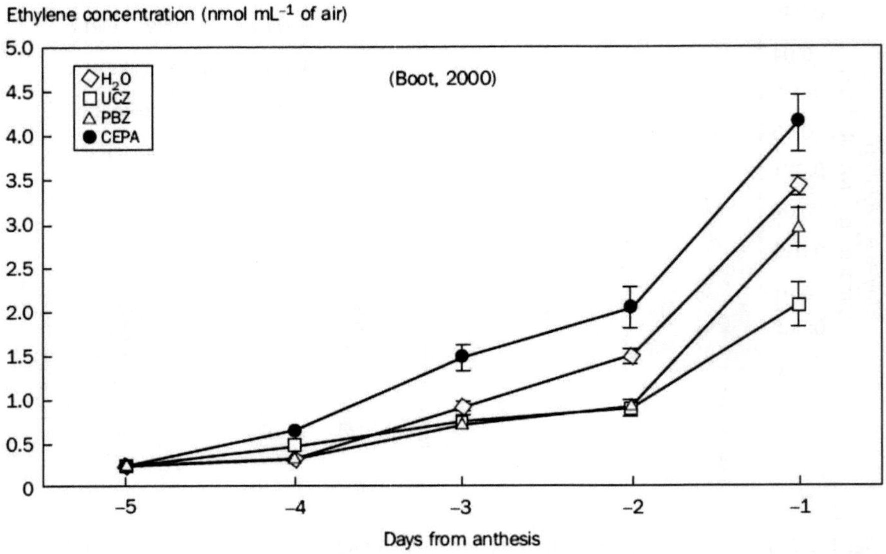

Fig. 11.32 The effect of ethylene regulating substances uniconazole (UCZ), paclobutrazol (PBZ), and 2-chloroethylphosphonic acid (CEPA) on ethylene production inside the boot of flag leaf sheath of rice cultivar Mahanadi in the pre-anthesis stage of panicle development. (Mohapatra and Mohapatra 2005)

part of the panicle. Mohapatra and Mohapatra (2005) suggested that ethylene emanating from the apical spikelets is stored in the boot of the flag leaf sheath, which perturbs the growth of un-emerged inferior spikelets (Fig. 11.32)

Boot ethylene Infringes Pericarp Functions

Boot ethylene has several pernicious effects on rice spikelet development. It accelerates structural degradation of the pericarp, lemma, and palea resulting in chlorophyll loss and enhancement of lipid peroxidation and peroxidase activity (Mohapatra and Mohapatra 2006) (Fig. 11.33).

Loss of seed coat function contributes directly to retardation of grain filling. The pericarp development in rice caryopsis has been co-ordinate with endosperm growth. This inference is possible because carbohydrate status (Weber et al. 1996) and activities of sugar metabolizing enzymes and transporters (Weschke et al. 2003) control partitioning of assimilates between the maternal (pericarp) and endosperm (filial) tissues and adverse effect on one is bound to have a consequential functional infringement of the others. Additionally, high ethylene production impacts male gametophyte development of inferior spikelets that contributes significantly to poor grain filling (Naik and Mohapatra 1999). Therefore suppression of ethylene action or synthesis or both, before anthesis through the application of specific inhibitors has been successful in optimizing grain filling and grain yield in rice (Mohapatra

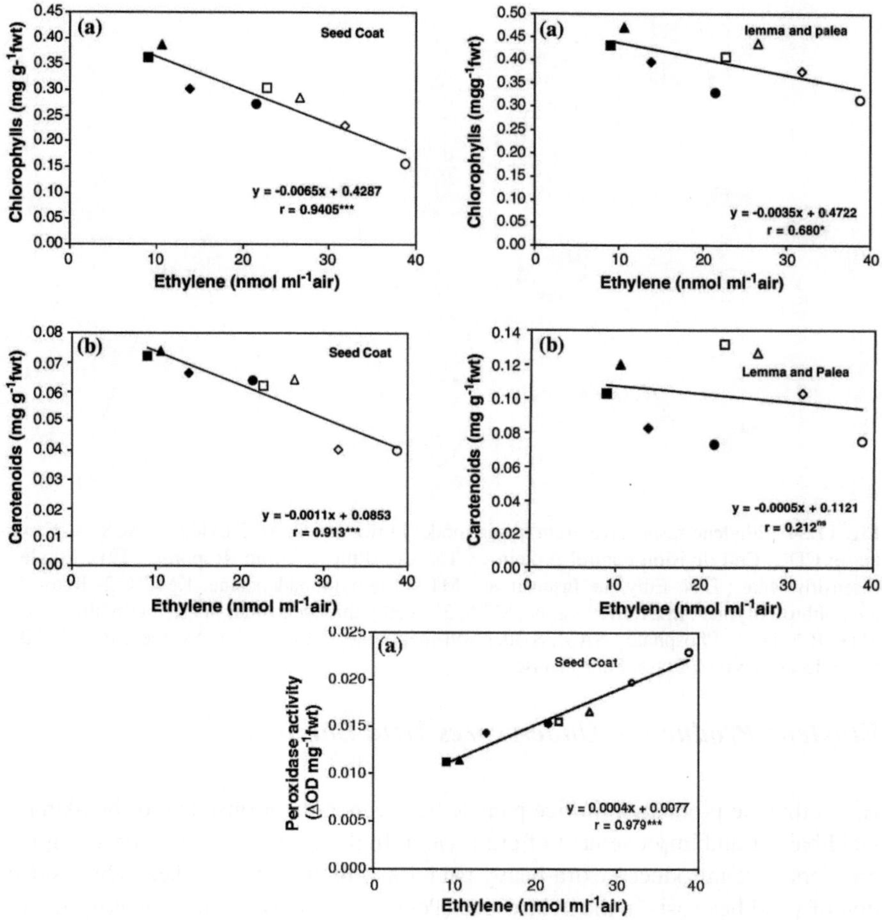

Fig. 11.33 Correlation between ethylene released inside flag leaf sheath at the time anthesis affected by treatments of various ethylene regulating substances, Control= rhombus, AVG (aminoethoxy vinyl glycine- ethylene synthesis inhibitor) = rectangle, AGNO3 (ethylene action inhibitor) = triangle, and CEPA (2-chloroethyl phosphonic acid- ethylene releasing substance) = circle with photosynthetic pigments and peroxidase activity of lemma-palea and seed coat of developing caryopsis at mid grain filling stage of two high yielding rice cultivars, Mahanadi (closed symbol) and Mahalaxmi (open symbol). Mohapatra and Mohapatra (2006)

et al. 2000, Rzewuski and Sauter 2008, Mohapatra et al. 2009, Zhang et al. 2014a). The researchers have used several ethylene inhibitors like silver, cobalt, 1-MCP, and AVG to block ethylene action/synthesis and thereby improve grain filling of inferior spikelets, often by injecting the chemicals into the boot of the flag leaf sheath. A model describing the possible mechanism of ethylene control of grain filling in rice is given in Fig. 11.34.

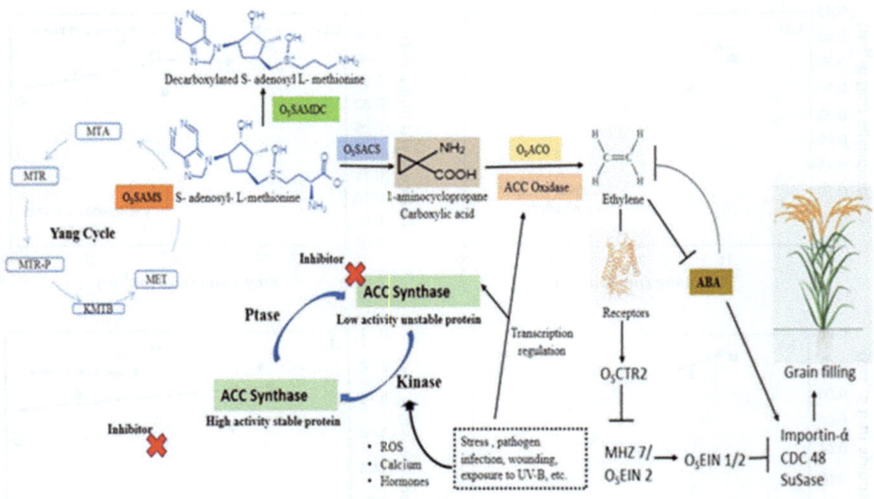

Fig. 11.34 Ethylene responsive grain filling model in rice. ACO, ACC oxidase ; ACS, ACC synthase; CDC, Cell division control protein ; CTR, Constitutive Triple Response ; EIL, Ethylene Insensitive like ; EIN, Ethylene Insensitive ; MTA, Methylthioadenosine; KMTB, 2- Keto- 4 – methylthio-butyrate ; MET, Methionine; MTA, 5'- methylthioadenosine; MTR, methylthioribose ; MTR-P, MTR-1- Phosphate ; SAM, S-adenosyl-metionine; SAMS, SAM Synthetase ; SAMDC, SAM decarboxylase; Ptase, Phosphatase

Ethylene Production Underscores Yield Barriers

High ethylene production in rice panicle has been a major obstacle for breaking of yield barrier and improvement of grain yield. In the quest for high grain yield, rice breeders have introduced extra-heavy panicles with numerous spikelets by a selection of yield heterosis in the hybrid rice (Peng et al. 1999) and new plant-type rice. In these large panicle cultivars, extra spikelets that contributed to the increase in number were accommodated on secondary branches close to each other. The compromise of inter spikelet distance for an increase of number made panicle architecture compact and stimulated ethylene production significantly. High ethylene production undermined the advantage of high spikelet numbers for the improvement of grain yield (Mohapatra and Mohapatra 2006; Panda et al. 2009). In particular, grain filling of inferior type spikelets located on secondary branches of the proximal part of panicle was decimated more on account of high ethylene production as evident in the comparison between lax- and compact-type panicle rice cultivars (Panda et al. 2009, Zhang et al. 2014b). It was shown that a high spikelet number of dense panicle rice stimulates overproduction of ethylene because of the rise in ACC-synthase and –oxidase activities and application of ethylene inhibitors down plays the activities (Panda et al. 2016- Fig. 11.35.

Ethylene Perception Induces Over-Expression of Its Signal Transducers

Ethylene production inducting vulnerability of spikelet growth has been elucidated further by exploring the expression of genes encoding receptor and transducers proteins in the downstream pathway of ethylene signal transduction following ethylene perception. Wuriyanghan et al. (2009) identified ethylene response2 (ETR2) receptor in endoplasmic reticulum membrane causing a delay in flowering and accumulation of starch in stem by down regulating the expression of a monosaccharide transporter gene. The antagonistic relationship between ethylene receptors and starch synthesizing enzymes has been proposed for rice grain development (Mohapatra and Panigrahi 2011). The whole-genome sequencing in rice has revealed the identity of five ethylene receptors (Pareek et al. 2006), and follow up reviews on the subject have proposed that the pathway of ethylene signal transduction has been similar to Arabidopsis (Rzewuski and Sauter 2008; Ma et al. 2010; Zhao and Guo 2011). The reviews have highlighted the nature of perception of ethylene in rice subjected to submergence but precluded discussion on the importance of the expression and activity of the receptors in endosperm cells during grain filling.

Sekhar et al. (2015b) with PCR kit identified expression of genes encoding ethylene receptors ERS1, ERS2, ETR2, ETR3, and ETR4 during the early period of grain filling of which *ERS1* gene was more prominently expressing than the others. Besides, all the receptor genes except *ETR4* over-expressed in inferior spikelets of compact-panicle rice compared to the lax-panicle type. In the pathway downstream of receptors five components of ethylene signaling, namely EIN2, EIN3, ERF2, ERF3, and EREBP5, were discovered in rice caryopsis at the very early stage of grain filling (Sekhar et al. 2015b) (Fig. 11.36). Four rice cultivars, two each belonging to compact- and lax-panicle types were used for the assessment of variation in expression of the genes encoding these proteins in both superior and inferior spikelets of the panicle. The magnitude of expression of the genes was cultivar specific as well as spikelet specific. The expression was much poorer in the high sterile compact-panicle cultivars compared to low sterile lax-panicle cultivars. High ethylene production in spikelets matched with over expression of ethylene receptor and signal transducer genes (Panda et al. 2016). It was noticed that grain filling of compact-panicle cultivars is more sensitive to ethylene, where over expression of components of signal transduction pathway by ethylene during the first few days of anthesis limits cell division of the juvenile endosperm that infringes up on reduced cell cycle activity, endo-reduplication of endosperm cells and starch biosynthesis. Such deleterious activities are neutralized and sink strength for grain filling is improved when ethylene action is blocked by inhibitors like 1-methyl cyclopropane (Panda et al. 2016).

Fig. 11.35 Effect of ethylene inhibitor 1-methyl cyclopropane (1-MCP) on ethylene evolution (**a**), ACC synthase activity (b), ACC-oxidase activity (**c**), and total chlorophyll concentration of flag leaf of apical (A) and basal (B) spikelets of dense panicle rice cultivar OR-1918 on 0 and 3 days after anthesis. (Panda et al. 2016).

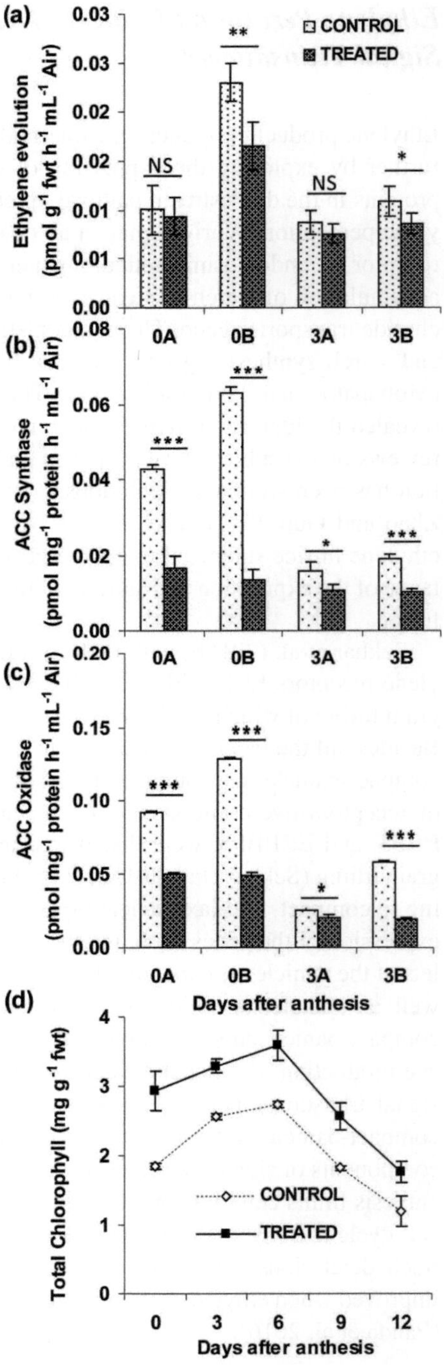

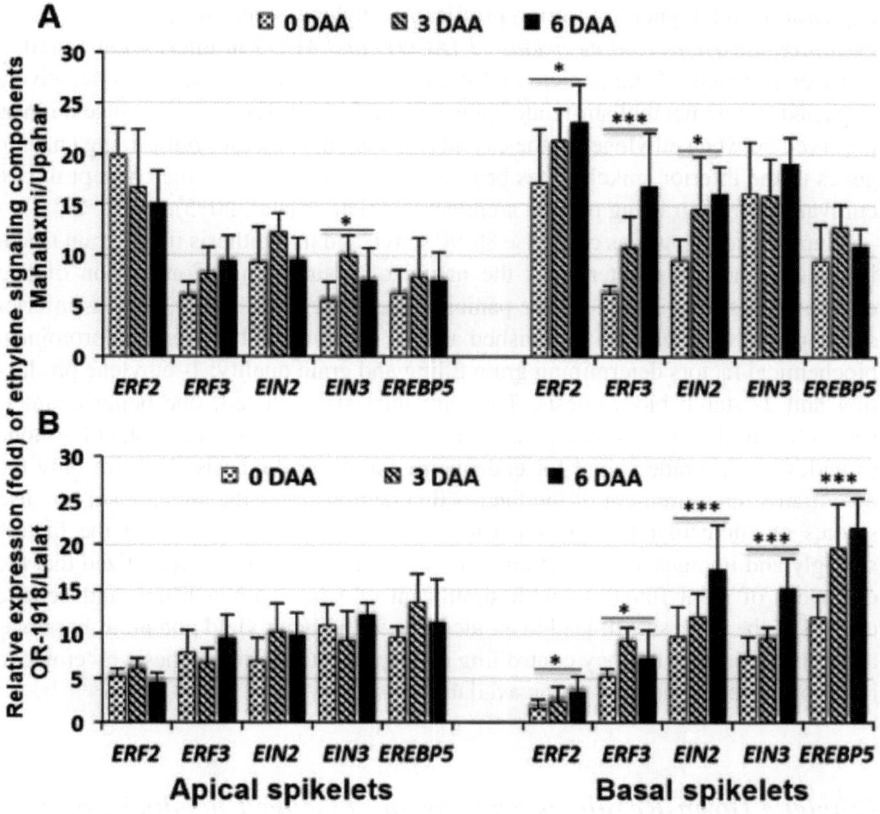

Fig. 11.36 Expression levels (relative levels in folds) of various ethylene signaling components in apical and basal spikelets of rice cultivar A: Mahalaxmi (dense panicle) over their respective counterparts in Upahar (lax panicle) and that in B: OR-1918 (dense panicle) over Lalat (lax panicle) on different days after anthesis (DAA) as determined by real-time PCR (qRT-PCR)- Sekhar et al. (2015b)

Ethylene is a Retrograde Signal for Endosperm Starch Biosynthesis

Production of ethylene in rice spikelets has been counter-productive for grain filling and starch synthesis (Table 11.2). Minimized action of ethylene has promoted endosperm cell division and sink efficiency for starch synthesis in these inferior spikelets otherwise disadvantaged for assimilate partitioning (Naik and Mohapatra 2000; Panda et al. 2009). There is a wealth of information available on high ethylene production limiting activities of endosperm starch synthesizing enzymes and resulting in partial grain filling of inferior spikelets of rice panicle (Naik and Mohapatra 2000; Yang and Zhang 2006; Zhang et al. 2009a; Zhang et al. 2009b). With help of DNA microarray and real-time PCR, Zhu et al. (2011) documented greater expression and higher transcript profile of ethylene synthesis genes *1- methyl cyclopropane-1-carboxylate oxidase 1 (ACO1) and ACO3* in inferior compared to superior spikelets of the panicle and greater expression of these genes negatively impacted starch metabolism-related genes in the former. The negative correlation as observed between ethylene synthesis and expression of starch synthesizing enzyme genes in the inferior spikelets has been corroborated by testing the concept in rice cultivars with contrasting panicle architectures (Panda et al. 2015).

Unpolished rice grains comprise 80-90 starch and its synthesis in the grain ripening period primarily determines the nature of grain filling. Comparison of two extreme types of spikelets of rice panicle concerning grain filling, such as inferior and superior spikelets, has established a close relationship between two prominent biochemical factors determining grain filling and grain quality; 1- ethylene production and 2- starch biosynthesis. They are intricately related, one being counterproductive to the other. In the other words, ethylene as an important plant hormone provides a retrograde signal for endosperm starch biosynthesis. It is not only the quantitative measurement of the factors that demonstrates the antagonism but also studies on molecular biology have refurbished the tradeoff between the factors strongly and immensely (Panigrahi et al. 2019). It has been conceptualized that the discovery of plant functions with significant effects on grain filling and genetic control of the process will lead to an ideotype with higher yield potential. In pursuance, the molecular biology controlling endosperm starch biosynthesis by ethylene holds promise for the best option available now.

Ethylene Down-Regulates Expression of Genes Encoding starch Biosynthesizing Enzymes

Starch synthesis of inferior spikelets of rice panicle is slow because of preferential movement of assimilates to superior spikelets; it contributes to poor sink efficiency of the former (Mohapatra and Sahu 1991; Mohapatra et al. 1993) and this type effect resulting in differential assimilate partitioning is subjugated to intrinsic ethylene evolution. It is discussed earlier that starch synthesis in rice endosperm cells begins with the entry of sucrose from embryonic apoplasm and concludes with

Table 11.2 Correlations of ethylene concentration (pmol h^{-1} g^{-1}fwt. mL^{-1} air) of apical and basal spikelets of rice cultivars Upahar (lax panicle type) and Mahalaxmi (dense panicle type) at anthesis with concentrations of starch (μg mg^{-1}dwt.) and soluble carbohydrates (μg mg^{-1}dwt.) of grains on day 28 after anthesis and grain filling rate (mg day^{-1}) between days 7 and 28 after anthesis (Sekhar et al. 2015b)

	Ethylene	Grain filling/rate	Starch
Grain filling rate	-0.786^{***}		
Starch	-0.882^{***}	0.953^{***}	
Soluble carbohydrates	0.889^{***}	-0.965^{***}	-0.991^{***}

***Significant at the 0.001 levels of probability (df = 22)

amylose or amylopectin formation in amyloplast for which co-ordinated actions of several enzymes is necessary. In this context, Mohapatra et al. (2009) investigated the effects of ethylene on important starch synthesizing enzymes sucrose synthase and AGPase during grain filling period of inferior and superior spikelets of three rice cultivars differing in grain size and weight. Ethylene concentration of spikelets at anthesis stage correlated inversely with cell division rate, grain filling rate, starch concentration, and activities of sucrose synthase and AGPase (Table 11.3)

Further, Panda et al. (2015) compared ethylene evolution and activity of expression of genes encoding various isoforms of starch synthesizing enzymes like sucrose synthase, AGPase, and soluble and granule bound starch synthase in superior and inferior spikelets of two groups of cultivars contrasting for grain filling/sterility percentage. In group one there were two cultivars, namely Upahar and CR-45. They had spikelets arranged in lax orientation, i.e. inter-spikelet distance is high. Spikelet fertility and grain filling percentage were also high. In contrast, group second consisted of Mahalaxmi and CR-29, both had high spikelet numbers and low fertility percentage. Inter-spikelet distance in this group was low implying dense packaging of the organs (Table 11.4). Inferior spikelets produce more ethylene (Fig. 11.37) and exhibit poor expression of genes encoding starch synthesis enzymes, suggesting an inverse relationship between the two functional processes. The genes encoding starch synthesizing enzymes always exhibited a greater expression profile and had higher transcript levels in superior spikelets compared to their inferior counterparts at the early and middle stages of grain filling (Fig. 11.38, 11.39 and 11.40).

CR-45 and CR-29 were recombinant inbred lines belonging to the medium duration group, whereas, Upahar and Mahalaxmi were of long duration group. The first cultivar of each group bears a lax panicle in contrast to the compact panicle of the second cultivar. Inter grain space = total length of primary and secondary branches of panicle/panicle grain number. Grain density = panicle length/panicle grain number. Filled grains with specific gravity ≥1.0 were considered for 1000 grain weight. Mean values ($n = 3$) followed by a common letter within a column shows no significant difference at ($P \leq 0.05$) in Duncan's multiple range test (DMRT)

Table 11.3 Correlation between ethylene concentration with cell division rate, grain filling rate, starch concentration and activities of sucrose synthase and AGPase of developing endosperm of apical and basal spikelets of three contrasting *indica* rice cultivars differing in grain weight and size [Jogesh (large), Udayagiri (medium) and Sidhanta (small)] following treatments of kinetin and cobalt nitrate at the time of anthesis compared to the control plants treated with water (Mohapatra et al. 2009)

	Ethylene	Grain filling rate	Cell division rate	AGP activity	SS activity
Grain filling rate	–0.9643***				
Cell division rate	–0.9268***	0.9628***			
AGP activity	–0.9262***	0.9137***	0.9669***		
SS activity	–0.8549***	0.8740***	0.9407***	0.9679***	
Starch	–0.8809***	0.8871***	0.9565***	0.9600***	0.9544***

***$P < 0.001$

Table 11.4 Yield-related parameters of panicle on the main shoot of compact- (Mahalaxmi and CR-29) and lax-panicle (Upahar and CR-45) cultivars of rice. Compact-panicle cultivars had more number grains/panicle compared to the lax-panicle cultivars, but most of the grains, particularly on the basal part were unfilled or partially filled (Panda et al. 2015)

Cultivars	Panicle length (cm)	Primary branch number	Secondary branch number	1000 Apical grain weight (g)	1000 Basal grain weight (g)	Length of primary and secondary branches (cm)	Grain number (Panicle^{-1})	Filled grain (%)	Inter grain space (cm)	Grain density (cm)
Mahalaxmi	25.3a	11.00a	35.33a	21.28b	10.86b	211.93a	252.67a	62.33b	0.84b	0.10b
Upahar	24.77a	10.67a	21.00b	25.63a	23.83a	165.63b	150.00b	86.00a	1.10a	0.17a
l.s.d. 5%	3.21	1.85	7.89	1.19	2.77	25.72	23.15	6.85	0.13	0.02
CR-29	29.40b	16.33a	64.3a	27.64a	21.13b	442.03a	473.33a	56.10b	0.93b	0.06b
CR-45	33.13a	15.33a	53.33b	28.45a	25.91a	379.80b	280.33b	91.07a	1.35a	0.12a
l.s.d. 5%	2.03	2.07	8.86	1.13	2.69	34.58	21.57	5.26	0.05	0.01

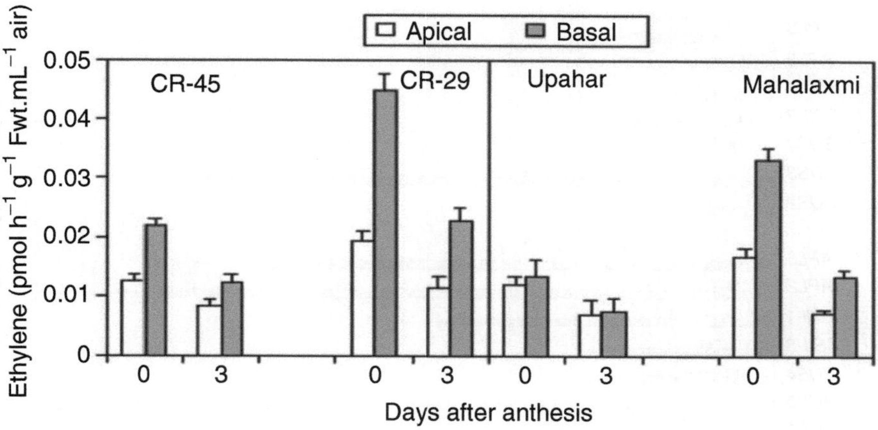

Fig. 11.37 Ethylene concentration of apical (superior) and basal (inferior) spikelets of the panicle on the main shoot of the compact- panicle rice cultivars (Mahalaxmi and CR-29) and lax- panicle rice cultivars (Upahar and CR-45) on different days after anthesis. Vertical bars represents' SD values ($n = 3$). The variance between apical and basal spikelets and between the cultivars of each group was significant at $P \geq 0.05$. (Panda et al. 2015).

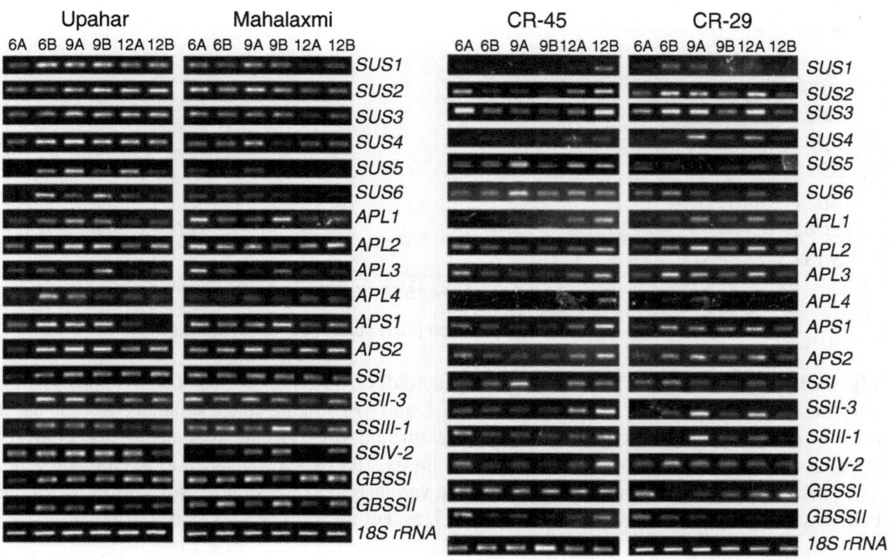

Fig. 11.38 Temporal and spatial expression of *sucrose synthase* (*SUS1*, *SUS2*, *SUS3*, *SUS4*, *SUS5*, *SUS6*), *AGPase* (*APL1*, *APL2*, *APL3*, *APL4*, *APS1*, *APS2*), and *starch synthase* (*SSI*, *SSII*, *SSIII*, *SSIV*, *GBSSI*, and *GBSSII*) enzyme genes in spikelets of rice cultivars, Upahar verses Mahalaxmi and CR-45 verses CR-29. An equal amount of the total RNA isolated from apical (A) and basal (B) spikelets of the rice cultivars on various days (6, 9, and 12) post-anthesis was reverse transcribed and the individual cDNA preparations were used to amplify genes using gene-specific primers on a thermocycler. 18S rRNA was taken as a loading control (Panda et al. 2015)

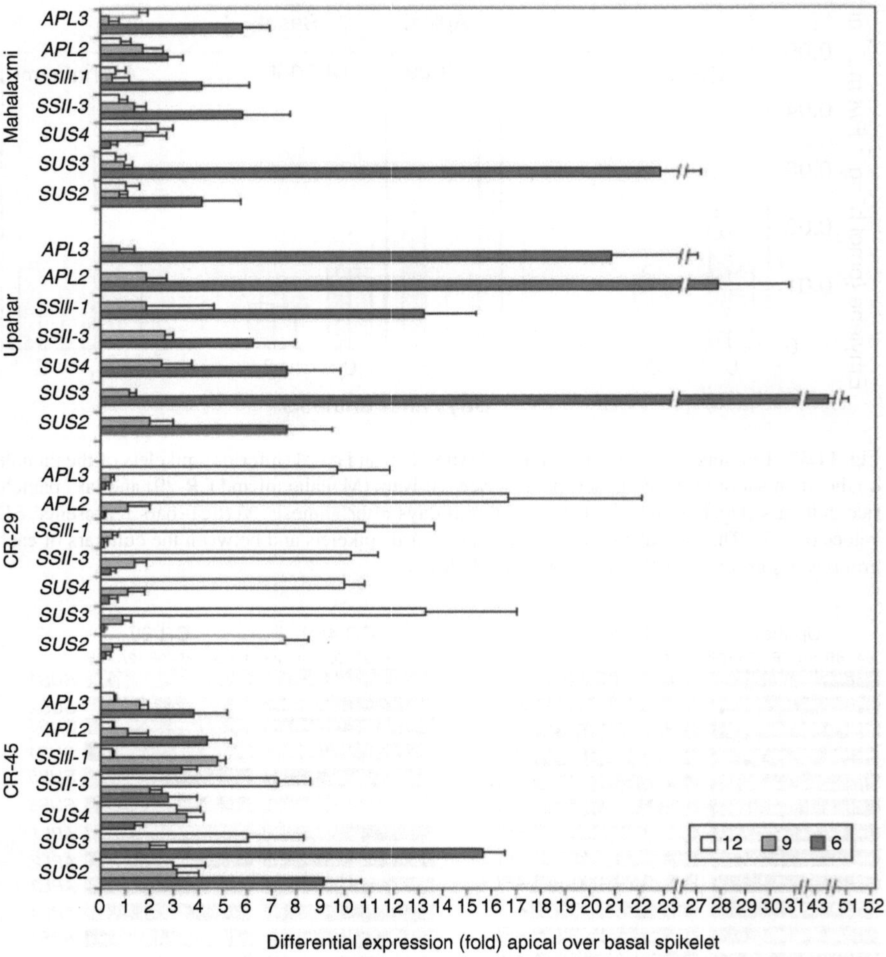

Fig. 11.39 The relative level of expression (in the fold) of various isoforms of starch synthesizing enzymes the *SUS2*, *SUS3*, *SUS4*, *SSII*, *SSIII*, *APL2*, and *APL3* in the apical spikelets over the basal ones in compact-panicle rice cultivars (Mahalaxmi and CR-29) and lax-panicle rice cultivars (Upahar and CR-45) on days 6, 9 and 12 post-anthesis. 18S rRNA was taken as a reference gene. The horizontal bars represent ± s.d. ($n = 3$). The variance in expression of isoforms between the cultivars of each group was significant at $P \leq 0.05$, except for *SSIII* (Panda et al. 2015)

Inverse Relationship of Ethylene with Starch Synthesis

The inverse relationship between ethylene and grain starch synthesis was further corroborated, when inferior spikelets disadvantaged for grain filling on panicle structure owing to spatial or genetic discrimination exhibited transcripts levels of ethylene receptors and signal transducers proteins for a longer time in the post-anthesis stage of grain development compared to the superior spikelets (Panda et al. 2016) (Fig. 11.41)

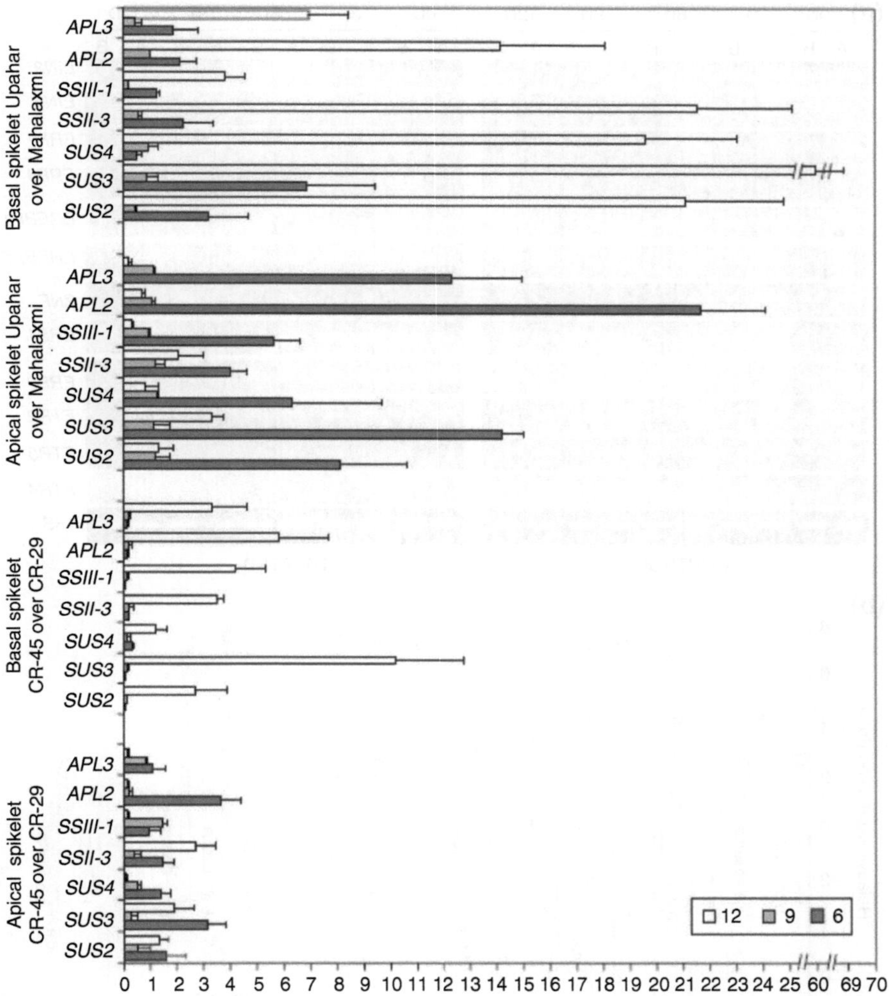

Fig. 11.40 The relative level of expression (in the fold) of various isoforms of starch synthesizing enzymes *SUS2*, *SUS3*, *SUS4*, *SSII*, *SSIII*, *APL2,* and *APL3* in the comparison between apical and basal spikelets of lax-panicle rice cultivars Upahar and CR-45 over the apical and basal spikelets of compact-panicle Mahalaxmi and CR-29 respectively, on days 6, 9 and 12 post-anthesis. 18S rRNA was taken as a reference gene. The individual horizontal bars represent mean ± SD ($n = 3$). (Panda et al. 2015)

In sharp contrast to ethylene expression, the inferior spikelets had under-expression of most of the starch synthesizing enzymes of the endosperm during grain maturation and this has been evident in cultivars with varying panicle architecture (Panda et al. 2015). This work revealed the omnipresence of ethylene perception factors in individual spikelets of panicle, but the expression pattern of the genes encoding them was different in the two types of spikelets and tissue-specific. The superior spikelets of panicle reach anthesis early and gain competence for

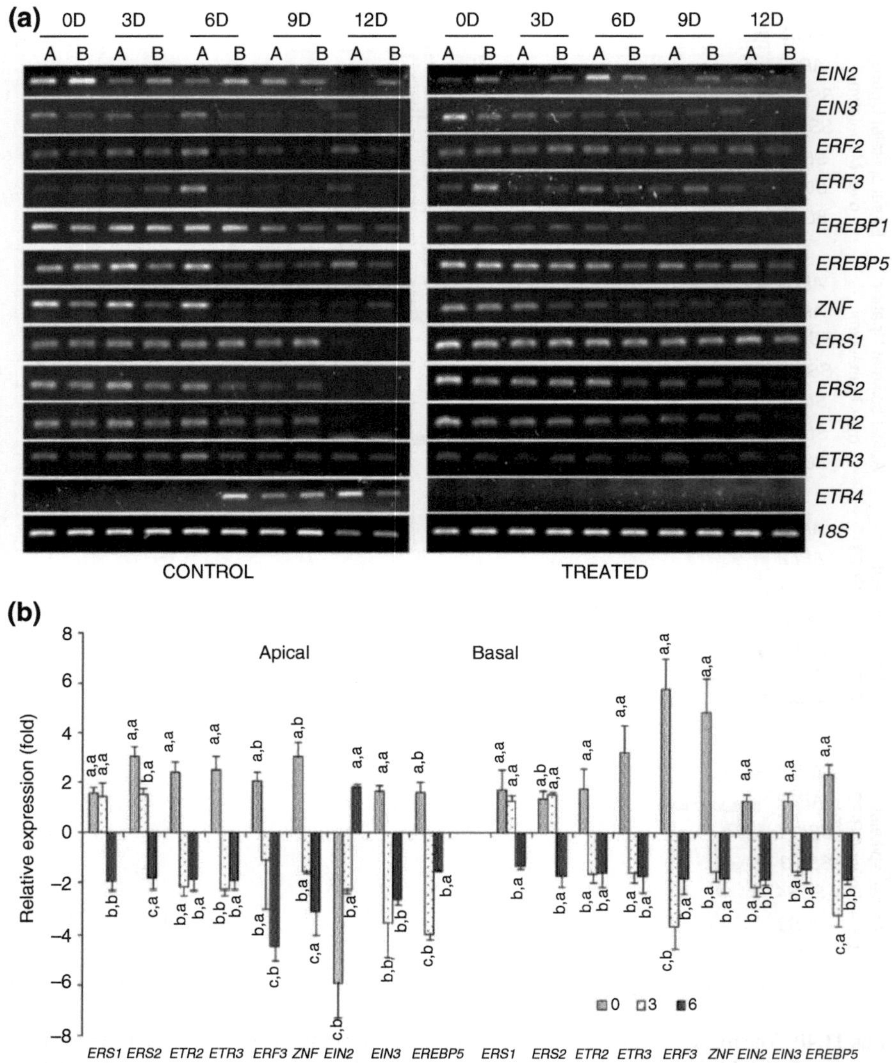

Fig. 11.41 The effect of 1-MCP treatment on expression of various ethylene receptor and signal transducer genes during kernel development of superior (apical) and inferior (basal) spikelets of rice cultivar OR-1918. a. RT-PCR products visualized in 2% agarose gel, b. changes in temporal expression (relative expression level in the fold) as determined by qRT- PCR of total RNA isolated from apical (**a**) and (**b**) spikelets of control and 1-MCP treated plants on 0,3 and 6 (D) days after anthesis. Common letters near (first) to the vertical bars among various days after anthesis of different genes and second common letters among apical and basal spikelets are not significantly different (Panda et al. 2016)

ethylene perception earlier than the inferior spikelets because of metabolic dominance and early release from the boot of flag leaf sheath. Response to adverse action of ethylene was short-lived in them compared to the long-lasting effects of a higher concentration of ethylene in the inferior spikelets. The enhanced perception of ethylene in superior spikelets implies that the hormone is necessary to complete various physiological functions including grain filling. In contrast, the persistence of ethylene response for a longer time in inferior spikelets is detrimental for starch synthesis and grain filling. Some of the high sterile dense panicle cultivars not only produce higher levels of ethylene but also exhibit greater perception of the hormone by overexpressing genes encoding ethylene signal transducers particularly in the inferior basal spikelets of the panicle. Ethylene not only down- regulates expression of genes for starch synthesis but also affects the synthesis of essential proteins for cell progression and division like importin-α, elongation factor 1-β and cell division control protein 48 (CDC 48) (Das et al. 2016) and some regulatory proteins (Sekhar et al. 2015a) that pre-empt reduction of grain sink size of inferior spikelets. The adverse effects of ethylene on starch synthesizing enzyme gene expression, grain filling, and cell number and size in inferior spikelets can be overwhelmed by exogenous application of ethylene inhibitors like 1-MCP ultimately lifting the sink size of the grain (Panda et al. 2016).

The interplay of Ethylene with Other Hormones

Recently Zhang et al. (2016) reported that poor grain occurrence of inferior spikelets results from the low intrinsic levels of zeatin+zeatin ribosides, IAA, polyamines, and ABA leading to low cell division rate of endosperm cells, fewer cell number, slower grain filling rate, and smaller grain weight. Similarly, Tamaki et al. (2015) found that chemicals that inhibit GA, IAA, and BR biosynthesis decrease grain filling whereas application of inhibitors of auxin perception and biosynthesis of jasmonic acid and ethylene improve grain yield. These reports, however, do not provide a comprehensive solution to the antagonism between the two types of growth regulators dictating the final yield of rice. It is surmised that ethylene appears to be the primary growth regulator for the process of grain development in rice with other growth regulators acting as suppressers of the ethylene effects. In a recent review, Deveswar et al (2020) have proposed a model for the interplay of ethylene with other hormones working in the control of tillering and panicle architecture (Fig. 11.42). Tamaki et al. (2015) reported that specific inhibition of ethylene or jasmonic acid biosynthesis by exogenous application of chemicals improved rice yield. In contrast, chemicals inhibiting auxin, GA, and brassinosteroid biosynthesis act otherwise (Table 11.5).

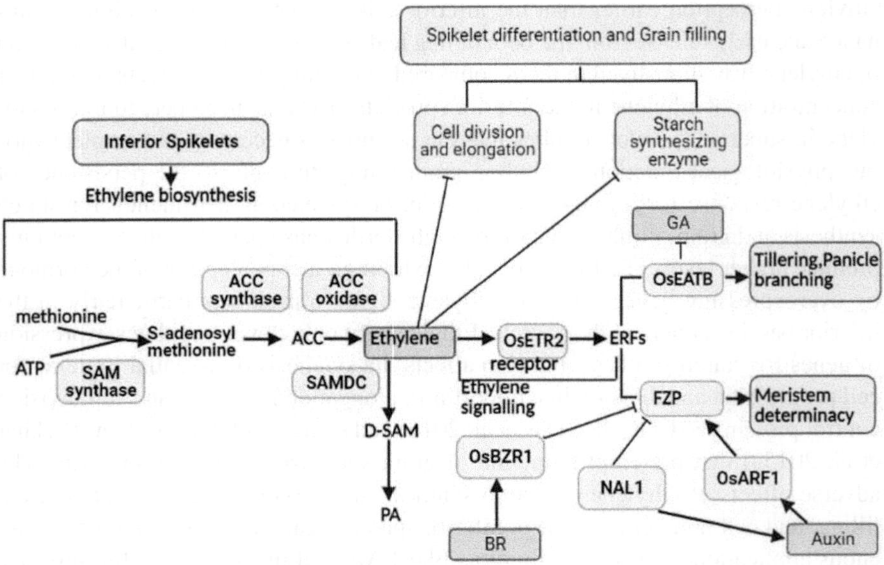

Fig. 11.42 Mode of action of ethylene in consortium with other hormones (pink hexagons) during panicle development and grain filling. Inferior spikelets accumulate ethylene at anthesis stage by promoting its biosynthesis. Polyamines (PA) compete and re- route ethylene metabolism, thus antagonizing the ethylene response.Meristem transition determination by FZP greatly affects the branching potential of panicle. FZP interacts with BR (brassinosteroids) and auxin pathway components. OsEATB co-ordinating with GA (gibberellic acid) affects tillering and panicle branching. D-SAM, decarboxylated S- adenosyl-methioine; SAMDC, SAM decarboxylase. Arrow-heads and bar-heads represent positive and negative reulations respectively. (Adopted from Deveswar et al. 2020)

Interaction of Ethylene with Other Hormones in Different Plant Organs

There are other examples like leaf abscission, underwater stem elongation, and leaf senescence; where the physiological events are regulated by a balance between ethylene as negative and other hormones as a positive regulator. For the control of leaf senescence, a balance between intrinsic ethylene and cytokinin is crucial, where ethylene promotes and cytokinin retards the destructive effects. Auxin application promotes ethylene synthesis in the abscission zone for the shedding of fruits, flowers, and leaves. In deep water rice ethylene stimulates internode elongation by increasing sensitivity and abundance of gibberellins by decreasing the level of ABA; ABA is antagonistic to gibberellins. In this case, ethylene is shown to be positively involved with gibberellins activity. The facts propose a similar balance between ethylene and the positive regulators controlling grain growth. The balance becomes crucial for grain occurrence when ethylene production in spikelets is stimulated on account of the disadvantage of spatial location on the panicle axis due to either proximal or high-density placement. A balance in favor of the negative regulator

Table 11.5 Relative expression of different genes responding to hormones upon inhibitor treatment. The primary targets of the inhibitors were as follows. NDGA (ABA synthesis), Paclobutrazol (GA synthesis), Imazalil (Brassinosteroid synthesis), L-AOPP (Auxin synthesis), AIB (Ethylene synthesis), $AgNO_3$ (Ethylene perception) (Adopted from Tamaki et al. 2015)

Chemicals	mM	ABA responsive gene *OsbZIP23*	GA responsive gene *GA20ox2*	BR responsive gene *DWF4*	Auxin responsive gene *OsIAA4*	Ethylene responsive gene *OsETR2*
NDGA	0.12	0.9 ± 0.04	**1.3 ± 0.07***	1.2 ± 0.09	0.9 ± 0.20	1.0 ± 0.08
	0.6	**0.7 ± 0.02***	**1.7 ± 0.23***	1.2 ± 0.11	0.9 ± 0.11	0.9 ± 0.12
	3	**0.4 ± 0.02***	**1.8 ± 0.32***	1.0 ± 0.23	**0.6 ± 0.03***	**0.7 ± 0.09***
Paclobutrazol	0.016	0.8 ± 0.1	1.2 ± 0.14	0.8 ± 0.10	0.8 ± 0.08	1.0 ± 0.06
	0.08	0.9 ± 0.03	1.2 ± 0.14	0.8 ± 0.10	1.1 ± 0.12	1.0 ± 0.03
	0.4	0.9 ± 0.05	**1.6 ± 0.06***	0.8 ± 0.18	0.8 ± 0.12	0.9 ± 0.07
Imazalil	0.04	0.8 ± 0.12	0.9 ± 0.07	**1.5 ± 0.10***	1.1 ± 0.24	0.9 ± 0.10
	0.2	0.8 ± 0.01	0.9 ± 0.07	**1.5 ± 0.08***	1.1 ± 0.25	1.0 ± 0.08
	1	**0.7 ± 0.15***	**2.5 ± 0.25***	1.4 ± 0.22	**2.0 ± 0.38***	1.0 ± 0.17
Propiconazole	0.04	0.8 ± 0.02	**1.6 ± 0.31 ***	0.8 ± 0.16	1.0 ± 0.07	0.9 ± 0.23
	0.2	0.8 ± 0.07	1.2 ± 0.12	0.9 ± 0.07	1.0 ± 0.06	0.9 ± 0.05
	1	**0.7 ± 0.14***	**1.8 ± 0.16***	1.0 ± 0.07	1.2 ± 0.10	1.0 ± 0.10
L-AOPP	0.04	**0.5 ± 0.02***	1.0 ± 0.07	**0.7 ± 0.07***	**0.7 ± 0.10***	1.1 ± 0.12
	0.2	**0.6 ± 0.14***	1.2 ± 0.10	**0.7 ± 0.11***	**0.7 ± 0.11***	1.2 ± 0.12
	1	**0.6 ± 0.11***	**1.8 ± 0.17***	**0.7 ± 0.12***	0.9 ± 0.12	1.3 ± 0.26
PEO-IAA	0.04	**0.7 ± 0.16***	1.1 ± 0.12	0.8 ± 0.11	0.8 ± 0.07	1.1 ± 0.18
	0.2	**0.6 ± 0.13***	1.3 ± 0.32	0.8 ± 0.13	**0.7 ± 0.08***	1.3 ± 0.16
	1	**0.6 ± 0.02***	1.4 ± 0.26	**0.7 ± 0.11***	0.9 ± 0.10	0.9 ± 0.15
AIB	0.2	0.9 ± 0.11	**1.8 ± 0.27***	1.1 ± 0.12	1.1 ± 0.25	**0.8 ± 0.03***
	1	1.0 ± 0.11	**1.5 ± 0.06***	**1.6 ± 0.15***	1.1 ± 0.25	**0.8 ± 0.09***
	5	1.0 ± 0.16	**1.6 ± 0.15***	**1.6 ± 0.32***	**1.6 ± 0.27***	1.1 ± 0.16
$AgNO_3$	0.12	**0.7 ± 0.05***	**1.6 ± 0.20***	1.0 ± 0.07	1.3 ± 0.16	**0.8 ± 0.07***
	0.6	1.0 ± 0.12	**2.1 ± 0.15***	**2.5 ± 0.50***	1.3 ± 0.04	0.9 ± 0.10
	3	1.3 ± 0.17	**4.6 ± 0.47***	**2.8 ± 0.33***	1.4 ± 0.20	**1.7 ± 0.08***

Values are Mean ± SE (n= 6). Relative gene expression value setting UTC as 1
*Indicates significance at $P \leq 0.05$

ethylene decimates grain filling and vice-versa. Several experiments conducted in recent years have corroborated the negative impact of ethylene on grain filling using biochemical intervention techniques with the application of positive and negative regulators ethylene synthesis and actions (Mohapatra et al. 2000; Panda et al. 2015; Panda et al. 2016; Sekhar et al. 2015b). Although factually illusive for a mechanism, Yang et al. (2004) first proposed that the antagonist reaction in the cross-talk between ethylene and ABA could be involved in mediating post-anthesis growth of rice caryopsis in plants subjected to moderate water stress. Under the circumstances, a high ratio of ABA to ethylene enhances grain filling significantly. The antagonistic relationship between the two types of stress hormones was confirmed further by

Table 11.6 Levels of ABA, ethylene, and ACC expressed based on dry weight and per grain in the superior and inferior spikelets during grain filling of rice (Adopted fromYang et al. 2006). Superior spikelets are those that flowered on the first two days and inferior ones are those that flowered on the last two days on a panicle. Each value represents the mean of the 15 measurements taken at 2, 4, 6, 8, 10, 12, 14, 16, 18, 20, 22, 24, 26, 28, and 30 DPA. ± values standard deviation of four replications for each measurement

Cultivar	Group of	Ethylene		ACC		ABA	
	spikelets	(pmol	(pmol g-	(pmol	(pmol g-	(pmol	(pmol g-
		grain-1h-1)	1DWh-1)	grain-1)	1DW)	grain-1)	1DW)
LY-9	Inferior	0.53±0.04	76.1±6.18	21.6±1.35	3097±242	3.97±0.28	269±18.5
	Superior	0.14±0.01	8.32±0.64	5.74±0.43	431±32.9	7.68±0.51	521±41.6
YD-4	Inferior	0.51±0.03	60.2±4.15	20.8±1.14	2435±137	5.14±0.42	365±27.3
	Superior	0.12±.01	8.53±0.76	4.64±0.32	343±25.8	10.6±0.57	735±56.8

intrinsic analyses in superior and inferior spikelets of rice panicle where high ABA concentration can overwhelm ethylene action and vice-versa (Yang et al. 2006) (Table 11.6).

Ethylene Interaction with ABA and IAA

Promotion of ABA synthesis in kernels of inferior spikelets of rice panicle by moderate soil drying at the post-anthesis stage has been reported to increase grain sink strength (Zhang et al. 2012), possibly by countering detrimental effects of ethylene. Zhu et al. (2011) extended the concept with evidence that exogenous application of ethylene releasing compound ethephon was more effective in downregulating the expression of most of the endosperm starch metabolism-related genes compared to ABA application. The interplay of antagonistic interaction between ethylene and ABA has not been explained with responses at the transcriptomic level for grain development of rice as of today. However, models have been proposed for the interaction for the regulation of cell elongation in coleoptile and mesocotyl and root growth of rice (Yin et al. 2017). For cell elongation, ethylene promotes *OsEIN2* and *OsEIL2* transcription. ABA acts upstream of ethylene to block the transcription and preclude coleoptiles elongation. For root growth inhibition ethylene action is correlated with ABA. In this case, the ethylene transduction signal pathway for growth inhibition involves activation of *OsEIN2* and *OsEIL1* transcription, which in turn promote ABA production. Heterogeneous architecture of panicle facilitates higher ethylene production in the inferior spikelets on basal primary branches because interaction is involved between IAA and ethylene transports (Sekhar et al. 2015b).

Ethylene is the Second Messenger for Auxin Signal

In correlative metabolic inhibition of inferior basal spikelets, it was surmised that IAA moving basipetal from the apical spikelets of rice panicle infringes export of IAA from the inhibited inferior spikelets. Because IAA from the dominant superior spikelet (apical bud) cannot reverse its direction and move upwards, it needs genesis of a secondary messenger for suppression of growth in the inferior basal spikelet (lateral or axillary bud), and ethylene is most suitable for the purpose. In reality, ethylene production is high in the basal spikelet, which is accountable for growth suppression. Co-ordinated antagonism in the action of IAA and ethylene has been the most plausible explanation for the correlative inhibition of inferior basal spikelets of rice panicle hitherto. IAA application to young rice panicle promotes the growth of apical spikelets making them more dominant in growth at the cost of the development of basal spikelets (Patel and Mohapatra 1992). Antagonism between ethylene and cytokinin actions on grain filling of rice kernel has also been evaluated (Mohapatra et al. 2009). Ethylene retards cell division at the early stage of grain development and limits sink size for assimilate partitioning and grain growth in inferior spikelets of rice panicle, but cytokinin application increases grain filling by improving sink strength (Table 11.7). This action of cytokinin is similar to that of ethylene synthesis inhibitor $Co(NO_3)_2$, which implies a possible antagonism between the two hormones.

Ethylene Interaction with Polyamines During Grain Filling

S-adenosyl-1-methionine (SAM) is the biosynthetic precursor for ethylene. Polyamines like spermidine and spermine also share SAM for biosynthesis (Ravanel et al. 1998). Spermidine and spermine are also derived from another polyamine putrescine. Because of the intricate relation, fluctuation in the level of either ethylene or polyamine chemical during biosynthesis in response to the environmental cue is bound to elicit a corresponding effect on the intrinsic level of the other. Both of the growth regulators have a significant influence on physiological processes controlling cell division, senescence, embryogenesis, fruit set, and growth and morphogenesis. In rice, studies undertaken hitherto have explored the influence of ethylene on seed germination, stress tolerance, and grain filling, but seldom efforts have been made to explore the interaction between them. For example, high ethylene production slackens post-anthesis development of caryopsis in the inferior spikelets of panicle disadvantaged in spatial distribution. However, synchronous changes in the polyamines levels corresponding to the fluctuation of ethylene levels and their interaction regulating grain filling have been less investigated.

Table 11.7 Effects of $Co(NO_3)_2$ (ethylene inhibitor) and kinetin on yield parameters of the main panicle of three *indica* rice cultivars differing in grain size and weight; Jogesh (large), Udayagiri (medium), Sidhanta (small) during the dry season of 2007 (Mohapatra et al. 2009)

Cultivar type and treatment	Panicle dry wt. (g)	Dry wt. of spikelets (g)	Total no. of spikelets	Average dry wt. of spikelet (g)	% of filled spikelets	% Unfilled spikelets	% of high density grains
Jogesh							
Dist. water	4.53 ± 0.27	4.32 ± 0.17	180.32 ±0 7.5	0.0240 ± 0.0012	70.12 ±5.82	29.80 ± 3.10	30.28 ± 3.20
Kin.	5.28 ± 0.12	5.05 ± 0.09	188.66 ± 10.5	0.0268 ± 0.0009	88.14 ± 4.15	11.83 ± 2.20	42.14 ± 4.80
Cob. nit.	5.36 ± 0.21	5.13 ± 0.11	190.00 ± 09.0	0.0270 ± 0.0011	91.40 ± 2.38	08.60 ± 1.29	44.20 ± 2.18
Udayagiri							
Dist. water	3.30 ± 0.36	3.15 ± 0.07	176.33 ± 11.5	0.0179 ± 0.0014	78.43 ± 4.89	21.57 ± 2.80	22.04 ± 2.88
Kin.	4.05 ± 0.23	3.89 ± 0.14	182.00 ± 07.0	0.0214 ± 0.0009	90.58 ± 5.01	09.42 ± 2.13	38.56 ± 1.20
Cob. nit.	4.19 ± 0.39	4.02 ± 0.08	184.00 ± 11.0	0.0218 ± 0.0013	91.82 ± 5.40	08.18 ± 2.59	41.06 ± 2.06
Sidhanta							
Dist. water	2.29 ± 0.18	2.18 ± 0.10	158.66 ± 08.0	0.0138 ± 0.0011	68.24 ± 3.40	31.75 ± 2.83	16.80 ± 1.20
Kin.	2.74 ± 0.14	2.61 ± 0.04	167.00 ± 07.0	0.0156 ± 0.0010	84.80 ± 4.20	15.20 ± 2.62	35.40 ± 2.60
Cob. nit.	2.81 ± 0.26	2.69 ± 0.11	169.00 ± 09.0	0.0159 ± 0.0008	89.26 ± 3.80	10.73 ± 3.47	38.26 ± 1.15
F-value/L.S.D							
Between treatments	$41.88^{**}/0.4075^{*}$	$40.97^{**}/0.4025^{*}$	$96.95^{***}/6.05^{*}$	$31.70^{**}/0.00184^{*}$	$55.75^{**}/8.68^{*}$	$56.19^{**}/8.63^{*}$	$64.76^{***}/14.87^{*}$
Between cultivars	$379.09^{***}/0.7628^{*}$	$356.80^{***}/0.7533^{*}$	$500.43^{***}/6.05^{*}$	$340.81^{***}/0.00344^{*}$	5.42^{NS}/_	5.46^{NS}/_	$12.81^{*}/4.80^{*}$

The ± values indicate standard deviation of the mean ($n = 3$). Level of significance for 2-way ANOVA test, $^{NS}P > 0.05$, $^{*}P < 0.05$, $^{**}P < 0.01$, $^{***}P < 0.001$

Chemical Structure of Polyamines

In sharp difference to ethylene action, free polyamines like spermidine and spermine have been found beneficial for grain filling of rice and wheat (Yang et al. 2008; Yang et al. 2014). Zhang et al. (2016) reported that the endogenous concentration of free putrescine, spermidine, and spermine increased briskly during the early part of grain filling and declined to a minimum level later; and the temporal pattern of fluctuation matched with that of the positive growth regulators like ABA, IAA, and cytokinins exhibiting significant correlation with endosperm filling and cell division rates. Similar to these hormones, the concentration of the polyamines was found lower in the inferior spikelets compared to the superior spikelets. Subjecting field-grown plants to moderate level drought or soil drying during panicle differentiation in rice, increases differentiated spikelet number and reduce spikelet degeneration and sterility, and these morphological responses are closely associated with increase and decrease in biosynthesis of polyamines and ethylene respectively (Zhang et al. 2017) (Fig. 11.43).

Chemical Structure of Polyamine

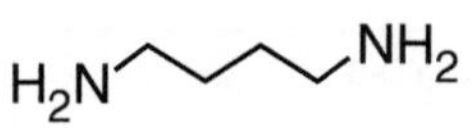

Putrescine

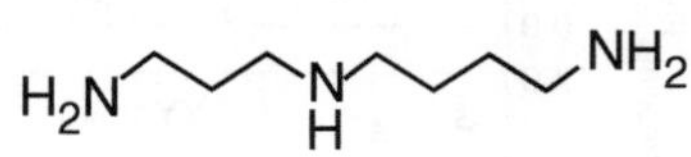

Spermidine

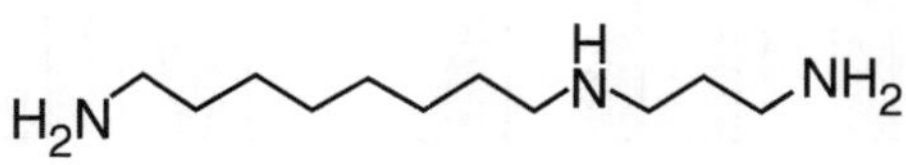

Spermine

These empirical shreds of evidence give credence to the antagonistic interaction between ethylene and polyamines during the grain development of rice. Ethylene biosynthesis is promoted and polyamine synthesis reduced when a spikelet is disadvantaged for space on the panicle axis or the plant is placed in an inappropriate environment and vice versa. Thus, utilization of SAM in polyamine and ethylene biosynthesis through the expression of genes encoding ACC synthase and polyamine synthase respectively is not constitutive and amenable for induction.

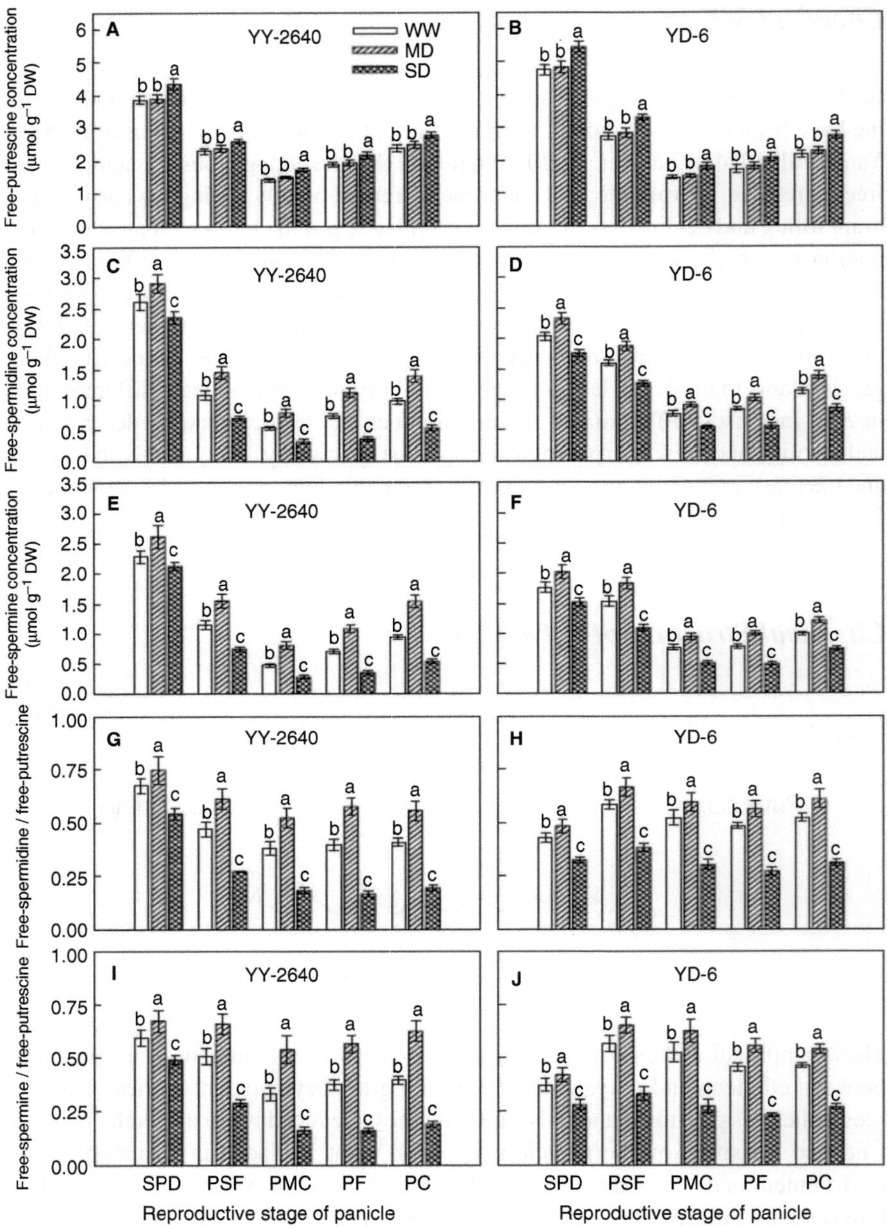

Fig. 11.43 Changes in free putrescine, spermidine, and spermine and ratios spermidine/ putrescine and spermine/putrescine in young panicles of rice cultivars YY -2640 and YD -6 under WW (well- watered), MD (moderate drought), and SD (severe soil drought) treatments. SPD, PSF, PMC, PF, and PC represent spikelet primordium differentiation, pistil and stamen formation, pollen mother cell meiosis, pollen filling, and pollen completion stages respectively. Vertical bars represent ±SEM (n=3) where these exceed the size of the symbol. Different letters above bars indicate a significant difference at P=0.05 within the same measurement stage. (Adopted from Zhang et al. 2017)

Polyamine Effects on Grain Filling

It was reported that treatment of moderate soil drying from panicle initiation to pollen completion stage of young rice panicle increases spikelet differentiation, spikelet number, grain number, and grain yield by decreasing the number of degenerated/sterile spikelets by increasing contents of polyamines like free spermine, spermidine, and putrescine against intrinsic levels of ethylene and ACC (Zhang et al. 2017). Because S-adenosyl-L- methionine (SAM) is a common precursor for both polyamines (spermine, spermidine) and ethylene, the increase of one becomes a causative factor for depression of the other. The synchronous changes in the chemicals and their interactive effects have been analyzed for possible encouragement of polyamine synthesis at the cost of ethylene during the reproductive stage of the rice plant to harness better grain yield (Zhang et al. 2017).

References

Abeles FB, Morgan PW, Saltveit Jr ME (1992) Ethylene in plant biology. Academic Press. San Diego. ISBN:9780120414512

Burg SP, Thimann KV (1959) The physiology of ethylene formation in apples. Proc Nat Acad Sci U S A 45: 335–344

Das K, Panda BB, Sekhar S, Kariali E, Mohapatra PK, Shaw BP (2016) Comparative proteomics of the superior and inferior spikelets at the early grain filling stage in rice cultivars contrast for panicle compactness and ethylene evolution. J Plant Physiol 202: 65–74

Deveswar P, Prusty A, Sharma S, Tyagi A (2020) Phytohormone –mediated molecular mechanism involving multiple genes and QTL govern grain number in rice. Frontiers in Genetics11:586462. https://doi.org/10.3389/fgene.2020.586462

Liu K, Ye Y, Tang C, Wang Z, Yang J (2008) Responses of ethylene and ACC in rice grains to soil moisture and their relations to grain filling. Front Agric China 2: 172–180

Ma B, Chen S, Zhang J (2010) Ethylene signaling in rice. Chinese Sci Bulletin 55: 2204–2210

Ma B, He S-J, Duan K-X, Yin CC, Chen H, Yang C, Xiong Q, Song QX, Lu X, Chen HW, Zhang WK, Lu TG, Chen SY, Zhang JS (2013) Identification of rice ethylene-response mutants and characterization of MHZ7/OsEIN2 in distinct ethylene response and yield trait regulation. Mol Plant 6: 1830–1848

Mohapatra PK, Sahu SK (1991) Heterogeneity of primary branch development and spikelet survival in rice in relation to assimilate contents of primary branches. J Exp Botany 42: 871–879

Mohapatra PK, Patel R, Sahu SK (1993) Time of flowering affects grain quality and spikelet partitioning within rice panicle. Aus J Plant Physiol 20: 231–241

Mohapatra PK, Naik PK, Patel R (2000) Ethylene inhibitors improve dry matter partitioning and the development of late-flowering spikelets on rice panicle. Aus J Plant Physiol 27: 311–323

Mohapatra PK, Mohapatra R (2005) Sink source relation and yield potential of rice- effect of ethylene on grain filling of late-flowering spikelets. “Rice is life: scientific perspectives for the 21st century”. Editors- K. Toriyama, K.L.Heong, and B. Hardy.CD published on the occasion of World Rice Research Conference, 4-7 November 2004, Tokyo/Tsukuba, Japan. International Rice Research Institute, Philippines, pp. 120–124. ISBN-971-22-0204-6

Mohapatra R, Mohapatra PK (2006) Ethylene control of seed coat development in low and high sterile semi-dwarf indica rice cultivars. Plant Growth Reg 50: 47–55

Mohapatra PK, Naik PK, Patel R (2000) Ethylene inhibitors improve dry matter partitioning and the development of late-flowering spikelets in rice panicle. Aus J Plant Physiol 27: 311–323

Mohapatra PK, Sarkar RK, Kuanar SR (2009) Starch synthesizing enzymes and sink strength of grains of contrasting rice cultivars. Plant Sci 176: 256–263

Mohapatra PK, Panigrahi R (2011) Ethylene control of grain development in inferior spikelet of rice panicle. In Advances in Plant Physiology, vol. 12. Editor- Hemantaranjan A, Scientific Publisher (India), Jaipur, pp. 79–89

Naik PK, Mohapatra PK (1999) Ethylene inhibitors promote male gametophyte survival in rice. Plant Growth Reg 28: 29–39

Naik PK, Mohapatra PK (2000) Ethylene inhibitors enhanced sucrose synthase activity and promoted grain filling of basal rice kernels. Aus J Plant Physiol 27: 997–1008

Panda BB, Kariali E, Panigrahi R, Mohapatra PK (2009) High ethylene production slackens seed filling in compact-panicled rice cultivars. Plant Growth Reg 58: 141–151

Panda BB, Badoghar A, Das K, Panigrahi R, Kariali E, Das SR, Dash SK, Shaw BP, Mohapatra PK (2015) Compact panicle architecture is detrimental for growth as well as sucrose synthase activity of developing rice kernels. Funct Plant Biol 42: 875–887

Panda BB, Badoghar AK, Sekhar S, Shaw BP, Mohapatra PK (2016) 1-MCP treatment enhanced expression of genes controlling endosperm cell division and starch biosynthesis for improvement of grain filling in a dense-panicle rice cultivar. Plant Sci 246: 11–25

Panigrahi R, Kariali E, Panda BB, Lafarge T, Mohapatra PK (2019)Controlling the trade-off between spikelet number and grain filling: the hierarchy of starch synthesis in spikelets of rice panicle in relation to hormone dynamics. Funct Plant Biol 46: 507–523

Pareek A, Singh A, Kumar M, Kushwaha HR, Lynn AM, Singla-Pareek SL (2006) Whole- genome analysis of Oryza sativa reveals the similar architecture of two-component signaling machinery with Arabidopsis. Plant Physiol 142: 380–397

Patel R, Mohapatra PK (1992) Regulation of spikelet development in rice by hormones. J Exp Botany 43: 257–262

Peng S, Cassman, KG, Virmani SS, Sheehy, J, Khush GS (1999) Yield potential trends of tropical rice since the release of IR8 and the challenge of increasing rice yield potential. Crop Sci 39: 1552–1559

Ravanel S, Gakière B, Job D, Douce R (1998) The specific features of methionine biosynthesis and metabolism in plants. Proc Nat Acad Sci U S A 95: 7805–7812

Rzewuski G, Sauter M (2008) Ethylene biosynthesis and signaling in rice. Plant Sci 175: 32–42

Sekhar S, Panda BB, Mohapatra T, Das K, Shaw BP, Kariai E, Mohapatra PK (2015b) Spikelet-specific variation in ethylene production and constitutive expression of ethylene receptors and signal transducers during grain filling of compact-and lax-panicle rice (Oryza sativa) cultivars. J Plant Physiol 179: 21–34

Sekhar S, Gharat SA, Panda BB, Mohaptra T, Das K, Kariali E, Mohapatra PK, Shaw BP (2015a) Identification and characterization of differentially expressed genes in inferior and superior spikelets of rice cultivars with contrasting panicle-compactness and grain-filling properties. PLoS ONE 10: e0145749

Tamaki H, Reguera M, Abdel-Tawab YM, Takebayashi Y, Kasahara H, Blumwald E (2015) Targeting hormone-related pathways to improve grain yield in rice: A chemical approach. PLOS ONE 10(6): e0131213. https://doi.org/10.1371/journal.pone.0131213

Weber H, Borisjuk L, Wobus U (1997) Sugar import and metabolism during seed development. Trends Plant Sci 2: 169–174

Weschke W, Panitz R, Gubatz S, Wang Q, Radchuk R, Wever H, Wobus U (2003) The role of invertases and hexose transporters in controlling sugar ratios in maternal and filial tissues of barley caryopses during early development. Plant J 33: 395–411

Wuriyanghan H, Zhang B, Cao W, Ma B, Lei G, Liu Y, Wei W, Wu HJ, Chen LJ, Chen HW, Cao YR, He SJ, Zhang WK, Wang XJ, Chen SY, Zhang JS (2009) The ethylene receptorETR2 delays floral transition and affects starch accumulation in rice. Plant Cell 21: 1473–1494

Yang J, Zhang J, Ye Y, Wang Z, Zhu Q, Liu L (2004) Involvement of abscisic acid and ethylene in the responses of rice grains to water stress during filling. Plant Cell Environ 27: 1055–1064
Yang J, Zhang JH (2006) Grain filling of cereals under soil drying. New Phytol 169: 223–236
Yang J, Zhang J, Wang Z, Liu K, Wang P (2006) Post-anthesis development of inferior and superior spikelets in rice in relation to abscisic acid and ethylene. J Exp Botany 57: 149–60
Yang J, Cao Y, Zhang H, Liu L, Zhang J (2008) Involvement of polyamines in the post-anthesis development of inferior and superior spikelets in rice. Planta 228: 137–149
Yang W, Yin Y, Li Y, Cai T, Ni YL, Peng DL, Wang ZL (2014) Interactions between polyamines and ethylene during grain filling in wheat grown under water deficit conditions. Plant Growth Reg 72: 189–201
Yang C, Lu X, Ma B, Chen S-Y, Zhang J-S (2015) Ethylene signaling in rice and Arabidopsis: conserved and diverged aspects. Mol Plant 8: 495–505
Zhang H, Tan G, Wang Z, Yang J, Zhang J (2009a) Ethylene and ACC levels in developing grains are related to the poor appearance and milling quality of rice. Plant Growth Reg 58: 85–96
Zhang ZC, Xue YG, Wang ZQ, Yang JC, Zhang JH (2009b) The relationship of grain filling with abscisic acid and ethylene under non-flooded mulching conditions. J Agric Sci 147: 423–436
Zhang H, Li H, Yuan L, Wang Z, Yang JC, Zhang J (2012) Post-anthesis alternate wetting and moderate soil drying enhance activities of key enzymes in sucrose-to-starch conversion in inferior spikelets of rice. J Exp Botany 63: 215–227
Zhang J, Zhu L, Yu S, Jin Q (2014a) Involvement of 1-methyl cyclopropane in plant growth, ethylene production, and synthase activity of inferior spikelets in hybrid rice differing in panicle architectures. J Plant Growth Reg 33: 551–561
Zhang Z, Zhao H, Tang J, Chen D, Lin W (2014b) A proteomic study on the molecular mechanism of poor grain-filling of rice (Oryza sativa L.) inferior spikelets. PLoS ONE 9: e89140
Zhang W, Cao Z, Zhou Q, Chen J, Xu G, Gu J, Liu L, Wang Z, Yang J, Zhang H (2016) Grain filling characteristics and their relations with endogenous hormones in large-and small- grain mutants of rice. PLoS ONE 11: e0165321
Zhang W, Chen Y, Wang Z, Yang J (2017) Polyamines and ethylene in rice young panicles in response to soil drought during panicle differentiation. Plant Growth Reg 82: 491–503
Zhao Q, Guo H-W (2011) Paradigms and paradoxes in the ethylene signaling pathway and interaction network. Mol Plant 4: 626–34
Zhu G, Ye N, Yang J, Pend X, Zhang J (2011) Regulation of expression of starch synthesis genes by ethylene and ABA in relation to the development of rice inferior and superior spikelets. J Exp Botany 62: 3907–3916

Brassinosteroids

Abstract Brassinosteroids are a class of polyhydroxy steroids and were recognized as the sixth class of plant hormones involved in many developmental processes like cell elongation, cell expansion in association with auxins, senescence, pollen tube formation and elongation, vascular tissue differentiation, and protection of plants subjected to environmental stresses. In regulating plant cell elongation, the action of brassinosteroids is closely synonymous with GA. Natural brassinosteroids identified from plant species include several biosynthetic congeners of brassinolide, such as cathasterone, teasterone, 3-dehydroteasterone, typhasterol, and castasterone as well as another series of 6-deoxoteasterone, 3-dehydro-6-deoxoteasterone, 6-deoxotyphasterol, and 6-deoxocastasterone. These compounds are derived from campesterol as the plant sterol progenitor. The BR-signaling pathway in rice was discovered with the identification of BR-deficient dwarf mutant d61, where a loss of function mutation of brassinosteroid insensitive1 homolog prevented internode elongation and bending of leaf lamina joint. There are several instances, where brassinosteroids have been reported to influence morphological features of rice, such as tiller number, leaf size, and leaf angle, plant height, grain size, flowering, senescence, and environmental adaptations. The hormone mobilizes a greater flow of assimilates to the reproductive parts for an increase of both grain number and size.

Keywords Brassinosteroids · Campesterol · Brassinolide · Cell elongation · Rice grain yield · Campesterol · Brassinolide · Cell elongation · Rice grain yield

Properties of Brassinosteroids

The role of steroidal hormones in the bio-regulation of living organisms was previously confined to animals and fungi. However, a new dimension was added with the discovery of brassinosteroids in plants. Plants use to produce a large variety of steroids and very often a function of some of them were postulated as hormones. However, it was not until 1979 a new steroidal lactone brassinolide was identified as having hormonal action. This brassinolide was the first brassinosteroid (BR) isolated from pollen extracts of oilseed rape *Brassica napus* (Grove et al. 1979) and shown to have biological activity in promoting stem elongation and cell division. In regulating plant cell elongation, the action of brassinosteroids is closely synonymous with GA. Since the discovery of brassinolide, structurally and functionally related compounds numbering more than 70 have been identified and they are collectively named brassinosteroids. Brassinosteroids are a class of polyhydroxy steroids and were recognized as the sixth class of plant hormones involved in all developmental aspects of life. Because brassinosteroids are widespread in the plant kingdom and exhibit various kinds of regulatory activities in growth and development, they are now considered promising compounds for application in agriculture. They can increase not only yield but also the quality of the produce (Prusakova et al. 1999). In recent times discovery of several biosynthetic and signaling mutants have confirmed BRs as plant hormone and identified its mechanism of signal transduction. The hormone is involved in many plant processes like cell elongation, cell expansion in association with auxins, senescence, pollen tube formation and elongation, vascular tissue differentiation, and protection of some plants subjected to environmental stresses (Clouse and Sasse 1998).

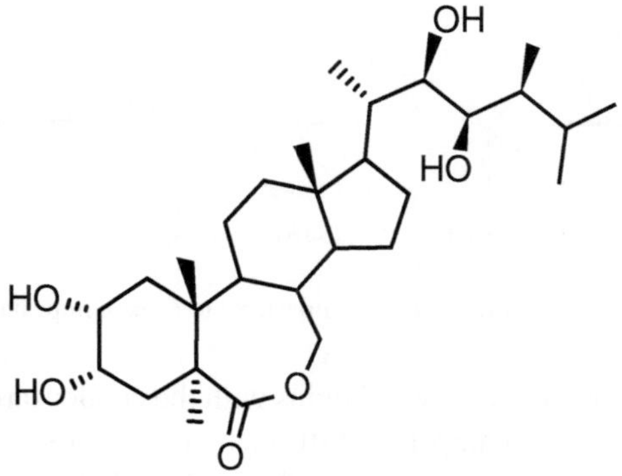

Brassinolide

Biosynthesis of Brassinosteroids

The site of BR synthesis is not specific to any plant organ. It is understood that all plant tissues produce BRs. The genes regulating biosynthesis and signal transduction are expressed in many different plant organs suggesting omnipresence of the substance in all tissues of the plant system. The BRs are not transported long distances but are utilized near the site of synthesizing cells. Plants respond to very low doses of brassinosteroid; the response is mediated by either direct action on the genome or through an epigenetic route. A comprehensive account of biosynthetic enzymes and signaling proteins that perceive and transduce the signal could be established by coupling molecular genetics with the biosynthetic pathway in mutants of Arabidopsis and pea embracing defects in BR synthesis or sensitivity. The BR-deficient and BR-insensitive mutants were all dwarfs and needed exogenous BR application for a gain of cell elongation and restoration of normal growth (Yokota 1997).

Brassinolide has been the most active brassinosteroid, showing high biological activity in plants. Its biosynthesis pathways have been elucidated by comparison of BR- biosynthetic and -insensitive mutants with wild type in plants Arabidopsis (*dwf*4; *cpd*, *cbb*3, *dwf*3; *bri*1, *cbb*2, *dwf*2), pea (*lka*), and tomato (*dpy*, *cu*-3) (Clouse and Sasse 1998; Noguchi et al. 2000). Natural brassinosteroids identified from plant species include several biosynthetic congeners of brassinolide, such as cathasterone, teasterone, 3-dehydroteasterone, typhasterol, and castasterone as well as another series of 6-deoxoteasterone, 3-dehydro-6- deoxoteasterone, 6-deoxotyphasteroI, and 6-deoxocastasterone (Fujioka and Sakurai 1997). Co-occurrence of these compounds with brassinolide strongly suggests their derivation from campesterol as the plant sterol progenitor. Besides similarities in biological activities and molecular structure of the intermediary compounds corroborates the origin of BR from campesterol. The overall pathways are tri-terpenoid pathways, where squalene 2,3-oxide generated from mevalonic acid is cyclized to cycloartenol and then to campesterol in multiple steps.

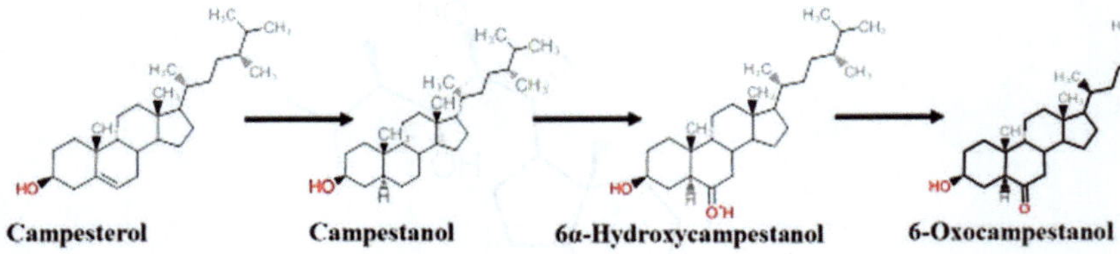

Fig. 11.44 The biosynthetic pathway from campesterol to 6-oxocampestanol

In the plant cells, BR biosynthesis begins with the condensation of mevalonate in the isoprenoid pathway to campesterol through compounds like isopentenyl pyrophosphate, geranyl pyrophosphate, farnesyl pyrophosphate, squalene, cycloartenol, and 24 methylene cholesterol (Clouse and Sasse 1998). The reduction of campesterol to campestanol is committed to the synthesis of brassinosteroids. When labeled campesterol is fed to cultured cells of *Catharanthus roseus* (Suzuki et al. 1995), campestanol, 6 hydroxycamapestanol, and 6 oxocampestanol appear as metabolites in a sequence (Fujioka and Sakurai 1997) (Fig. 11.44). There were two types of parallel C6-oxidation pathways, early and late, leading to biosynthesis of brassinolide from campesterol (Noguchi et al. 2000).

Early Oxidation Pathway

In the early C6-oxidation pathway, campesterol after reduction to campestanol is oxidized further to 6-oxocampestanol and thereafter it is converted to cathasterone. In Arabidopsis, the *DWF4* gene encodes an enzyme for C22 hydroxylation of 6-oxocampestanol to cathasterone and the reaction is considered a potential rate-limiting step for BR- biosynthesis. In a chain of steps, cathasterone is converted to brassinolide through intermediary compounds teasterone, 3-dihydrosteasterone, typhasterol, and castasterone in a sequence. Castasterone is finally converted to brassinolide. The conversion through the intermediaries has been demonstrated by feeding labeled intermediary compounds (Fujioka and Sakurai 1997) (Fig. 11.45).

Late Oxidation Pathway

With several examples shown in rice, tomato, zinnia, pea, and tobacco, it was concluded that early and late C-oxidation pathways are common in the plant kingdom (Fujioka and Yokota 2003). The late C6-oxidation pathway co-exists with the early-oxidation pathway. Akira and Shozo (1997) discussed the possible blocked steps of biosynthesis in brassinosteroid deficient mutants of Arabidopsis and pea and proposed the existence of the two pathways in a wide variety of plants. In the late C6-oxidation pathway, campestanol instead of being converted into 6-oxocampastenol goes through a series of intermediaries like deoxocathasterone, deoxoteasterone, 3-dehydro-6-deoxo-teasterone, 6-deoxotyphasterol, and

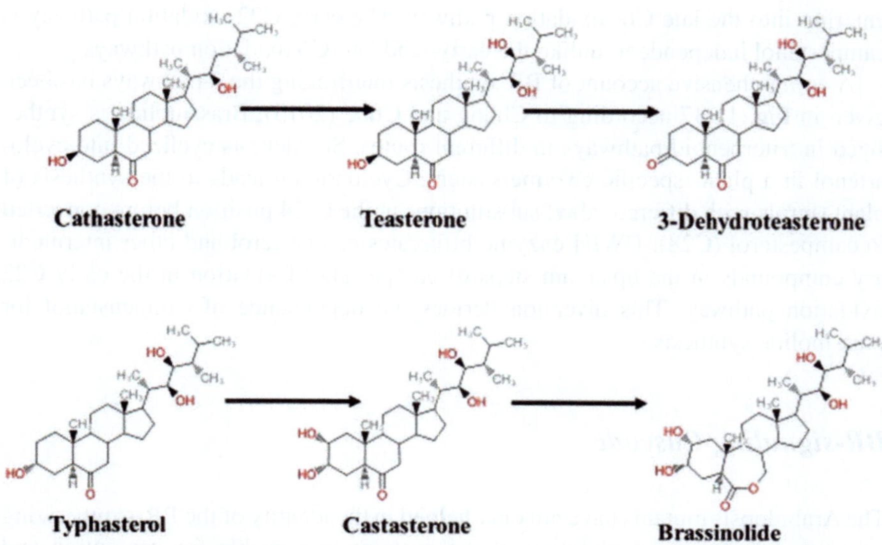

Fig. 11.45 Early C6-oxidation pathway

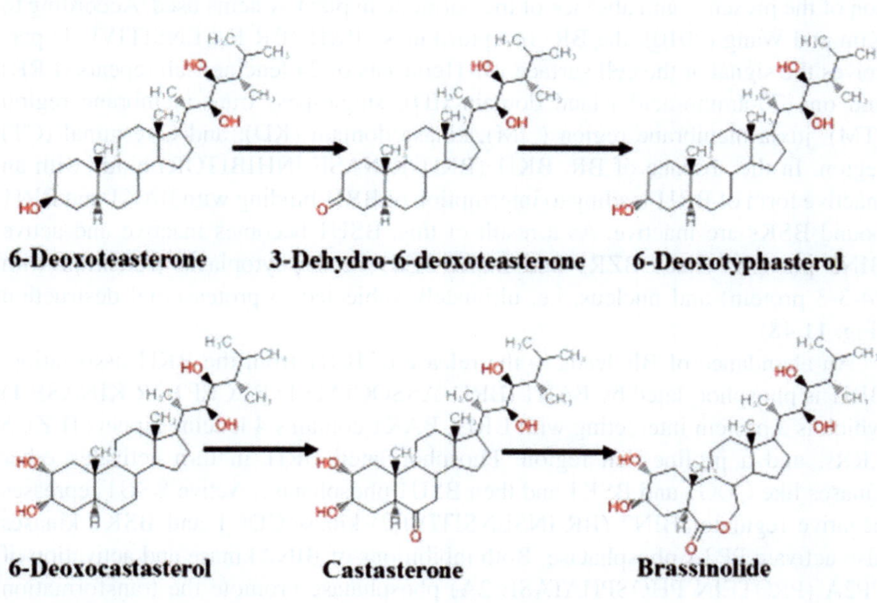

Fig. 11.46 Biosynthesis of brassinolide in late C6-oxidation pathway

6-deoxocastasterone. 6-deoxocastasterone is converted to castasterone that is the precursor of brassinolide, similar to the early C6-oxidation pathway (Fig. 11.46).

Step-wise metabolic experiments using campesterol as the primary precursor, a third BR-biosynthesis pathway has been demonstrated (Fujioka and Yokota 2003). In this case (early C22 oxidation pathway), campesterol is hydroxylated by DWF4 enzyme to 22- hydroxy-campesterol, which forms several intermediaries before entering into the late C6- oxidation pathway. The early C22 oxidation pathway is campestanol independent, unlike the early- and late-C6 oxidation pathways.

A comprehensive account of BR synthesis interlinking the 3 pathways has been given in Fig. 11.47 according to Chung and Choe (2013). Brassinolide is synthesized in triterpenoid pathways in different routes. Squalene is cyclized into cycloartenol in a plant- specific enzyme system. Cycloartenol leads to the synthesis of plant sterols with different alkyl substitutions at the C-24 position before converted to campesterol (C28). DWF4 enzyme bifurcates campesterol and other intermediary compounds in the upstream steps of campestanol formation in the early C22 oxidation pathway. This diversion derides the dependence of campenstanol for brassinolide synthesis.

BR-signaling Cascade

The Arabidopsis mutants have not only helped in the identity of the BR-synthesizing enzymes but also an exploration of the proteins responsible for perception and transduction of the signal. The BR-signaling cascade has been explored by comparison of the presence and absence of the hormone in plant systems used. According to Kim and Wang (2010), the BR- receptor kinase BRI1 (BR INSENSITIVE 1) perceives the signal at the cell surface. BRI1consists of 24 leucine-rich repeats (LRR) and one 70-amino-acid island domain (ID), single-pass trans-membrane region (TM), juxta-membrane region (JM), kinase domain (KD), and C-terminal (CT) region. In the absence of BR, BKI1 (BRI1 KINASE INHIBITOR) binds with an inactive form of BRI1 leading to interruption of BRI1 binding with BAK1 and BRI1 bound BSKs are inactive. As a result of this, BSU1 becomes inactive and active BIN2 phosphorylates BZR1 and BZR2/BES1 in the cytoplasm (retention with 14-3-3 protein) and nucleus, i.e. ultimately subjected to proteasomal destruction (Fig. 11.48).

An abundance of BR leads to the release of BRI1 from the BKI1 association. BRI1 is phosphorylated by BAK1 (BRI1-ASSOCIATED RECEPTOR KINASE 1) which is a protein interacting with BRI1. BAK1 contains 4 leucine zippers (LZ), 5 LRRs, and a proline-rich region. Phosphorylated BRI1 in turn activates other kinases like CDG1 and BSK1 and then BSU1 phosphatase. Active BSU1 represses negative regulator BIN2 (BR-INSENSITIVE2) kinase.CDG1 and BSK1 kinases also activate PP2A phosphatase. Both inhibitions of BIN2 kinase and activation of PP2A (PROTEIN PHOSPHATASE 2A) phosphatase promote the transformation of phosphorylated BZR1 (BRASSINAZOLE RESISTANT 1) and BES1 (*bri1* SUPPRESSER 1) to dephosphorylated forms that migrate to the nucleus and form various transcriptional complexes to regulate the expression of BR-responsive genes (Fig. 11.49).

The level of the hormone in Arabidopsis is regulated by controlled expression of biosynthetic genes by nuclear transcription factor BZR1 (BRASSINAZOLE RESISTANT1) and its homolog BES (BR11-EMS-SUPPRESSOR1)/BZR2. High-level cellular brassinosteroids induce binding of BZR1 to the promoter region of brassinosteroid biosynthetic genes and repress their expression. In contrast low

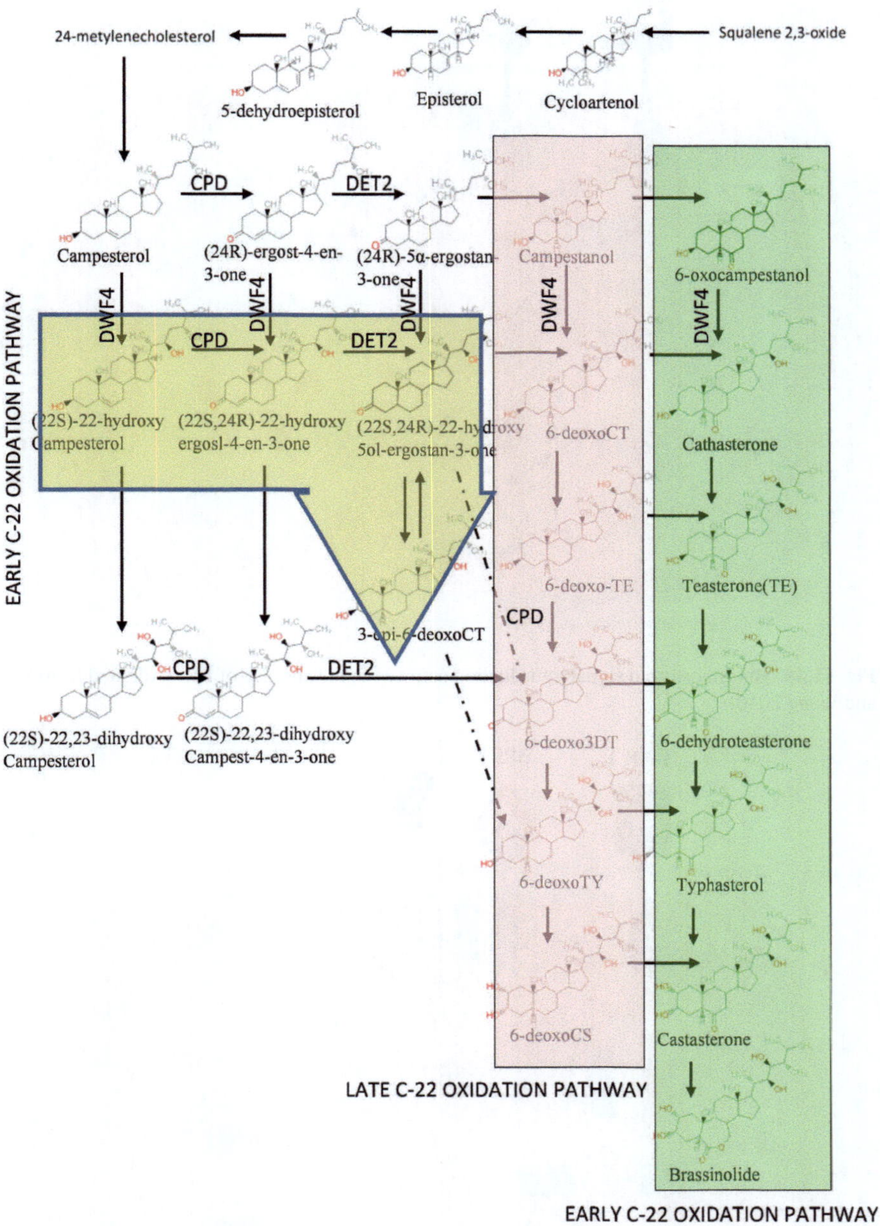

Fig. 11.47 Biosynthesis pathways for Brassinolide synthesis in plants adopted from Chung and Choe (2013), Box 1. Early C22 oxidation pathway, Box 2. Early-C6 oxidation pathway, Box 3. Late-C6 oxidation pathway. Enzymes controlling the reaction are indicated on top of the arrows.

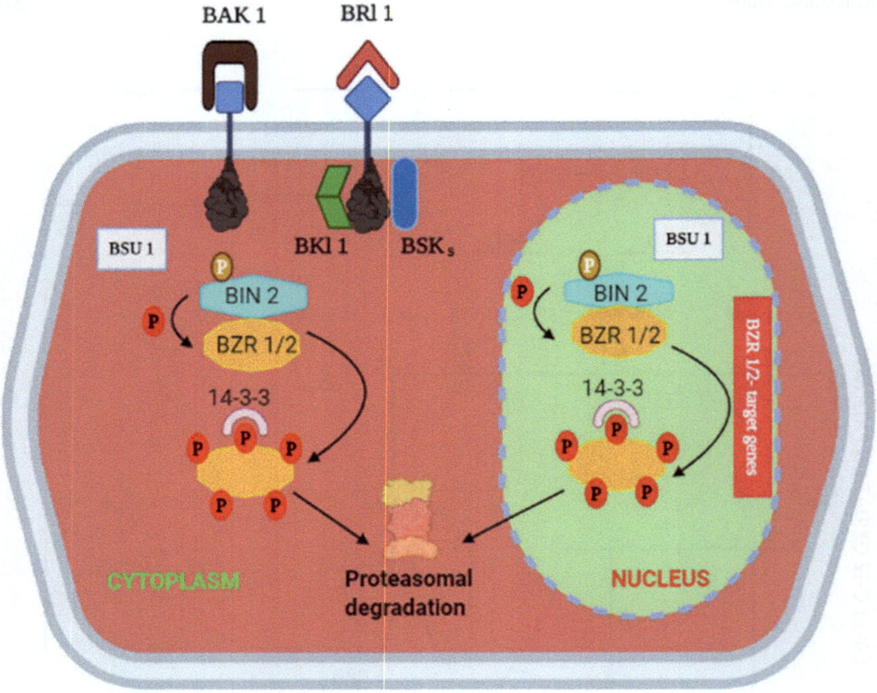

Fig. 11.48 Brassinosteroid signal transduction pathway in the absence of BR. (Adopted from Kim and Wang 2010)

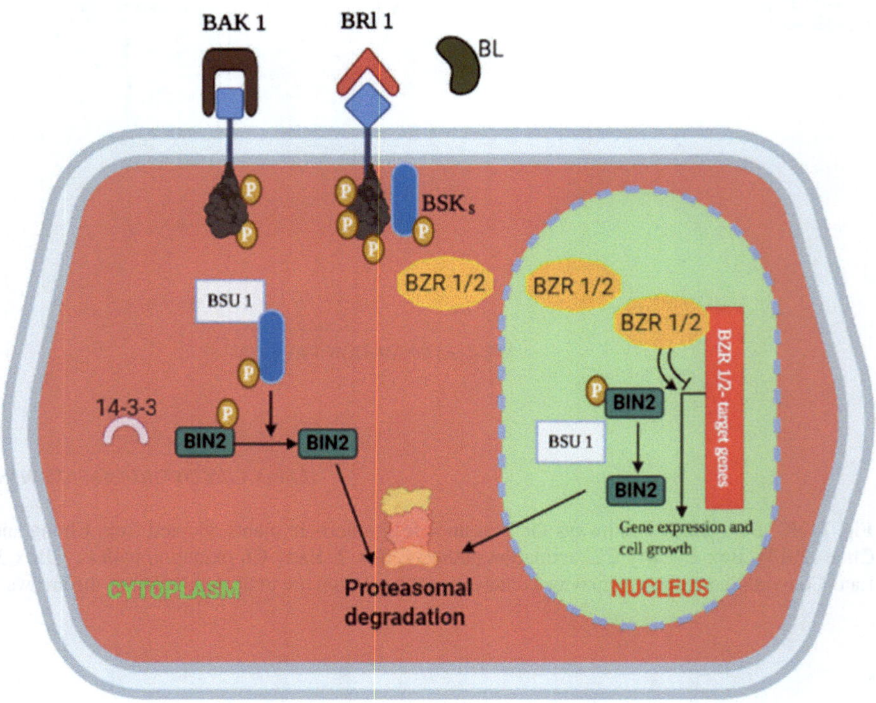

Fig. 11.49 Brassinosteroid signal transduction pathway in the presence of BR. (Adopted from Kim and Wang 2010)

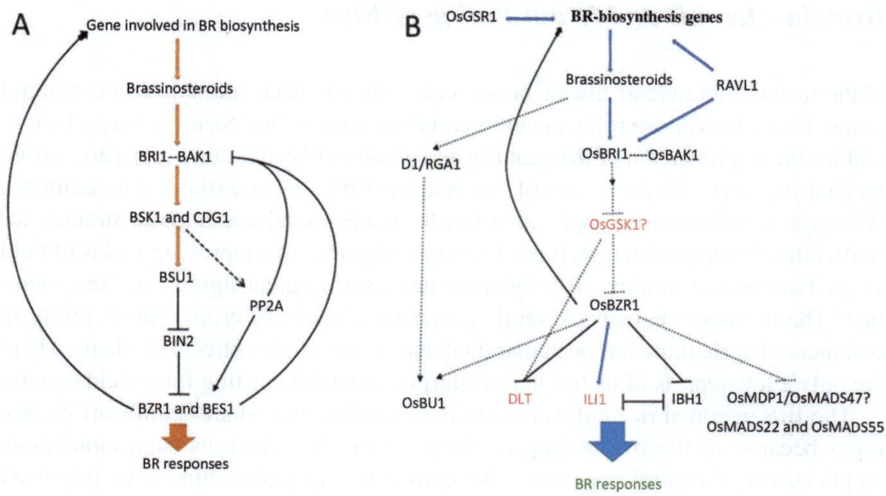

Fig. 11.50 A comparative account of BR-signal transduction between Arabidopsis and rice has been illustrated in the figure. Integrated diagrams for BR-signaling A- Arabidopsis, B- Rice. Arrows show promotion actions and bar ends show inhibitory action. Solid lines show direct regulation, dotted lines indicate hypothetical regulation, and curved lines indicate feedback regulation. (Adopted from Tong and Chu 2012)

cellular level brassinosteroids render BZR1 inactive through phosphorylation vide action of protein kinase BR INSENSITIVE2 (BIN2). GSK1-like kinase also derepresses the expression of the biosynthetic genes. Similarly, auxin induces brassinosteroid biosynthesis by inhibiting the binding of BZR1 to the DWF4 promoter. This signaling pathway has also the complication of some feedback actions. *BRI1* gene expression is negatively regulated by BZR1 and BES1 at transcription and PP2A at translation levels. Also, BZR1 and BIN2 are not omnipresent in the system and are targeted by proteasome-mediated degradations.

The BR-signaling pathway in rice was discovered with the identification of BR-deficient dwarf mutant *d61*, where a loss of function mutation of *brassinosteroid insensitive1* homolog prevented internode elongation and bending of leaf lamina joint (Yamamuro et al. 2000). The *OsBRI1* gene is closely linked to the *d61* locus. The gene functions in various growth and development processes of rice plants. It controls internode elongation, bending of lamina joint, and skotomorphogenesis. The signaling system discovered is mostly similar to that of Arabidopsis. The BR-signaling components of the plant were traced further by forward and reverse genetics (Tong and Chu 2012). The forward genetics have identified expression of *OsBRI1, OsGSK1, DLT, D1/RGA1, ILI1/IBH1,* and *RAVL1* genes, while reverse genetics revealed expression *OsBZR1, OsMDP1/OsMADS47, OsMADS22, OsMADS55, OsGSR1, OsBU1, OsBAK1,* and some other BR-response genes. A comparative account of BR signaling pathways in Arabidopsis and rice is given in Fig. 11.50.

Brassinosteroids and Grain Filling of Rice

Hitherto the commercial use of brassinosteroids has been limited to horticultural crops. Using brassinosteroids in commercial production has been a strategy beneficial for the improvement of the quantity and quality of horticultural crops apart from its contribution to the protection of plants against stresses in unstable environments. Although brassinosteroid application has been a practical and hopeful strategy for horticultural crop production, it has been sparsely used for improving yields of field crops. However, as of today, BR-signaling has assumed great significance in agriculture. The hormone regulates several agronomic features of crops, but its utility in commercial agriculture has been marginal due to its complex effects on plants. There are only a few reports of its use in rice, despite potential existing for a yield benefit.

The IR8-parented rice cultivars used in Green Revolution are semidwarf phenotypes because of the pyramiding of the *sd1* gene that slackens stem elongation. Applications of natural hormones like cytokinin and gibberellins have improved biomass accumulation and seed filling rate in the panicle. In particular, the hormone applications stimulated the development of inferior spikelets on the basal part of the panicle (Patel and Mohapatra 1992). Although brassinosteroids have the potential for rescuing dwarf mutants for tallness in Arabidopsis, there have been few efforts for an investigation similar to rice. There are several instances, where brassinosteroids have been reported to influence morphological features of rice, such as tiller number, leaf size, and leaf angle, plant height, grain size, flowering, senescence, and environmental adaptations (Sakamoto et al. 2006; Morinaka et al. 2006; Tong and Chu 2018). In a more recent work bearing agronomic importance for rice, Zhang et al. (2019) elucidated the BR-mediated effect of soil drying on spikelet differentiation and degeneration. The elevated level of BR in panicle under moderate soil drying was proposed to promote spikelet development by enhancing inflorescence meristem activity, ascorbic acid recycles, and non-structural carbohydrate partitioning to the developing organs (Fig. 11.51). The identification and elucidation of specific BR-receptors and downstream signal transducers and their relationship with rice phenotype in recent years have revealed feasibility for the pyramidization of new desirable traits into the breeding system for yield benefits.

Gene Manipulation for Brassinosteroid Synthesis in Rice

It is assumed that there is a great potential of the hormone in improving plant growth and grain yield. Wu et al. (2008) introduced a transgene encoding sterol C-22 hydroxylase that controls brassinosteroid synthesis using a promoter active only in stem, leaves, and roots to rice. A rise in brassinosteroid level increased tiller number and seed filling in the inferior spikelets of the panicle. These phenotypic changes in the transgenic plant increased seed yield by 15-44% compared to the control wild-type plants. The transgenic plants mobilized a greater flow of assimilates to the reproductive parts to increase both grain number and size. This was possible by enhanced CO_2-assimilation that increased the pool of carbon assimilates of the flag

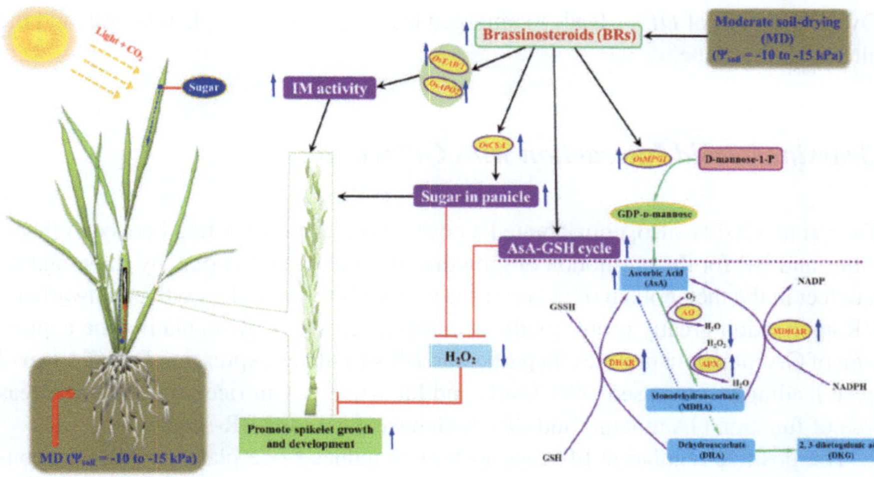

Fig. 11.51 Diagrammatic presentation of BR functions in spikelet growth and development under moderate soil drying (MD) treatment during the early part of panicle development. MD does not impair photosynthesis to interrupt carbon supply from leaf to panicle for meristem activities. Ascorbic acid (AsA) synthesis and recycle (AsA-GSH cycle) get enhanced in young panicle by an increase of BR level to protect spikelet growth and development from deleterious effects H2O2. Black arrow = promotion, red arrow = inhibition, upward blue arrow = up-regulation, down-ward blue arrow = down-regulation under MD treatment. (Adopted from Zhang et al. 2019)

leaf and greater use of assimilates in sink activity of the developing grains, particularly grains located on the basal part of inflorescence by increasing conversion of glucose to starch. In later studies, direct regulation of grain growth of rice by brassinosteroids was illustrated. Che et al. (2015) reported that grain length associated with QTL, *GL2* has the potential to improve rice grain weight and total yield by 27.1 and 16.6% respectively. The QTL is directly involved in brassinosteroid signaling in young rice panicles. It is allelic to *OsGRF4* (Growth Regulating Factor 4) and contains a mutation in the miRNA396 targeting sequence and this mutation induces brassinosteroid response by up-regulating several brassinosteroid-induced genes for promotion of grain development. OsGRF4 belongs to a family of transcription factors that have been involved in the regulation of cell proliferation and cell size and the genes encoding growth-regulating factors (GRF) are targeted by miRNA396. Additionally, these workers found that the central negative regulator of rice brassinosteroid signaling pathway GSK2 directly interacts with OsGRF4 and inhibits its ability for transcription activity to mediate in elongation of grain length by the hormone. Over-expression of *GL2* activates brassinosteroid response as well and this response was reported to be mediated through the consequential promotion of GA biosynthesis genes including *GA20ox-1, GA20ox-2,* and *GA3ox-2* and inactivation of *GA2ox-3*. The role of negative regulator GSK2, which targets DLT, the grass family protein for BR-response has been elucidated further with the identification of Ovate family Protein1 (OFP1) (Xiao et al. 2017). OFP1 is an additional regulator of BR-response and its influence on rice plant architecture depends on DLT function.

Overexpression of *OFP1* leads to enlarged leaf angle, reduced plant height, and the altered grain shape.

Brassinosteroid Interaction with Gibberellins

Tong et al. (2014), also, corroborated a positive correlation existing between gibberellins and BR for the promotion of growth and development in rice, by showing that a defect in the metabolism of either of the two leads to stunted growth and dwarfism. BR application greatly promotes the accumulation of GA by regulating the expression of GA metabolism genes. In particular, BR stimulates expression *D18/GA3-ox-2* gene leading to increased GA1 levels and bioactive GA in rice seedlings, whereas loss of function GA mutants and *d18*, both have decreased BR-sensitivity.

The positive regulation of brassinosteroids-induced rice plant growth was confirmed further. Wu et al. (2015) reported that Dof (DNA binding with one finger) proteins, a class of plant-specific transcription factors involved in plant development stress responses. Overexpression of *OsDof12* in transgenic rice leads to shortened plant height and reduced panicle size with smaller numbers of primary and secondary branches and these features are controlled by brassinosteroid hyposensitivity of the plant. *OsDof12* overexpression downregulates two positive brassinosteroid signal regulators like *OsBR11* and *OsBZR1* for suppression of brassinosteroid action. In conformity, Qin et al. (2018) found a brassinosteroid deficient mutant named *ltbsg1* showing a longer top panicle branch and shorter grain size from *indica* rice cultivar Zhenong 34 developed by ethyl methane sulfonate mutagenesis. In the mutant rice amino acid variation (Glycine-Arginine) in the FAD (Flavin adenine dinucleotide)-binding domain of delta (24)-sterol reductase precludes brassinosteroid biosynthesis by preventing the conversion of 24-methylenecholesterol (24-MC) to campesterol (CR) in the early stage. This observation reveals the complementary nature of BR response with GA in cell elongation of rice because a defect in either of the two results in reduced cell elongation and dwarfism (Tong et al. 2014).

Incidentally, it is construed that at normal physiological level BR induces GA biosynthesis, inhibits its inactivation, and stimulates its synthesis for promotion of cell elongation. GA inhibits BR overproduction and action. Cell elongation is inhibited and GA remains inactive at a high BR level, whereas high GA promotes cell elongation by increasing BR-signaling. The complimentary talk between BR and GA has been explored further (Ross and Quittenden 2016). It is suggested that the hormones interact at the signaling level where BZR1 is the positive regulator for BR-response. DELLA protein, the negative regulator of GA-signaling physically interacts with BZR1 and this interference reduces growth extension. DELLA proteins are degraded by an abundance of bioactive GAs. Also, Wang et al. (2017) have found abnormal brassinosteroid and gibberellins activation in *OsMADS1* transgenic rice reduces both stem and panicle size. Overexpression of *OsMADS1* could have induced excess of BR accumulation leading to GA deficiency in stem by up-regulation of *GA2ox-3* gene restricting internode elongation. Conversely, up-regulation of BR negative regulators might have scaled-down BR level in panicle, and overexpression of *GA2ox-1* in this situation could have resulted in gibberellins deficiency to detriment of meristematic growth. In a nutshell, the BR-signaling system in rice has been summarized by Tong and Chu (Tong and Chu 2018), as follows.

In rice, the BR-signal is transduced through a pathway intended to destabilize the GSK2 kinase that targets various substrates including several transcription factors regulating BR- responsive genes. The BR-signal is perceived by the OsBRI1 receptor and its co-receptor OsBAK1, transduced to OsBSK3 kinase and then to an unidentified BSU1 phosphatase, and further represses GSK2 kinase to release repression of OsBZR1 transcription factor (TF), which regulates BR response genes to implement BR responses. GSK2 and its homologs belonging to the GSK2kinase-like family play central negative roles in BR-signaling. In addition to OsBZR1, the kinases also target several other TFs including DLT, LIC, RLA1/SMOS1, and OFP8 to inhibit their functions and modulate BR-responses. *DLT* encodes a GRAS family protein and positively regulates BR responses. LIC interacts with and antagonizes OsBZR1 to negatively regulate BR responses. High BR levels tend to induce *LIC* expression through OsBZR1 to attenuate BR signaling and LIC will, in turn, suppress OsBZR1 expression to further inhibit BR responses. RLA1/SMOS1 as an integrator interacts with both OsBZR1 and DLT to modulate their activity. OFP8 also positively regulates BR responses and its nuclear localization is modulated by GSK2 phosphorylation. By contrast, OFP19 negatively regulates BR signaling by forming a complex with DLT and OSH1.

References

Akira S, Shozo F (1997) Studies on Biosynthesis of Brassinosteroids. Bioscience, Biotechnology, and Biochemistry 61: 757–762

Anderson JL, Cook JC (1979) Brassinolide, a plant growth-promoting steroid isolated from Brassica napus pollen. Nature 281: 216–217

Che R, Tong H, Shi B, Liu Y, Fang S, Liu D, Xiao Y, Hu B, Liu L, Wang H, Zhao M, Chu C (2015) Control of grain size and rice yield by GL2-mediated brassinosteroid responses. Nat Plants 2:15195. https://doi.org/10.1038/nplants.2015.195.

Chung Y, Choe S (2013) The regulation of brassinosteroid biosynthesis in Arabidopsis. Critical Rev Plant Sci 32: 396–410

Clouse SD, Sasse JM (1998) Brassinosteroids: essential regulators of plant growth and development. Ann Rev Plant Physiol Plant Mol Biol 49: 427–451

Fujioka S, Sakurai A (1997) Biosynthesis and metabolism of brassinosteroids. Physiol Plantarum 100: 710–715

Fujioka S, Yokota T (2003) Biosynthesis and metabolism of brassinosteroids. Ann Rev Plant Biol 54: 137–164

Grove MD, Spencer GF, Rohwedder WK, Mandava N, Worley JF, WarthenJr, JD, Steffens GL, Flippen- Kim T-W, Wang Z-Y (2010) Brassinosteroid signal transduction from receptor kinases to transcription factors. Ann Rev Plant Biol 61: 681–704

Morinaka Y, Sakamoto T, Inukai Y, Agetsuma M, Kitano H, Ashikari M, Matsuoka M (2006) Morphological alteration caused by brassinosteroid insensitivity increases the biomass and grain production of rice. Plant Physiol 141: 924–931

Noguchi T, Fujioka S, ChoeS, Takatsuto S, Tax FE, Yoshia S, Feldmann KA (2000) Biosynthetic pathways of brassinolide in Arabidopsis. Plant Physiol 124: 201–209

Patel R, Mohapatra PK (1992) Regulation of spikelet development in rice by hormones. J Exp Botany 43: 257–262

Prusakova L, Ezhov M, Salnikov A (1999) The use of emistim, epibrassinolide, and uniconazole to overcome the quality difference of buckwheat grains. Agrarian Russia 1: 41–44

Qin R, Zeng D, Yang C, Akhter D, Alamin M, Jin X, Shi C (2018) LTBSG1, a new allele of BRD2, regulates panicle and grain development in rice by Brassinosteroid biosynthetic pathway. Genes 2018 9: 292. https://doi.org/10.3390/genes9060292

Ross JJ, Quittenden LJ (2016) Interactions between brassinosteroids and gibberellins: synthesis or signaling? Plant Cell 28: 829–832

Sakamoto T, Morinaka Y, Ohnishi T, Sunohara H, Fujioka S, Ueguch-Tanaka M, Mizutani M, Sakata K, Takatsuto S, Yoshida S, Tanak H, Kitano H, Matsuoka M (2006) Erect leaves caused by brassinosteroid deficiency increase biomass production and grain yield in rice. Nat Biotechnol 24: 105–109 Suzuki H, Inoue T, Fujioka S, Saito T, Takatsuto S, Yokota T, Murofushi N,, Yanagisawa X, Sakurai A (1995) Conversion of 24-methylcholesterol to 6-oxo-24-methylcholestanol, a putative intermediate of the biosynthesis of brassinosteroids, in cultured cells of Catharanthus roseus. Phytochemistry 40: 1391–1397

Tong H, Chu C (2012) Brassinosteroid signaling and application in rice. J Genetics Genomics 39: 3–9

Tong H, Chu C (2018) Functional specificities of brassinosteroid and potential utilization for crop improvement. Trends Plant Sci 23 https://doi.org/10.1016/j.tplants.2018.08.007

Tong H, Xiao Y, Liu D, Gao S, Liu L, Yin Y, Jin Y, Qian Q, Chu C (2014) Brassinosteroid regulates cell elongation by modulating gibberellin metabolism in rice. Plant Cell 26: 4376–4393

Wang L, Zeng X-Q, Zhuang H, Shen YN, Chen H, Wang ZW, Long JC, Ling YH, He GH, Li YF (2017) Ectopic expression of OsMADS1 caused dwarfism and spikelet alteration in rice. Plant Growth Reg 81: 433–442

Wu Q, Li D, Li D, Liu X, Zhao X, Li X, Li S, Zhu L (2015) Overexpression of OsDof12 affects plant architecture in rice (Oryza sativa L.). Front Plant Sci 6: 833

Wu C-Y, Trieu A, Radhakrishnan P, Kwok SF, Harris S, Zhang K, Wang J, Wan J, Zhai H, Takatsuto S, Matsumoto S, Fujioka S, Feldmann KA, Pennell RI (2008) Brassinosteroids regulate grain filling in rice. Plant Cell 20: 2130–2145

Xiao Y, Liu D, Zhang G, Tong H, Chu C (2017) Brassinosteroids regulate OFP1, a DLT interacting protein, to modulate plant architecture and grain morphology in rice. Front Plant Sci 8:1698. https://doi.org/10.3389/fpls.2017.01698

Yamamuro C, Ihara Y, Wu X, Noguchi T, Fujioka S, Takatsuto S, Ashikari M, Kitano H, Matsuoka M (2000) Loss of function of a rice brassinosteroid insensitive1 homolog prevents internode elongation and bending of the lamina joint. Plant Cell 12: 1591–1605

Yokota T (1997) The structure, biosynthesis, and function of brassinosteroids. Trends Plant Sci 2: 137–143

Zhang W, Sheng J, Xu Y, Xiong F, Wu Y, Wang W, Wang Z, Yang JC, Zhang J (2019) Role of brassinosteroids in rice spikelet differentiation and degeneration under soli-drying during panicle development. BMC Plant Biol 19:409 https://doi.org/10.1186/s12870-019-2025-2

Chapter 12
Effects of Environmental Stresses on Grain Filling of Rice Panicle

Abstract Rapid population growth and climate change-induced disturbances in rice agro-ecosystem undermine rice production and accentuate the demand for food. Rice plants demand as much as twice of water consumed by other cereal crops. However, the prospects of rice for climate resiliency are greater than the others. Under the emerging circumstances of the rise in global temperature irrigated rice cultivation has been a challenge because of a scarcity of water. Production losses of rice worldwide consequential to the two climate change scenarios like temperature increases and drought incidence are found high on account of either spikelet sterility or reduced accumulation of assimilates for grain filling. Water stress coinciding with the reproductive stage affects the number of secondary branches initiated on panicle and post-development of the branches causing spikelet abortion. The incidence of heat episodes concurrent with drought stress at anthesis or before anthesis leads to increased sterility of spikelets, poor seed setting, and loss of grain quality. In case of climate change-induced, high rainfall and consequential flooding and submergence, non-structural carbohydrates (NSC) have been an important source for providing the energy necessary for the maintenance of cell metabolism during submergence and recovery of the plant after de-submergence in rice. Stress-sensitive hormones like ethylene and ABA interact in mediating effects challenging environments on grain filling of rice.

Keywords Climate change · ABA · Ethylene · Heat episode · Submergence · Spikelet sterility · NSC

Climate Change and Food Security

Rice production must increase substantially to match the demand of explosive growth of rice consumers worldwide. Apart from population growth, climate change-induced disturbances in rice agro-ecosystem undermine rice production and accentuate the demand for food. Increased frequencies of extreme climatic events in rice-growing areas, such as typhoons, scarcity of rains for drought, heavy rainfall resulting in flooding, and rising sea level causing risk of submergence and salinity, are the inclement factors that destabilize rice crop pattern. Given the sensitivity of rice plants to climate change, thanks to the pernicious effects of high temperature, uncertain weather patterns involving cyclones, drought, and flooding, and the

P. K. Mohapatra, B. B. Sahu, *Panicle Architecture of Rice and its Relationship with Grain Filling*, https://doi.org/10.1007/978-3-030-67897-5_12

emergence of new stresses in the biosphere food security is at stake for the present and succeeding generations of human beings on earth. Therefore management of routes for an increase of yields and total production through greater tolerance of physicochemical stresses has been given priority in rice science to alleviate plant injury in stress-prone environments. Soil moisture regime is the most crucial factor for rice crop production. Rice plants demand as much as twice of water consumed by other cereal crops (Peng et al. 2006). Under the emerging circumstances of the rise in global temperature irrigated rice cultivation has been a challenge because of a scarcity of water. A paucity of irrigation water has been a major limiting factor for flooded rice production now a day. Tuong and Bouman (2003) reported that about two million hectares of Asia's irrigated rice land may face a serious shortage of water by the year 2025. The factor limiting rice production next to scarcity of water is the instability of ambient temperature. Production losses of rice worldwide consequential to the two climate change scenarios like temperature increases and drought incidence are found high on account of either spikelet sterility or reduced accumulation of assimilates (Korres et al. 2017). To minimize the deleterious impact of a high-temperature regime on grain filling agronomic operations should involve early showing. On the other hand, rice germplasm from exceedingly warm environments can be used for the selection of traits that are most appropriate for the development of high-temperature tolerant rice. Although the development of heat stress-tolerant varieties has been one of the best strategies to maintain productivity and make rice crop climate-resilient, obstacles are many. Heat stress tolerance is a multigenic trait and the candidate genes for the purpose are poorly investigated (Kilasi et al. 2018). To mitigate drought stress, water saving technology like alternate wetting and drying should be implemented; the technique conserves irrigation water and reduces methane emission by a margin of 48% compared to continuous flooding.

Stress Effects on Panicle Morphology of Rice

To cope with the yield losses under water-limited environments, Quinones et al. (2017) recommended retention of key physiological traits like globular-shaped floral meristem as a marker for optimum spikelet number and increased panicle size and panicle diameter and panicle nonstructural carbohydrates content for the avoidance of yield penalty under challenging environment. On a closer look, the yield penalties are reportedly imposed on the plant by the adverse influence of stress on non-structural carbohydrate (NSC) partitioning to the panicle and limitation of sink size with restriction of spikelet number and differentiation. In case of submergence tolerance, NSC before and after submergence becomes an important source for providing the energy necessary for the maintenance of cell metabolism during submergence and recovery of the plant after de-submergence in rice (Das et al. 2005). The underwater elongation ability, of course, is associated more with NSC consumed during submergence but not with NSC level before submergence. The adverse effects of the stress are developmental stage-specific for the plant. During the early stage of reproductive development immediately after panicle initiation, primary and secondary branches are formed and they bear pedicellate-spikelets in rice inflorescence. Water stress coinciding with this stage affects the number of secondary branches initiated and post-development of the branches causing spikelet abortion (Kato et al. 2008; Kato & Katsura 2010) and consequential loss of grain yield. It is construed that any reduction in plant water status under drought conditions directly correlates with pre-flowering spikelet abortion. The stress reduces grain yield by altering the floral meristem development restricting sink size for grain filling (Quinones et al. 2017). The panicle diameter at the neck node is reduced significantly showing a morphological shift towards slender shape, while non-structural carbohydrate contents of the panicle decline significantly precluding chances of supporting grain growth at the post-anthesis stage of grain development when the rice plants are shifted from continuous flooding to drier fields. In the post spikelet differentiation phase, the young panicle could be a potential target of drought or heat stress because of reduced transpiration driven rise in canopy and spikelet tissue temperature (Jagadish et al. 2011b, a). Water deficit alone and in combination with heat episode significantly reduced peduncle elongation trapping 32 and 55% of spikelets inside the flag leaf sheath boot that in turn reduced spikelet fertility by a margin of 27% compared to control (Jagadish et al. 2011b). Individually, each stress, either heat episode or water deficit, up-regulates expression of chaperonic heat shock transcripts and proteins, but the effect becomes additive under combined exposure. The incidence of heat episodes concurrent with drought stress at anthesis or before anthesis leads to increased sterility of spikelets, poor seed setting (Jagadish et al. 2014; Shi et al. 2015), and loss of grain quality (Lyman et al. 2013).

Generally, the anthesis stage of spikelet development is most sensitive to temperature stress for seed settings in rice. In general, the success of seed setting depends on pollination and fertilization, but the events become vulnerable to high temperature because of the adverse impact of the stress on male gametophyte development, while pistil remains insensitive. High temperature affects critically the anther dehiscence and impedes pollination. It is proposed that the presence of longer stigma and large anthers contribute to heat tolerance at anthesis in rice (Matsui and Omasa 2002). Moisture limited soil environment increases tissue temperature (Fig. 12.1) and slackens non-structural carbohydrates content (Fig. 12.2) of panicle at the early reproductive stage (Quinones et al. 2017) accruing yield loss. The pernicious effects of abiotic stresses on grain yield studied so long have been the whole panicle centric not being particular to the staggered distribution of spikelet development concerning time and space on the panicle rachis.

Genes/QTLs for Stress Tolerance

Significant improvement for stress resistance has been achieved in developing tolerant varieties through the identification of genes/alleles through screening of substantial genetic variability in grain yield under different stresses (GRiSP 2013). The new generations of stress-tolerant varieties are expected to sustain plant growth

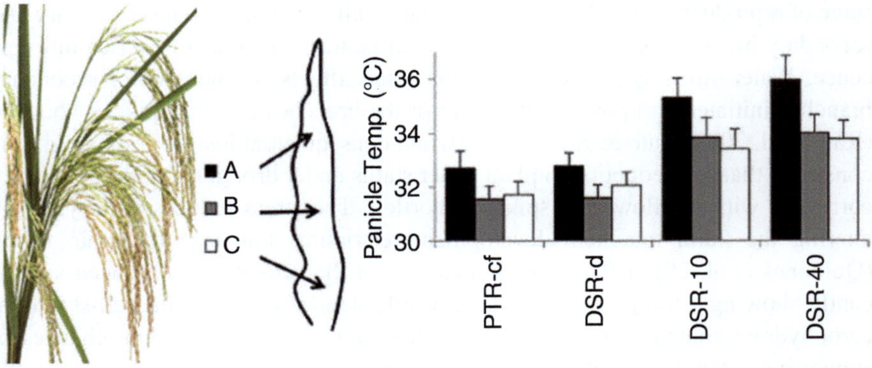

Fig. 12.1 The effect of soil moisture regime on panicle tissue temperature of rice (Adopted from Quinones et al. 2017). *PTR-cf* continuously flooded and transplanted rice, *DSR-d* Dry seeded rice with daily irrigation, *DSR-10* Dry seeded rice irrigated at 10 kPa soil moisture tension, *DSR-40* Dry seeded rice irrigated at 40 kPa soil moisture tension

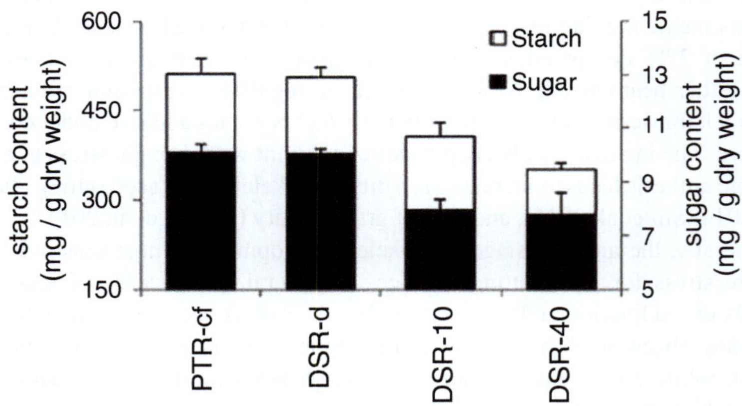

Fig. 12.2 The effect of soil moisture regime on panicle non-structural carbohydrate accumulation (Adopted from Quinones et al. 2017). *PTR-cf* continuously flooded and transplanted rice, *DSR-d* Dry seeded rice with daily irrigation, *DSR-10* Dry seeded rice irrigated at 10 kPa soil moisture tension, *DSR-40* Dry seeded rice irrigated at 40 kPa soil moisture tension

under the challenging environments like drought, flooding, and salinity. Development of rice variety with inherent tolerance against these stresses by pyramiding appropriate QTLs helped achieving sustained rice production under sub-optimal environments. The exploration of genes/qtls alleviating stress injuries is being updated temporally by rice scientists for novelty. In pursuance, several efforts have been made to introgress major effects QTLs such as *qDTY1.1*, *qDTY2.1*, *qDTY3.1*, and *qDTY6.1*for drought, *saltol* for salinity, and *Sub1* for flood tolerance, through marker-assisted breeding approach (Muthu et al. 2020).

Drought Tolerance

Drought stress has been one of the major challenges for sustain rice production in major rice-growing nations of the world impending risks of climate change-induced sub-optimal Henceconditions. Rice plant needs a semi-aquatic agro-ecosystem for production. Hence, the soil moisture regime has been the most critical factor that decides plant development and final yield. Rich biodiversity and diffusion widespread over contrasting growth habitats have contributed to significant variation in adaptation to drought, where several genes and quantitative trait loci related to drought tolerance are involved. Most of these traits have been identified from landraces and wild relatives of rice growing under water-limited environments (Vinod et al. 2019). For drought stress five QTLs for higher yield under drought were identified and were introgressed into popular cultivars through marker-assisted backcrossing. Kumar et al. (2007) detected a QTL on chromosome one adjacent to *sd1* that accounted for 32% of the genetic variation in yield under severe drought conditions induced by draining the field before flowering. Another QTL (*qtl12.1*) with a large effect on grain yield under drought stress was detected on chromosome 12 (Bernier et al. 2007). Grain yield apart, this locus increased harvest index, biomass yield, plant height besides reducing the number of days to flowering under reproductive stage drought stress in upland rice.

Venuprasad et al. (2009) using bulk segregant analysis of 490 recombinant inbred lines of a cross Apo2*/Swarna detected QTLs with large effects on grain yield in lowland drought stress. They are *DTY2.1* and *DTY3.1* and had a significant effect on grain yield under severe lowland drought stress. A model describing drought perception and mechanism of physiological and biochemical responses of rice plant leading to adverse impact on yield has been given in Fig. 12.3. In a field trial experiment, Swain et al. (2017) tested a set of genotypes known to be tolerant of drought at the reproductive stage for identification of breeding lines showing resilience to stress at both vegetative and reproductive stages and observed differences in stress tolerance mechanism between the two types. The genotypes that performed best under vegetative or vegetative plus reproductive stage drought incidence were different from the varieties doing best under reproductive stage drought. It was concluded that the development of improved genotypes with combined stress tolerance at multiple growth stages will help farmers in rainfed rice-growing regions.

Flooding Tolerance

Both deep water and some lowland rice cultivars survive stagnant flooding, although many of them die within the first week of exposure to the stress. For deepwater rice two ERF DNA binding proteins namely SNORKEL1 and SNORKEL2, allow the plant to adapt to submergence through the promotion of stem elongation matching level of water in the medium (Hattori et al. 2009); ethylene generated in plant organs

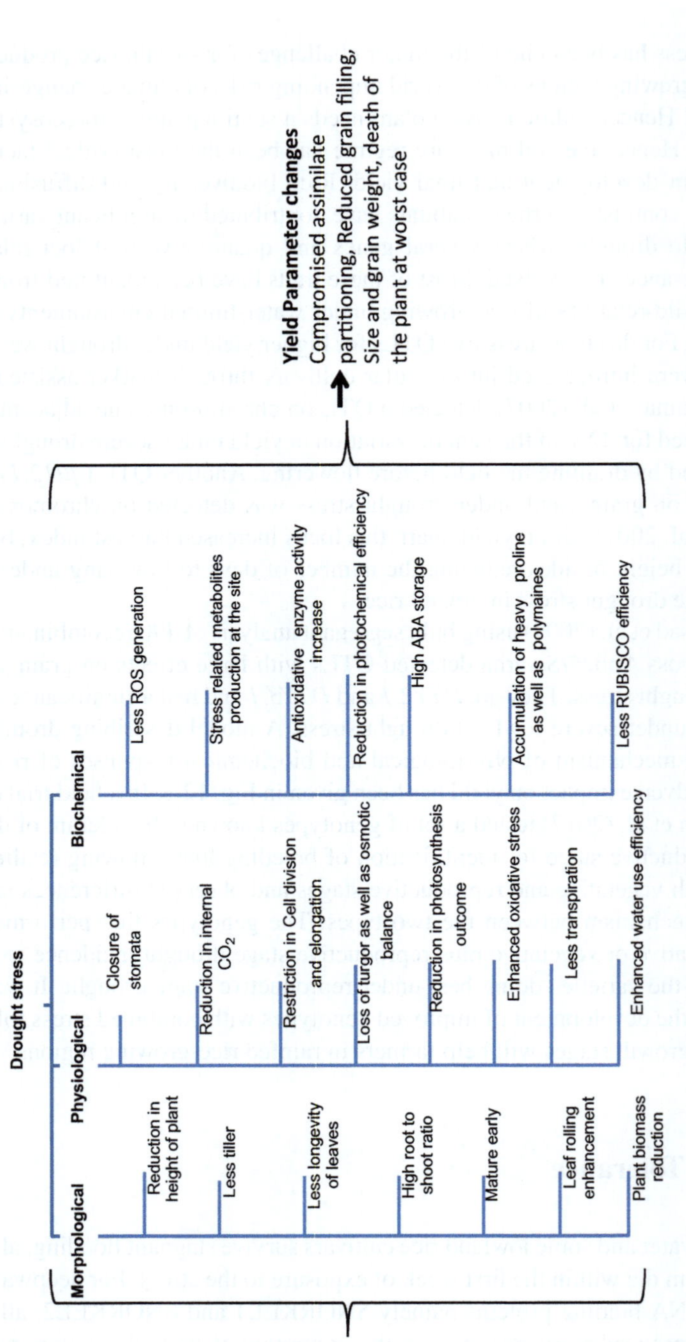

Fig. 12.3 A model for morphological, physiological, and biochemical responses governing yield attributes of rice plant under drought stress. (Adopted from Oladosu et al. 2019)

under submergence counters ABA response permitting GA action for stem elongation. Conversely, the mechanism of flood tolerance of low land rice is different. Another ERF located at *SUBMERGENCE1A* (*SUB1A*), also gene symbol synonym *Sub1A*, belonging to trait class stress tolerance has been identified. *Sub1* polygenic locus restrains stem elongation for conservation of carbohydrates, otherwise used for stem growth, for development of new organs upon de-submergence. Sub1 has been a major quantitative trait locus that has a cluster of three ERF-like genes. In flood intolerant japonica rice the locus has two *ERF* genes like *Sub1B* and *Sub1C*. In comparison, the locus has one additional *ERF* gene *Sub1A* in flood-tolerant *indica* rice like FR13A. It has several alleles like *Sub1A-1, Sub1A-2, Sub1A-3*. *Sub1A* confers submergence tolerance in domesticated rice by regulating metabolic and developmental acclimation (Xu et al. 2006, Fukao et al. 2006, 2009). The submergence tolerance conferred by *Sub1A* is mediated by GA signaling repressors SLR1 (Slender rice 1) and SLRL1 (SLR like1) for restriction of gibberellins response in rice (Fukao and Bailey-Serres 2008). It slackens ethylene production restricting GA-responsiveness for stem elongation. As discussed, ethylene production in rice stems is instrumental in controlling resilience to flooding, although the mechanism of one is opposite to the other. Transfer of *Sub1* QTL that confers resistance to flooding has been introgressed in many popular rice cultivars using marker-assisted selection for yield advantage of more than 1 t/ha over the original cultivar. Such introgression into mega rice cultivars like Swarna and Savitri, however, has not been beneficial for a yield advantage under stagnant flooding (Kuanar et al. 2019). Subjected to stress the QTL promotes genesis of ethylene in the panicle at anthesis that slackens activities of starch synthesizing enzymes leading to accumulation of soluble carbohydrates in developing panicle and poor filling of grains (Fig. 12.4).

Salinity Tolerance

Salinity is widespread in coastal plains, where rice crop suffers because of seawater ingress during high tides. In inland areas, salinity arises from salt deposits collected from repeated irrigation or runoff rainwater. Rice plants growing in coastal areas are home to a large number of salt-tolerant landraces. In general, the rice plant is sensitive to the incidence of salinity stress at two developmental stages, i.e., early seedling and reproductive stages. The impact of the stress on physiological functions of the plant are not similar at different stages of growth (Moradi and Ismail 2007), and hence, gene expression for coping with salt tolerance could be development stage-specific (Razzaque et al. 2019). Mapping of QTLs related to vegetative stage salinity tolerance carried out by interval mapping identified two QTLs, *qST1* and *qST3* on chromosomes 1 and 3 respectively conferring stress tolerance at the young seedling stage (Lee et al. 2007). In Genome-wide association studies, a major QTL named *Saltol* was mapped on chromosome 1 (Kumar et al. 2015). *Saltol*, which controls salinity tolerance at the seedling stage, was detected as a major association with the Na^+/K^+ ratio measured at the reproductive stage. Also, Bonilla et al. (2002) reported

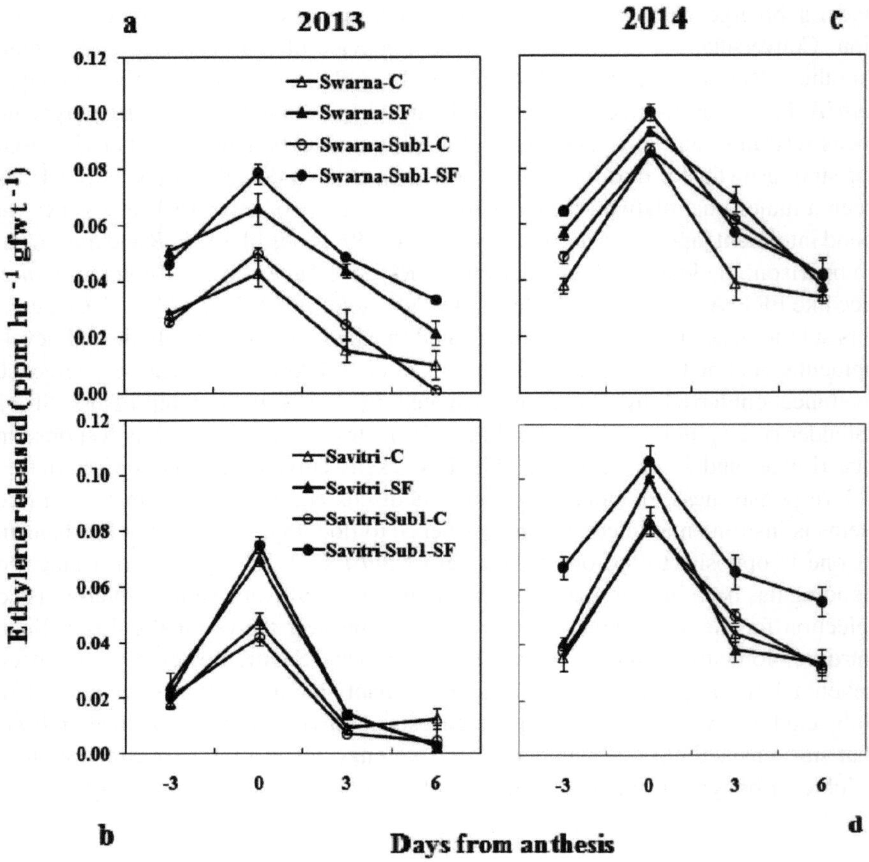

Fig. 12.4 The effect of stagnant flooding (SF) on ethylene concentration of panicle on the main shoot of rice cultivars Swarna and Savitri with and without *Sub1* introgression. (Kuanar et al. 2019)

that *saltol* was linked to major QTLs for Na^+ and K^+ uptake and Na^+/K^+ ratio on chromosome 1. Quantitative trait loci mapping approach using bi-parental population has led to the identification of several major and minor QTLs on different chromosomes in addition to *saltol*. Ammar et al. (2009) found 25 QTLs for ion concentration of K^+, Na^+, and Cl^- in leaf tissues at the reproductive stage on rice chromosomes 1, 2, 3, and 8. There was one QTL for seedling salt injury score, 9 for Na^+, three for K^+, and four for Cl^- concentrations. Combining QTL mapping and transcriptome profiling of bulked RILs for identification of functional polymorphism, Pandit et al. (2010) detected 8 QTLs for salt ion concentrations on rice chromosomes 1, 8, and 12 and Cheng et al. (2012) reported a total of 47 salt tolerant QTLs with strong genetic background effects using two sets of reciprocal introgression lines derived from a cross between salinity tolerant *japonica* and salinity susceptible *indica* Breading lines. QTL characterization for reproductive stage salt tolerance was extended further by Tiwari et al. (2016) by using bulk segregant analysis of bi-parental recombinant inbred lines. Genotyping of the parents and RIL bulk made based on salt sensitivity index for grain yield revealed 6068 polymorphic SNPs and 21 QTLs exhibiting homogeneity of contrasting alleles in the two bulks. Using bi-parental mapping population, about 70 salt-related QTLs have been identified in rice (Hu et al. 2012; Zang et al. 2008). In another approach, Kumar et al. (2015) conducted genotyping of 220 *indica* genotypes by using a custom-designed 6 K genic SNP array to successfully detect SNPs associated with the number of phenotypic traits at the reproductive stage salinity stress. SNP trait associations were stronger under salinity stress in comparison to the control condition. The study accounted for the distribution of 6000 SNP chips in known biotic and abiotic stress responses and unknown expressed rice genes across 12 chromosomes. Functional annotation of chips revealed 73.5% correspondence to abiotic stress-responsive genes followed by 16.3% biotic stress-related genes, unknown expressed genes 6%, and 4% hypothetical protein genes. The study identified *saltol* QTL located on chromosome 1. The QTL controls the salinity tolerance of rice with a major association with the Na^+/K^+ ratio at the reproductive stage and exerts influence on yield components.

The novel methods used for rapid scanning and characterization of QTLs for complex agronomic traits in rice have opened up avenues for the development of new ideotypes for stress tolerance. QTLs introgressed to major high-yielding cultivars through marker-assisted backcrossing resulted in significant yield benefit in the stress-prone environment. However, information is scant on the impact of stress on spikelet development and grain growth. The spikelet-specific impact of stress on grain development could be dependent on its coincidence with the stage of individual grain growth, the margin of sensitivity of which to the stress varies temporally. Being dependent on the time of exposure, stress effects on panicle growth do not follow a definite pattern, although the inferior spikelets are always vulnerable to the stress more than the superior ones.

Hormone Homeostasis Under Perturbed Environment

In rice panicle, grain filling properties of individual spikelets are not identical because of discrimination in assimilate partitioning, where inferior type spikelets deficient in positive growth regulators (Wang et al. 2006) lack sink strength for assimilate utilization in growth (Panigrahi et al. 2019). However, demand for assimilates is higher in compact panicle rice; grain filling becomes source limited when source activity is disturbed (Wang et al. 2006). Low sink strength of inferior spikelets increases vulnerability to environmental stress. It was noted that exposition to moderate level soil drying in the field at post-anthesis period improves the rate of grain filling, more in inferior than superior spikelets of panicle, by increasing activities of starch synthesizing enzymes because of the concomitant rise of ABA concentration (Zhang et al. 2012a). Zhang et al. (2012b) noted that overexpression of polar auxin transport protein gene *OsPIN3t* leads to up-regulation of *OsDREB2A* and *OsAP37* genes for increased drought tolerance of rice. A condition attributing to auxin deficiency, therefore, could be responsible for stress sensitivity. Stress-sensitive hormones like ABA and ethylene interact in mediating effects of water stress on grain filling. Levels of water stress, moderate or severe, increase significantly ABA and ethylene concentration of developing rice grains in the post-anthesis stage. Although the rise in ABA concentration is favorable for grain filling, the beneficial effects are out-performed by the high evolution of ethylene under severe stress. But under moderate level soil drying ethylene release is not sufficient to infringe the positive effects of ABA (Yang et al. 2004) (Fig. 12.5).

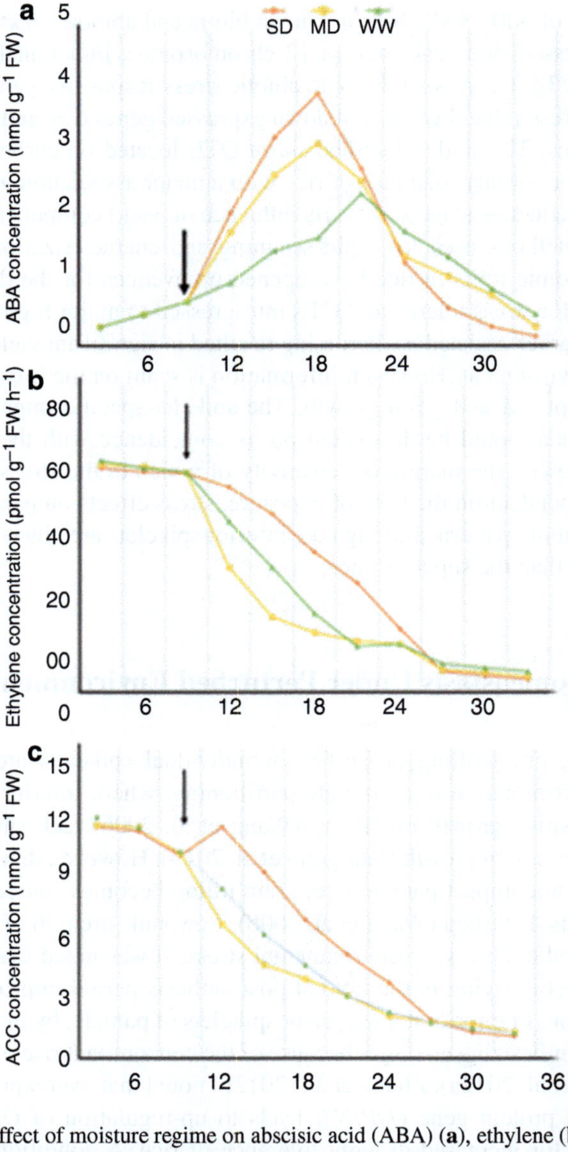

Fig. 12.5 The effect of moisture regime on abscisic acid (ABA) (**a**), ethylene (**b**), 1-aminocyclop ropane-1-carboxylic acid (**c**) in the developing grains of rice. The data are derived from Yang et al. (2004). *WW* well-watered, *MD* moderate water-stressed and *SD* severe water-stressed treatments. Arrows indicate the start of with-holding water

References

Ammar MHM, Pandit A, Singh RK, Sameena S, Chauhan M, Singh A, Sharma P, Gaikwad K, Sharma T, Mohapatra T, Singh NK (2009) Mapping of QTLs controlling Na^+, K^+ and Cl^- ion concentrations in salt-tolerant *indica* rice variety CSR27. J Plant Biochem Biotechnol 18:139–150

Berner J, Kumar A, Ramaiah V, Spaner D, Atlin GN (2007) A large-effect QTL for grain yield under reproductive-stage drought stress in upland rice. Crop Sci 47:507–516

Bonilla P, Dvorak J, Mackill D, Deal K, Gregorio G (2002) RFLP and SSLP mapping of salinity tolerance genes in chromosome 1 of rice (*Oryza sativa* L.) using recombinant inbred lines. Philipp J Agric Sci 85:68–76

Cheng L, Wang Y, Meng L, Hu X, Cui Y, Sun Y, Zhu L, Ali J, Xu J, Li Z (2012) Identification of salt-tolerant QTLs with strong genetic background effect using two sets of reciprocal introgression lines in rice. Genome 55:45–55

Das KK, Sarkar RK, Ismail AM (2005) Elongation ability and non-structural carbohydrate levels in relation to submergence tolerance in rice. Plant Sci 168:131–136

Fukao T, Bailey-Serres J (2008) Submergence tolerance conferred by Sub1A is mediated by SLR1 and SLRL1 restriction of gibberellin responses in rice. Proc Nat Acad Sci USA 105:16814–16819

Fukao T, Harris T, Bailey-Serres J (2009) Evolutionary analysis of the Sub1 gene cluster that confers submergence tolerance to domesticated rice. Ann Bot 103:143–150

Fukao T, Xu K, Ronald PC, Bailey-Serres J (2006) A variable cluster of ethylene response factor-like genes regulates metabolic and developmental acclimation responses to submergence in rice. Plant Cell 18:2021–2034

GRiSP (Global Rice Science Partnership) (2013) Rice almanac 4th edition. International Rice Research Institute, Los Baños, Philippines

Hattori Y, Nagai K, Furukawa S, Song X-J, Kawano R, Sakakibara H, Wu J, Matsumoto T, Yoshimura A, Kitano H, Matsuoka M, Mori H, Ashikari M (2009) The ethylene response factors SNORKEL1 and SNORKEL2 allow the rice to adapt to deep water. Nature 460. https://doi.org/10.1038/nature08258

Hu S, Tao H, Qian Q, Guo L (2012) Genetics and molecular breeding for salt-tolerance in rice. Rice Genomics Genet 3:39–49

Jagadish KS, Cairns JE, Kumar A, Somayanda IM, Craufurd PQ (2011a) Does susceptibility to heat stress confound screening for drought tolerance in rice? Funct Plant Biol 38:261–269

Jagadish KS, Craufurd P, Shi W, Oane R (2014) A phenotypic marker for quantifying heat stress impact during microsporogenesis in rice (*Oryza sativa* L.). Funct Plant Biol 41:48–55

Jagadish SK, Muthurajan R, Rang ZW, Malo R, Heur S, Bennett J, Crauford PQ (2011b) Spikelet proteomic response to combined water deficit and heat stress in rice (*Oryza sativa* cv. N22). Rice 4:1–11

Kato Y, Kamoshita A, Yamagishi J (2008) Preflowering abortion reduces spikelet number in upland rice (*Oryza sativa* L.) under water stress. Crop Sci 48:2389–2395

Kato Y, Katsura K (2010) Panicle architecture and grain number in irrigated rice, grown under different water management regimes. Field Crop Res 117:237–244

Kilasi NL, Singh J, Vallejos CE, Ye C, Jagadish SVK, Kusolwa P, Rathinasabapathi BR (2018) Heat stress tolerance in rice (*Oryza sativa* L.): identification of quantitative trait loci and candidate genes for seedling growth under heat stress. Front Plant Sci. 01 November 2018. https://doi.org/10.3389/fpls.2018.01578

Korres N, Norsworthy J, Burgos N, Oosterhuis D (2017) Temperature and drought impacts on rice production: an agronomic perspective regarding short-and long-term adaptation measures. Water Resour Rural Dev 9:12–27

Kuanar SR, Molla KA, Chattopadhyay K, Sarkar RK, Mohapatra PK (2019) Introgression of *Sub1 (SUB1) QTL* in mega rice cultivars increases ethylene production to the detriment of grain-filling under stagnant flooding. Sci Rep 9:18567. https://doi.org/10.1038/s41598-019-54908-2

Kumar R, Venuprasad R, Atlin GN (2007) Genetic analysis of rainfed lowland rice drought tolerance under naturally-occurring stress in eastern India: heritability and QTL effects. Field Crop Res 103:42–52

Kumar V, Singh A, Mithra SVA, Krishnamurthy SL, Parida SK, Jain S, Tiwari KK, Kumar P, Rao AR, Sharma SK, Khurana JP, Singh NK, Mohapatra T (2015) Genome-wide association mapping of salinity tolerance in rice (*Oryza sativa*). DNA Res 22:133–145

Lee SY, Ahn JH, Cha YS, Yun DW, Lee MC, Ko JC, Lee KS, Eun MY (2007) Mapping QTLs related to salinity tolerance of rice at the young seedling stage. Plant Breed 126:43–46

Lyman NB, Jagadish KS, Nalley LL, Dixon BL, Siebenmorgen T (2013) Neglecting rice milling yield and quality underestimates economic losses from high-temperature stress. PLoS One 8:e72157

Matsui T, Omasa K (2002) Rice (*Oryza sativa* L.) cultivars tolerant to high temperature after flowering: anther characteristics. Ann Bot 89:683–687

Moradi F, Ismail AM (2007) Responses of photosynthesis, chlorophyll fluorescence, and ROS-scavenging systems to salt stress during seedling and reproductive stages in rice. Ann Bot 99:1161–1173

Muthu V, Abbai R, Nallathambi J, Rahman H, Ramasamy S, Kambale R, Thulasinathan T, Ayyenar B, Muthurajan R (2020) Pyramiding QTLs controlling tolerance against drought, salinity, and submergence in rice through marker-assisted breeding. PLoS One 15(1):e0227421. https://doi.org/10.1371/journal.pone.0227421

Oladosu Y, Rafii MY, Samuel C, Fatai A, Magaji U, Kareem I, Kamarudin ZS, Muhammad I, Kolapo K (2019) Drought resistance in rice from conventional to molecular breeding: a review. Int J Mol Sci 20:3519. https://doi.org/10.3390/ijms20143519

Pandit A, Rai V, Bal S, Sinha S, Kumar V, Chauhan M, Gautam RK, Singh R, Sharma PC, Singh AK, Gaikwad K, Sharma TR, Mohapatra T, Singh NK (2010) Combining QTL mapping and transcriptome profiling of bulked RILs for identification of functional polymorphism for salt tolerance genes in rice (*Oryza sativa* L.). Mol Gen Genomics 284:121–136

Panigrahi R, Kariali E, Panda BB, Lafarge T, Mohapatra PK (2019) Controlling the trade-off between spikelet number and grain filling: the hierarchy of starch synthesis in spikelets of rice panicle in relation to hormone dynamics. Funct Plant Biol 46:507–523

Peng S, Bouman B, Visperas RM, Castañeda A, Nie L, Park H-K (2006) Comparison between aerobic and flooded rice in the tropics: agronomic performance in an eight-season experiment. Field Crop Res 96:252–259

Quinones C, Mattes N, Faronilo J, Yadav S, Jagadish KS (2017) Drought stress reduces grain yield by altering floral meristem development and sink size under dry-seeded rice cultivation. Crop Sci 57:2098–2108

Razzaque S, Elias SM, Haque T, Biswas S, Jewel GMNA, Rahman S, Weng X, Ismail AM, Walia H, Juenger TE, Seraj ZI (2019) Gene expression analysis associated with salt stress in a reciprocally crossed rice population. Sci Rep 9(2019):8249. https://doi.org/10.1038/s41598-019-44757-4

Shi WT, Ishimaru T, Gannaban RB, Oane W, Jagadish SVK (2015) Popular rice (*Oryza sativa* L.) cultivars show contrasting responses to heat stress at gametogenesis and anthesis. Crop Sci 55:589–596

Swain P, Raman A, Singh SP, Kumar A (2017) Breeding drought tolerant rice for shallow rainfed ecosystem of eastern India. Field Crops Res 209:168–178

Tiwari S, Krishnamurty SL, Kumar V, Singh B, Rao AR, Mithra SVA, Rai V, Singh AK, Singh NK (2016) Mapping QTLs for salt tolerance in rice (*Oryza sativa* L.) by bulked segregant analysis of recombinant inbred lines using 50K SNP chip. PLoS One 11(4):e0153610. https://doi.org/10.1371/journal.pone.0153610

Tuong T, Bouman B (2003) Chapter 4: Rice production in water-scarce environments. In: Kijne JW, Barker R, Molden D (eds) Water productivity in agriculture: limits and opportunities for improvement. CAB International, pp 53–67

Venuprasad R, Dalid CO, Del Valle M, Zhao D, Espiritu M, Sta Cruz MT, Amante M, Kumar A, Atlin GN (2009) Identification and characterization of large-effect quantitative trait loci for grain yield under lowland drought stress in rice using bulk-segregant analysis. Theor Appl Genet 120:177–190

Vinod KK, Krishnan SG, Tribhuvan R, Singh AK (2019) Genetics of drought tolerance, mapping QTLs, candidate genes, and their utilization in rice improvement. In: Rajpal V, Sehgal D, Kumar A, Raina S (eds) Genomics assisted breeding of crops for abiotic stress tolerance, Vol. II. Sustainable development and biodiversity, vol 21. Springer, Cham. https://doi.org/10.1007/978-3-319-99573-1_9

Wang F, Cheng F, Zhang G (2006) The relationship between grain filling and hormone content as affected by genotype and source-sink relation. Plant Growth Regul 49:1–8

Xu K, Xu X, Fukao T, Canlas P, Maghirang-Rodriguez R, Heuer S, Ismail AM, Bailey-Serres J, Ronald PC, Mackill DJ (2006) Sub1A is an ethylene-response-factor-like gene that confers submergence tolerance to rice. Nature 442:705–708

Yang J, Zhang J, Ye Y, Wang Z, Zhu Q, Liu L (2004) Involvement of abscisic acid and ethylene in the responses of rice grains to water stress during filling. Plant Cell Environ 27:1055–1064

Zang JP, Sun Y, Wang Y, Yang J, Li F, Zhou Y, Zhu L, Jessica R, Hosein FM, Xu J, Li Z (2008) Dissection of genetic overlap of salt tolerance QTLs at the seedling and tillering stages using backcross introgression lines in rice. Sci China C Life Sci 51:583–591

Zhang H, Li H, Yuan L, Wang Z, Yang J, Zhang J (2012a) Post-anthesis alternate wetting and moderate soil drying enhance activities of key enzymes in sucrose-to-starch conversion in inferior spikelets of rice. J Exp Bot 63:215–227

Zhang Q, Zhang W, Yan S, Wang R, Zhao J, Li Y, Qi Z, Sun Z, Zhu Z (2012b) The putative auxin efflux carrier *OsPIN3t* is involved in the drought stress response and drought tolerance. Plant J 72:805–812

Chapter 13
Designer Rice in Quest of High Grain Filling

Abstract The gain of specific leaf weight (SLW) and leaf area index (LAI) has enabled better light interception through higher radiation use efficiency and the associated increase of SPAD to reduce K values in the newly released rice cultivars for higher grain yield compared to the vintage. Despite this evolution in the shaping of the plant architecture of rice, there has been a yawning gap between the potential and actual grain yields. Lack of perfection in crop architectural design and agroecosystem management is the most probable factor responsible for this great failure. The ideal plant architecture proposed for new plant type rice, embodying prospects for higher grain yield, includes traits like low tiller number, no unproductive tillers, heavy panicle bearing numerous grains, stronger stem with dark green leaves, and robust root system. However, the yield advantage of the large panicle has been under-utilized because of a deficiency of information on the molecular basis of ideal plant architecture. Inadequate attention for the molecular basis of the genes controlling panicle architecture and the pleiotropic characterization of the relationship between high spikelet number and other yield-related features have not been investigated. Three crucial events of meristematic activity during inflorescence ontogeny collectively control the course of development and final panicle architecture of rice; such as (a) timing of rachis meristem abortion, (b) the transition of branch meristem to the terminal spikelet meristem, and (c) program shift of lateral meristem identity on a rachis branch in acropetalous fashion. Because the timing of these events is not identical, there is no single regulatory physiological mechanism common to all. It increases complexities for manipulation of different type's genetic expressions of meristem identity prevalent within the system. The physiological factors (hormones) and genes that control the phenotype of dense panicle rice bearing numerous spikelets with potential for high grain number are ideal targets for scrutiny in pursuance of efforts for developing designer rice panicle. In designer rice, there should be more emphasis on spikelet placement on primary than on secondary branches and conversion of the determinate panicle to an indeterminate pattern. An extra-large panicle possessing spikelets in compact orientation is not always an ideal option for high grain filling because of coincidental loss of apical dominance. The study of some rare genetic mutants has also opened up avenues for designing a panicle bearing multi-floret spikelets with 2–3 grains instead of one per spikelet.

Keywords Designer rice · Dense panicle · Genetic mutants · Meristem identity · Multi-floret spikelet

P. K. Mohapatra, B. B. Sahu, *Panicle Architecture of Rice and its Relationship with Grain Filling*, https://doi.org/10.1007/978-3-030-67897-5_13

The Gap Between Potential and Actual Yields

Research in rice science has contributed significantly to the improvement of grain yield as the yield potential has jumped from 4 to 10 tons per hectare per crop. It has been responsible for more than doubling overall world rice yield from 260 to over 700 million tons in the past 50 years. Improvement of phenotype attributes in the new plant-type rice lines, although potentially capable of harnessing greater yield (Wang and Li 2008) has not been of utility to raise the yield beyond the genetic yield ceiling. Of course, the evolution of growth phenotypes within rice cultivars cultivated over 70 years in the predominantly rice-growing areas of the Yangtze river basin of China has enabled newer cultivars for more efficient light energy utilization in achieving better yield compared to the vintage (Zhu et al. 2020). With progression in the year of release of the cultivars SPAD readings, SLW (specific leaf weight) and LAI (leaf area index) increased, while light extinction coefficient (K) declined to accrue gain of grain yield in new cultivars (Figs. 13.1 and 13.2). The gain of SLW in newer cultivars enabled better light interception through higher radiation use efficiency, which was associated with increased SPAD to reduce K values.

Despite this evolution in the shaping of plant architecture of rice, there has been a yawning gap between the maximum yield currently obtained in the farm and that could be achieved with improved farm management and the adoption of new high-yielding cultivars. Yoshida (1981) estimated the theoretical yield potential of rice at 15.9 t/ha and it is much higher than the best farm yield obtained as of now. Muralidharan et al. (2019) verified genetic gain for yields of rice genotypes tested in 11 rice ecosystems of India from 1995 to 2013 and found no evidence for either a genetic gain or loss of grain yields of genotypes developed for any of the agro-ecosystems when environmental effects were eliminated. The workers confirmed the potential yield of rice at 15–16 t/ha and that actual yields could be achievable to an extent not beyond 10 t/ha in the risk-free irrigated and shallow lowland ecosystems. Closing the yield gap of about 6.0 t/ha through corrective technological and policy interventions has not been successful anywhere. Similarly, van Wart et al. (2013) have acknowledged a yield gap existing between crop yield potential and current farm yields in both developed and developing nations. The farm yields plateau when they reach 75 to 85% level of estimated national yield potential. The authors advised for improved protocols in an appropriate crop model to obtain robust, transparent, and reliable estimates for yield potential at different geographical locations and make correct predictions for incidence of yield plateau. The stagnation of yield growth in major grain production areas has increased pressure on various rice-growing nations to accelerate either yield growth rate or expand the cultivated area to make up the difference between supply and demand. As discussed before, rice yield plateau for two reasons, (i) lack of perfection in crop and agroecosystem management techniques and, (ii) diminishing marginal yield benefit with additional input of resources as yield approaches the genetic ceiling. These pieces of evidence emphasize the need for a renewed consideration for genetic gain for

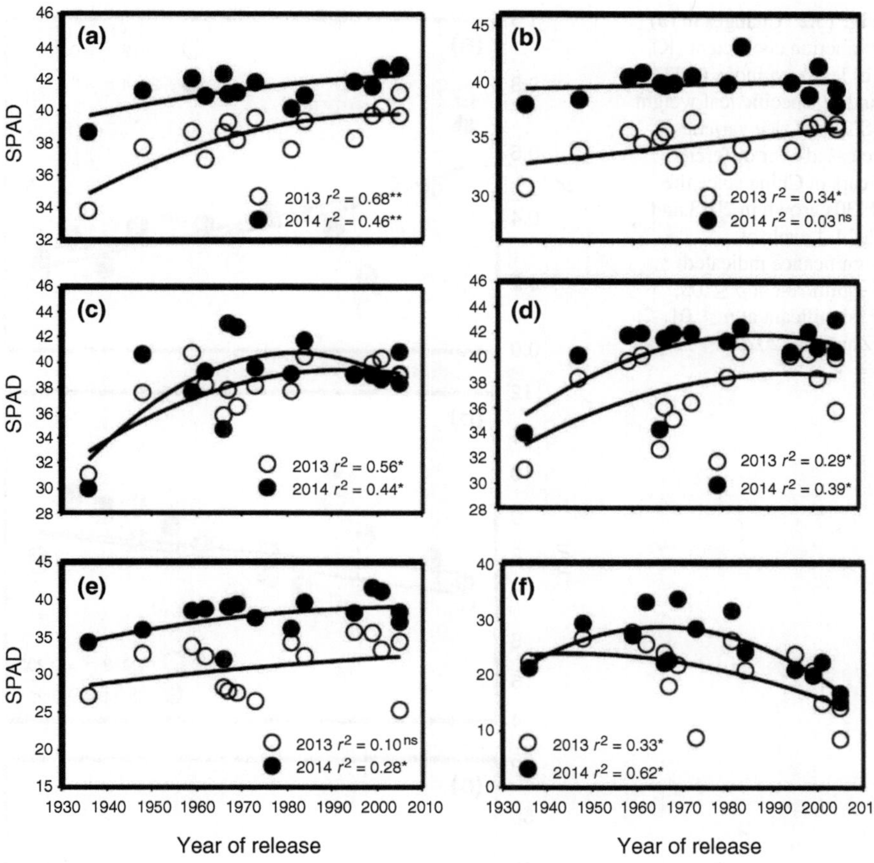

Fig. 13.1 Variation in SPAD value at various growth stages of rice varieties released over different years in China since the 1930s grown in 2013 and 2014. (**a**) Tillering stage; (**b**) panicle initiation stage; (**c**) heading stage; (**d**) 10 days after heading (DAH); (**e**) 20 DAH; (**f**) maturity stage (MT). (Zhu et al. 2020)

yields and ensure rice availability thereby matching the demand of an increasing number of rice consumers.

Manipulation of Panicle Architecture

The ideal plant architecture proposed for new plant type rice, embodying prospects for higher grain yield, includes factors like low tiller number, no unproductive tillers, heavy panicle bearing numerous grains, stronger stem with dark green leaves, and robust root system (Donald 1968; Khush 1995; Virk et al. 2005). However, the yield advantage of the new plant type has been under-utilized because of several

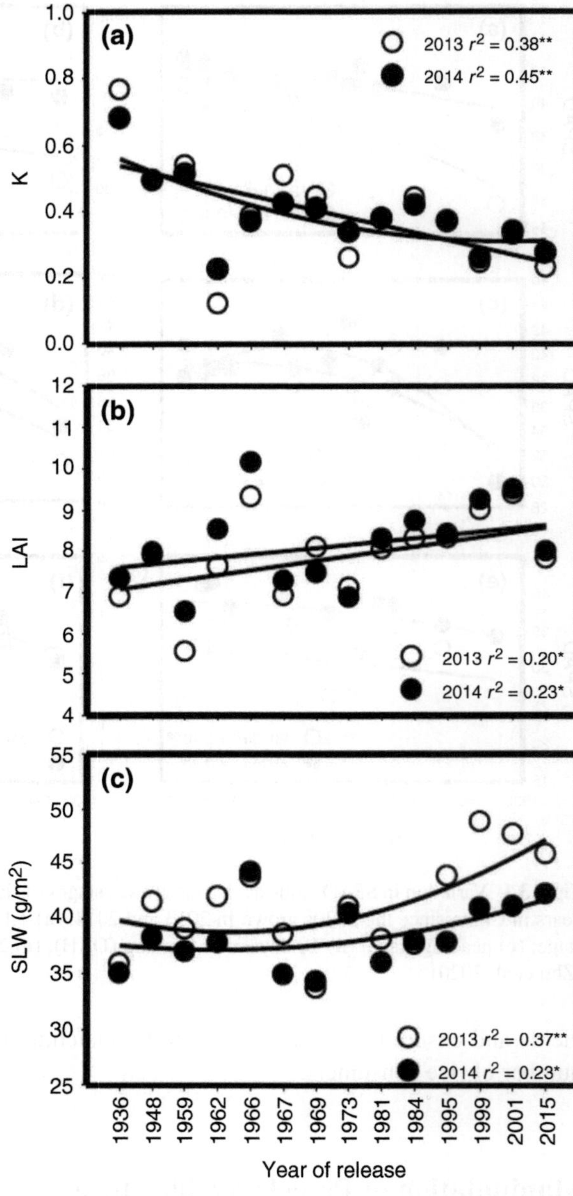

Fig. 13.2 Changes in (**a**) extinction coefficient (K), (**b**) leaf area index (LAI), and (**c**) specific leaf weight (SLW) of rice varieties released over different years in China since the 1930s grown in 2013 and 2014. Levels of significance indicated: *significant at $p \leq .05$, **significant at $p \leq .01$. (Zhu et al. 2020)

physiological constraints that need urgent solutions. Lack of adequate information on the molecular basis of ideal plant architecture is a real hindrance to genetic manipulation. It has been shown that *IPA1* (*IDEAL PLANR ARCHITECTURE1*) acts as a key regulator of the genetic network that finalizes plant architecture of new plant type rice for high grain yield. IPA1 is a functional transcriptional activator to which 1067 and 2185 genes are associated in shoot apices and young panicles respectively. *IPA1* interactions with QTLs like *DEP1*, *OsTB1,* and microRNAs miR156 and miR529 determine its developmental or cellular context-specific expression, in which plant height, tiller number, and panicle length are decided (Lu et al. 2013) (Fig. 13.3).

This concept has been extended further in a model proposed for breeding super hybrid rice by rational genetic design using *DEP1* and related genes in a cross between *japonica* sterile plant bearing desired alleles of *DEP1* and *GS3* and *indica* restorer plant with favorable alleles of *IPA1* and *Gn1a* (Xu et al. 2016) (Fig. 13.4).

Although heavy panicle feature of new plant type rice bearing numerous spikelets is most essential for high grain yield, lack of attention for molecular characterization of the genes controlling the feature, i.e. panicle architecture has precluded success of the new plant type rice as of today. Moreover, the pleiotropic characterization of the relationship between high spikelet number and other yield-related features have not been investigated correctly. Khush (1995) identified the close association between high total spikelet number with low tiller number and this

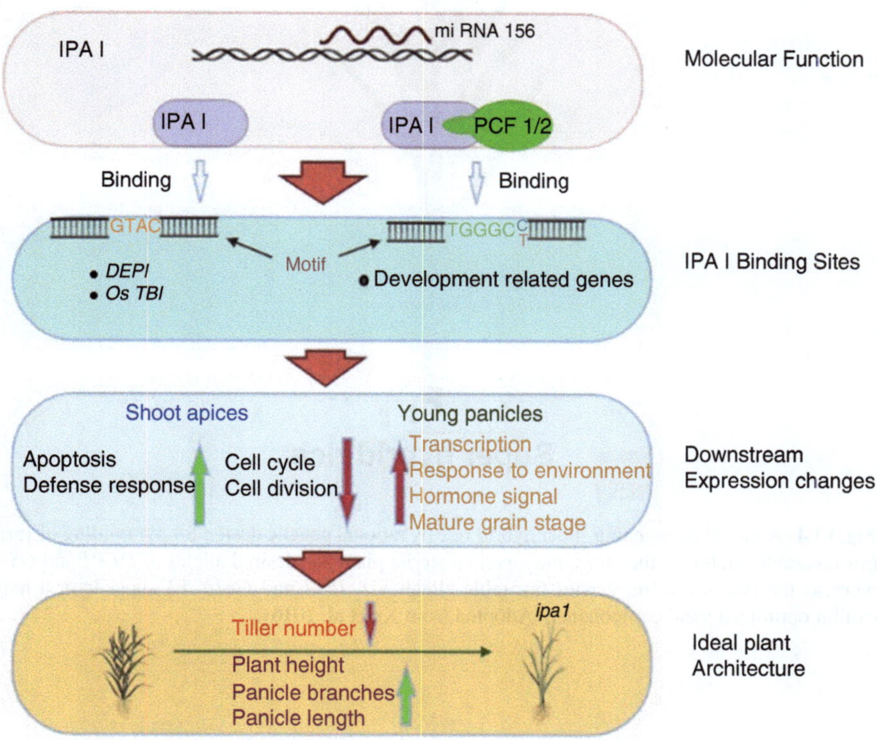

Fig. 13.3 Rice plant architecture regulation by *IPA1*. Direct binding of IPA1 to GTAC regulates plant architecture regulators like *DEP1* and *OsTB1*. On the other hand, it binds indirectly to the TGGGCC/T motif via PCF1 and PCF2 in modulating the development-related genes. As a result of the above interactions, ideal plant architecture is obtained from alteration in the expression of genes that govern various developmental processes *viz.* cell cycle, apoptosis, stress tolerance, other signaling of hormones. (Adopted from Lu et al. 2013)

Fig. 13.4 A model on breeding super hybrid rice by rational genetic design for pyramiding of various favorable alleles. In the cross, the *japonica* sterile plant has desired alleles of *DEP1* and *GS3*, whereas the *indica* restorer carries favorable alleles viz. *IPA1* and *Gn1a*. F1 super hybrid may exhibit optimized yield components. (Adopted from Xu et al. 2016)

inference was consolidated further by Sasaki et al. (2017). Thus the tradeoff between one agronomic trait with another has been a real impediment to revision of rice yield potential hitherto.

Spikelet-Specific Grain Filling

Yield potential is underscored when inferior spikelets discriminated against on account of disadvantage in the distribution of time and space during ontogeny, do not fill properly. The success of filling, hence, is spikelet specific and depends on the panicle architectural design. Ten thousand-year-old rice antiquity has revealed a change of panicle architecture in favor of adaptation to pernicious environments and for better grain yield. The display of a broad range of panicle structures not only contributes to plant architectural changes but also the reproductive success of rice under perturbed as well as stable environments; greater is the diversity larger is the reproductive success. Therefore architecture of rice inflorescence has been of great importance in agronomy. The wide diversity of architecture present in rice also contributes immensely to the developmental biology of other economically important grasses like wheat, barley, and maize because they are bereft of such advantages for adaptation. As discussed before, indices for panicle architecture estimation of rice are the length of rachis, the number, and length of primary and secondary branches, and the number of spikelets borne on the branches (Yu et al. 2017). In an enlarged spectrum, panicle architecture also includes grain shape defined by grain length, breadth, and length-width ratio (Bai et al. 2010), which contributes to yield benefit (Wang et al. 2019).

Change of Panicle Architecture in Rice Antiquity

Various dimensions of panicle architecture have changed considerably in favor of a compact and dense orientation in recent years when more spikelets are accommodated through innovative breeding efforts to increase grain number with marginal comprise of panicle size, but there has been no commensurate increase of grain yield always. The concept of increase of grain density to harness a yield benefit is a legitimate step because the lax orientation of spikelets in the panicle was considered incompatible for achieving greater yield in crop evolution, and domestication favored a transition from spread to compact type arrangement of spikelets in rice antiquity (Zhu et al. 2013). A single nucleotide polymorphism-6 in the *OsLG1* region has resulted in a change over to compact panicle architecture. For cultivated rice, an erect plant stature with a compact panicle embodies several advantages, such as high-density planting, improved harvest efficiency, more number of grains on panicle branches, and greater photosynthetic efficiency of sub-ordinate leaves spaced below the flag leaf. Yang et al. (2007) reported that the panicle of hybrid rice has more number spikelets with a yield advantage of 11–14% than the inbred *indica* or NPT rice, but such coincidence of yield benefit and spikelet number was not seen in many more large panicle rice cultivars bearing numerous spikelets. In new plant type rice, an increase of spikelet density changed panicle architecture and improved spikelet uniformity and homogeneity but exacerbated competition of individual spikelets for nutrition and other essential physiological resources for growth and development resulting in wide variation in fertility, weight, and chemical composition of individual grains. The lack of grain uniformity is unfavorable for milling and cooking qualities. As of today, there has been no clue to break the tradeoff between panicle grain number and quality and develop designer rice.

Designer Rice Panicle

Thematics for Architectural Changes

All panicle structures of rice are generated from inflorescence meristems. The mechanistic control of their development has attracted great attention in the quest for new panicle architecture. As discussed previously, there are three crucial events of meristematic activity during inflorescence ontogeny that collectively control the course of development and final panicle architecture; such as, (a) timing of rachis meristem abortion, (b) the transition of branch meristem to the terminal spikelet meristem, and (c) program shift of lateral meristem identity on a rachis branch in acropetalous fashion (Ikeda-Kawakatsu et al. 2009). The *APO1* (*ABERRANT PANICLE ORGANISATION1*) gene encoding an F-box protein defers precocious conversion of rachis branch meristems to spikelets to ensure generation of a certain number of spikelets and conserve panicle phenotype specificity. Because the timing of these events is not identical, there is no single regulatory physiological mechanism common to all. This observation increases complexities for manipulation of different types of meristem identity prevalent within the system. However, Kyozuka et al. (2014) has proposed a unique solution that merits some consideration. It was proposed that shoot apical meristem switches identity to inflorescence meristem at the onset of the reproductive period in rice and the latter generates branch meristems on its structure. Branch meristems produce several next order meristems with the progress of time, in which early ones acquire branch meristem identity and produce secondary branches, while the later ones gain spikelet meristem identity. After producing some branches and spikelets, branch meristems themselves transform into spikelet meristems. In this system, branch meristem identity has been considered indeterminate in contrast to the determinate spikelet meristem.

Spikelet Development Pre-Ordained in the Early Reproductive Phase

The transition of branch meristem to spikelet meristem identity in the meristematic phase, therefore, stands for floral organ development and conclusion of panicle organo-genesis. The specification of spikelet meristem identity precludes the scope of the ontogeny of new organ structures. The branch meristems otherwise are homologous to vegetative structures and have the potential for the genesis of new organs. Later the spikelet meristem identity more branches and more spikelets are expected in the panicle. This concept seems logical even though the ontogeny of rice panicle structures and their development have been classified into nine successive stages (Ikeda-Kawakatsu et al. 2009) and efforts are made to identify transcriptional changes at each of the stages. However, it is more prudent to classify the nine different stages broadly into two phases. In the first phase, meristematic activities define position and number branches, and spikelet differentiation on the branches provides a license for the panicle architecture. In the second phase, the branches elongate and spikelets develop on the branches. The nature of spikelet development and extent of branch elongation is pre-ordained in the first phase by a specification of meristematic identity staggered on the panicle branch nodes concerning the position of the organs. Thus exploration of genetic changes embodying phenotypic changes of the first phase is more important in the quest of designer rice.

Changes of Hormonal Balance in Transitional Stages

Gene cloning and QTL analyses have been useful for the identification of the genetic basis of yield-related traits. So long the genes/QTLs affecting panicle architecture have been characterized for the first phase in commensuration to attempts intended for increasing spikelet and branch numbers and finally grain yield. This has been done by using the F2 population, recombinant inbred lines, and double haploid populations in different studies. However, the utility of the genes has not been exploited in designer rice hitherto, because the events controlling various stages of panicle development in rice are far from simple and mired with complexities. Of course, the potential for manipulation is not out of bounds and needs a thorough revision of the concept for designer rice. Very recently, Xu et al. (2020) has predicted the presence of some genes related to panicle structure and grain shape in major QTLs of rice. The workers detected as many as 24 QTLs related to panicle structure and grain shape in the rice genome except chromosomes 8 and 11. In general, any change over of meristematic tissue identity in plant development has been considered a fate accomplished of switch in hormonal balance. The first and foremost signal that can be envisaged for the transition to spikelet meristem identity is a shift in the intrinsic ratio between cytokinin and auxin. In callus tissue differentiation, a higher concentration ratio of cytokinin to auxin promotes shoot development, an equivalent ratio enhances proliferation of undifferentiated callus and a lower ratio supports root development (Jing and Strader 2019). This notion would be most valid for organ differentiation in rice panicle also. Root meristem cell proliferation and cell elongation are promoted in rice when cytokinin contents are reduced (Wang et al. 2020).

Hormones for Organ Differentiation

In pursuance, we discussed the expression of *FON1–4, RCN1–2, TERMINAL FLOWER1, IPT, LP, WFP, LOG,* and *APO1* QTLS/genes and their role in the definition of rice panicle development earlier. The expression of all of them has been linked to a physiological signal thanks to a balance favorable for cytokinins as against auxins. The higher concentration ratio of cytokinin to auxin favored the indeterminacy of primary branch meristems more than the rachis meristem resulting in the initiation of spikelets beyond a sustainable limit, as seen in the dense panicle cultivars bearing many more spikelets than that taken to maturity. Greater indeterminacy of primary against rachis meristem increased secondary branch number, but not that of primary branch number. The closer spacing of the branches made the architecture compact. It is not understood, why the high concentration ratio of cytokinins to auxins promoted indeterminacy of the former but not the latter. Ontogenetic characterization in the meristematic phase of panicle development reveals that the mode of termination of rachis meristem activity on a time scale does not coincide with that of the primary branch meristem. After producing several branch primordial, rachis meristem loses cell proliferation activity and ends with a degenerative point. While the rachis meristem closes activity early, the branch meristems continue to produce secondary branch meristems until transformation into spikelet initial. Because of differences in temporal articulation and structural specification, it is implied that a common physiological signal cannot be the decisive factor for the endpoint activity of rachis and primary branch meristems. It is logically inferred that the role of hormones other than cytokinin could be also crucial for designer rice panicle.

Manipulation of Hormones

The role of hormones in the post-flowering stage of rice panicle development, although mired with mystery, is not as much complex as that of the pre-flowering stage. The information available is more explicit fore bearing a valid suggestion for designer rice panicle. The expression of genes controlling synthesis and activity of different hormones during grain filling of rice panicle has been discussed. Also, the antagonism that exists between positive regulators like cytokinins, gibberellins, brassinosteroids, and auxin and their negative counterpart jasmonic acid and ethylene has been elucidated. Tamaki et al. (2015), in particular, recommended inhibition of ethylene and/or jasmonic acid activity at the early stage of grain filling for improved grain yield because of the positive influence of such manipulations on source activity and carbon assimilate partitioning to grains in the post-anthesis stage. In addition to source activity, the sink strength of the spikelets for assimilate utilization has been considered crucial for grain development and yield (Mohapatra et al. 2011). This concept gains strength because appropriate doses of ethylene inhibitors like 1-methylcyclopropene (1-MCP), 2-aminoisobutyric acid (AIB), Co(NO3)2 administered at the crucial time of development, preferably before anthesis, depressed ethylene action within the first week after anthesis, when rapid cell cycle activity of the fertilized ovule fixes the final cell number and sink size of the kernel (Mohapatra et al. 2009; Panda et al. 2016). Contrary to ethylene, cytokinins have a positive effect on panicle development. The expression of genes regulating cytokinin action in rice panicle has been development stage-specific. Yamburanko et al. (2017) identified a sub-network of genes expressed at different stages of early panicle development that could modulate inflorescence architecture substantially. Ethylene has been found counterproductive to cytokinin action in underscoring endosperm cell division rate and number (Mohapatra et al. 2009).

In this context, the physiological factors that control the phenotype of dense panicle rice bearing numerous spikelets with potential for high grain number are ideal targets for scrutiny in pursuance of efforts for developing designer rice panicle. In these cultivars, the objective for high grain yield failed because of the correlative inverse relationship between spikelet number and grain filling percentage (Panigrahi et al. 2019). Other than the level of ethylene production at anthesis, there has been no other hormonal signal found instrumental for the inverse relationship. In rice panicle, the magnitude of ethylene production has been spikelet specific with higher level resulting in ill-developed spikelets. The demonstration of the influence of ethylene level in grain development through the hormone specific promoter/inhibitor intervention techniques as well as through comparative genomics and proteomics study of genotypes of varied panicle architecture has opened up possibilities of selectively improving grain occurrence of inferior spikelets in designer rice (Mohapatra et al. 2011) (Fig. 13.5).

FZP Controlling Meristem Identity

In this context, the genes/QTLs controlling synthesis and activity of gibberellins, ethylene, auxins, ABA, and brassinosteroids were assessed with their role in meristematic activity as well as identity. The promotion of meristem activity of inflorescence and/or branches, either primary or secondary, and appropriate delay in spikelet meristem specificity could help to increase spikelet number and panicle size. The most distinguished among them is the identification *FRIZZY PANICLE (FZP)* gene that represses panicle branching and positively influences floral meristem identity through the expression of an ethylene-responsive factor (ERF) (Komatsu et al. 2001,

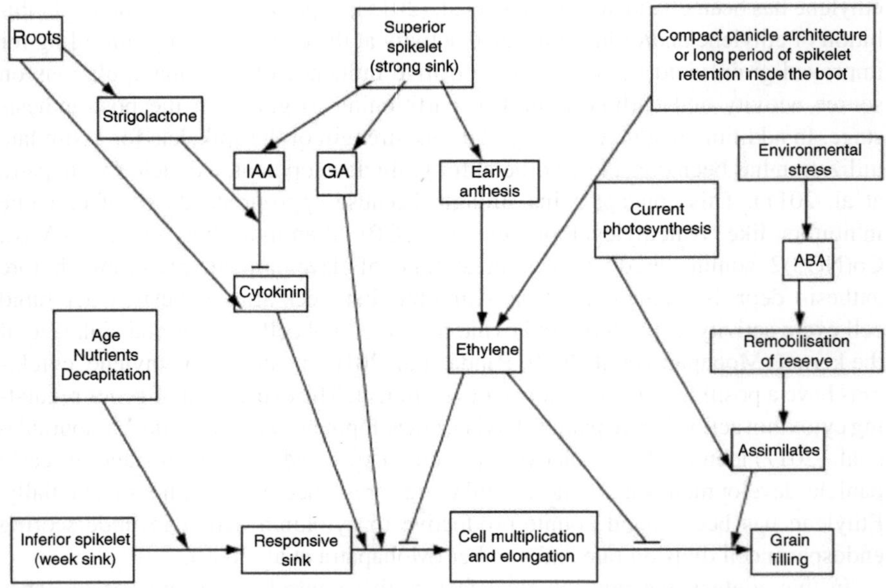

Fig. 13.5 Schematics of factors influencing the development of superior and inferior spikelets of rice panicle. (Mohapatra et al. 2011)

2003). In comparison, transcriptional dynamics of other hormones related to inflorescence meristematic tissue specification or proliferation are scantly studied. To elucidate the relationship existing between different floral identity genes, Bai et al. (2016) have monitored the roles of *FZP* and *RFL/APO2* (*ABERRANT PANICLE ORGANISATION2*) genes in the panicle meristems. *FZP* is a major negative regulator of *RFL/APO2* and its expression is crucial for the transition of panicle branching to spikelet formation. Superimposition of *FZP* on expression of QTLs/genes controlling panicle branching of rice panicle (*GN1a, DEP1, RCN1, LAX1, RFL/APO2*) at two stages of reproductive development i.e. very early stage with panicle length of 1.5–2.5 mm and later stage with length more than 1 cm revealed transcription of *RFL/APO2* high in the *fzp-11* mutant, medium in wild type and low in *FZP*-overexpressed (*OX-FZP-(fzp-11)* line) at both the stages panicle development. Compared to wild type, panicle branch number was higher in *fzp-11*, but the number declined in *OX-FZP-(fzp-11)* (Table 13.1).

In this situation probably downregulation of ERF expression in *fzp-11* allowed elevation of cytokinin pool of the meristematic tissue for the production of more panicle branches by modification of ABCDE floral organ identity genes; particularly *FZP* overexpression favorably regulated B-class genes, *AGL6* (*OsMADS6, OsMADS17*) and E-class genes (*OsMADS1, OsMADS7, OsMADS8*) in the floral meristem. Transcriptional silencer-mediated *FZP* expression increased spikelet number per panicle in rice (Bai et al. 2017). This group of researchers has corroborated that repressed expression of *FZP* along with five *SEPALLATA*-like genes

Table 13.1 Phenotype values of rice plants *fzp-11*, Wt(DJ), and *OX-FZP-(fzp-11)*

Traits	Panicle length (cm)	Flag leaf width (cm)	Flag leaf length (cm)	No. of roots	No. of primary branch	No. of secondary branch	Spikelets per panicle
***fzp*-11(M ± SD)**	18.0 ± 3.0	1.31 ± 0.10	39.0 ± 14.8	51 ± 15	9.9 ± 2.2	38 ± 9.1	0
Wt(DJ) (M ± SD)	18.3 ± 2.7	1.29 ± 0.15	37.2 ± 16.0	48 ± 13	10.3 ± 2.0	23.3 ± 5.6	100 ± 6
OX-*FZP*-(*fzp-11*) (M ± SD)	8.5 ± 1.1[a]	1.48 ± 0.16[a]	21.3 ± 4.7[a]	34 ± 11[a]	9.8 ± 2.1	5.2 ± 2.7[a]	53 ± 19[a]

[a]Significantly different between the plants of *OX-FZP-(fzp-11)* and wild type (DJ) and between *OX-FZP- (fzp-11)* and *fzp-11*. (Adopted from Bai et al. 2016)

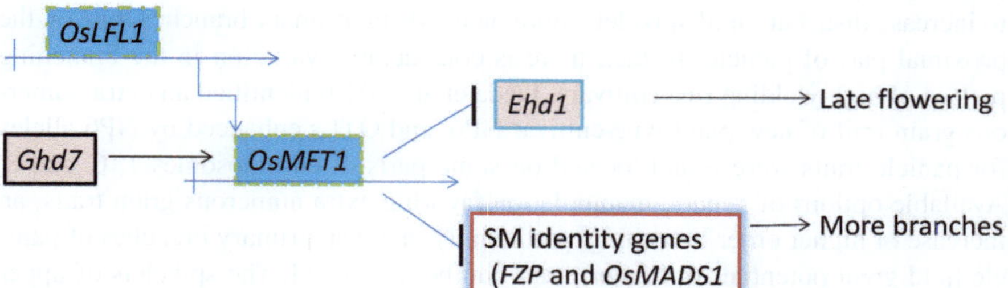

Fig. 13.6 A working model proposed for OsMFT1. OsLFL1 and GHD7 are transcription factors. *Ghd1(Grain number, plant height, and heading date1)*, *Ehd1(Early heading date1)*. (Adopted from Song et al. 2018)

would delay the transition of the branch to floral meristem identity and increase panicle size (Song et al. 2018). They identified a gene *MOTHER OF FT AND TFL1* (*OsMFT1*), which down-regulates expression of *FZP* and five *SEPALATTA like* (*OsMADS1, OsMADS5, OsMADS7, OsMADS8, OsMADS34*) genes and thereby delays heading date and promotes branch and spikelet production. The *SEPALLATA* like MADS-box genes belongs to lass E of floral organ determinacy gene and specification of organs like sepals, petals, stamens, and carpels (Cui et al. 2010). The over-expression of *OsMFT1* increases spikelet number as much as by two times. Accordingly, a model for *OsMFT1* regulation has been proposed by Song et al. (2018) (Fig. 13.6). OsLFL1 and GHD7 are transcriptional factors necessary for the activation of *OsMFT1* expression. OsLFL1 directly binds to OsMFT1 to activate its expression. *OsMFT1* negatively regulates flowering down-stream of *Ghd1* and up-stream of *Ehd1*. *OsMFT1* represses the expression of genes for floral meristem identity and thereby prolongs branch differentiation to increase branch number. Hitherto it is too premature to speculate how ethylene would regulate meristematic activity and specification flower organs of rice panicle in the reproductive stage with co-ordinated action of cytokinin.

Promotion of Higher-Order Branching

Assimilate partitioning in spikelets is dependent on the expression of genes for the endosperm starch synthesizing enzymes and the genes express poorly in the inferior spikelets because of high ethylene synthesis and enhanced transduction of its signal (Sekhar et al. 2015). This trait is genetically determined by a double recessive allele where a corresponding dominant allele can overwhelm the linkage between ethylene and grain starch synthesis successfully (Das et al. 2018). Ethylene apart, there is no information available on any physiological/hormonal factors targeting control of assimilate partitioning and spatial distribution of grain filling attributes of spikelets on various branches of rice panicle. In general, ethylene generated at the pre-anthesis stage accumulates in the boot and proximal spikelets get exposed more to it than the apical ones. Hence, designing panicle architecture should give attention to increase distribution of spikelets more on the distal primary branches than on the proximal part of panicle. In fact, there is considerable variation in the branching pattern of high yielding rice cultivars. Ikeda et al. (2010) identified an extra numerous grain trait of new plant type cultivar NP6, and QTLs enhanced by NP6 alleles for panicle traits were found located on some parts of chromosomes 1, 6, and 8. Available options of genetic manipulation favoring extra numerous grain traits, an increase of higher order branching, particularly in apical primary branches of panicle hold great potential for higher grain number and yield. The spikelets of upper branches emerge early from the boot, they escape the deleterious effects of ethylene for the promotion of grain filling. Fujishiro et al. (2018) identified a QTL *qSrn7* responsible for the increase of secondary branching with no concomitant effect on primary branch number. Association of this QTL with *FZP* promoted higher order branching in the upper primary branches of the panicle. Hence, a combination package of *qSrn7/FZP* assuring higher order branching on the distal primary branches could be a useful strategy for designer rice.

Change Over to the Indeterminate Pattern

In Chap. 9 it is observed that inferior spikelets of the proximal part of the panicle do not fill grains on account of metabolic dominance of superior spikelets on the apical part and the dominance contributes significantly to the heterogeneous architecture of the panicle amounting to the formation of several ill-developed grains. During ontogeny cell division activity of rachis meristem could be a potential determinant for branching ability of primary branch meristem in producing secondary branches and spikelets (Yamaki et al. 2010); fast decline in activity of rachis meristem successively upwards reduces secondary organ number of primary branches in acropetal fashion resulting in a conical shaped primitive type panicle. In this situation role of *TFL1* (*TERMINAL FLOWER1*) a member of PEBP family genes of cereal plants (Chardon and Damerval 2005) becomes crucial. *TFL1* regulates transition to flowering, where another member of PEBP gene family *FT* has an antagonist influence. *TFL1* inhibits apical meristem achieving floral identity and accounts for indeterminate growth of inflorescence producing more numbers of primary branches. In comparison, FZP (Bai et al. 2016) and TAW1 (Yoshida et al. 2013) regulate the transition of branch meristem to spikelet identity fixing the number of secondary branches. Either way manipulation of primary or secondary branching in the meristematic phase increasing indeterminacy can improve the spikelet number of the panicle. A moderate reduction of primary branch meristem activity along the panicle axis encourages secondary branching ability of primary branches at the distal part of the panicle with the cost of reduction of the numbers on the proximal part, amounting to truncated conical or cylindrical shaped panicle (Yamaki et al. 2010). In the new large panicle rice cultivars released in the late 1980s (Peng et al. 1999), the spikelet number per panicle increased thanks to an increase of secondary or higher-order branching (Yamaki et al. 2010), instead of any significant improvement of primary branching. The activity of inflorescence meristem declines progressively and degenerates finally to fix the final number of primary branches.

Homogeneity in Secondary Branching of Primary Branches

The increase of secondary as against primary branches made the architecture compact increasing competition for resources between the closely packed spikelets. As expected, there was no improvement in yield despite a rise in spikelet/grain number per panicle. Hence, the targeted grain number was scaled down to 150 from more than 250 in the second version of new plant type rice (Peng et al. 2008). Given this anomaly, i.e. the tradeoff between spikelet number and grain filling, designer rice panicle must embrace alleles for reduced dominance, although not absolute homology in development that is not possible because of the temporal pattern of genesis and primigenic dominance of the older over newer organs in any plant system (Mohapatra et al. 2011). In rice panicle homogeneity of spikelet development could be achieved by reducing the gradient in the number of secondary branches/spikelets between the apical and proximal parts of the panicle. Improving the genesis of these organs on the apical part of the panicle would present a cylindrical pattern of spikelet distribution, which is considered evolutionarily advanced and sustainable for grain filling as against the conical pattern considered primitive and low yielding. Comparison of development, spikelet number, and grain filling percentage between old traditional cultivar Akul and relatively modern cultivars Jaya and IR36 have indicated that the difference between the potential and the actual number of grain decreases in all cultivars from base to apex of panicle and this difference widens with an increase of spikelet number (Mohapatra 2013) (Fig. 13.7).

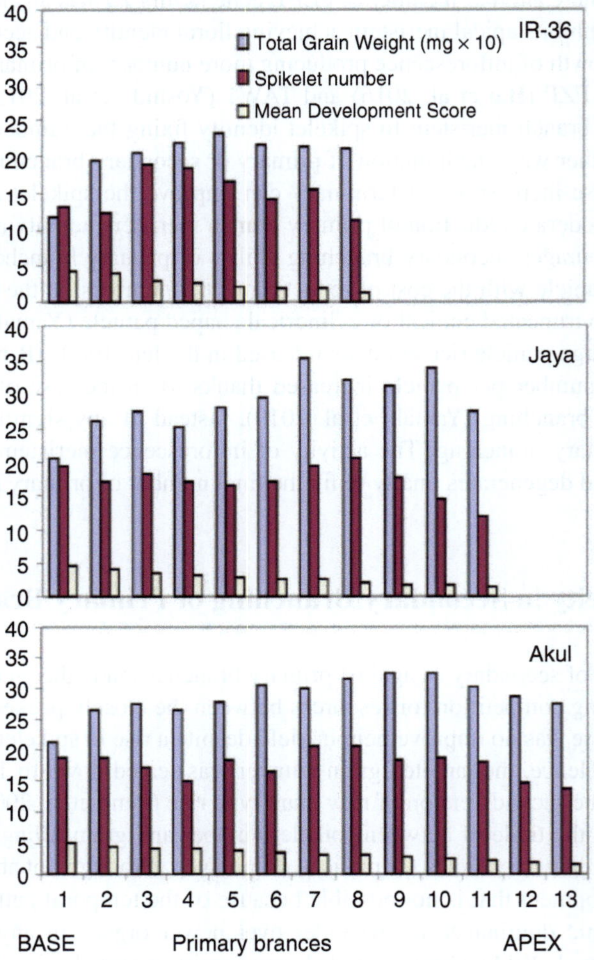

Fig. 13.7 Grain weight, spikelet number and mean developmental score of primary branches of two semidwarfs (IR36 and Jaya) and one tall *indica* (AKUL) rice cultivars differing in panicle size. A high developmental score indicates less development and vice versa. (Mohapatra 2013)

Cylindrical Panicle Architecture for Designer Rice

It is surmised that rice cultivars developed so long have not achieved a perfect cylindrical-shaped panicle structure, although at best they could be truncated conical. The newly developed large panicle rice has failed in achieving the objective of high grain filling for not being able to embody a cylindrical pattern of spikelet distribution. It is stated in Chap. 8 that rice panicle architecture, being defined by primary and secondary branch numbers, is determined by activities of rachis and branch meristems. The activity of meristem stronger in rachis than that of a lateral primary branch would increase spikelet distribution more on primary branches compared to the secondary branches in designer rice for the formation of a cylindrical pattern. In this changed scenario, there will be a preference for more primary branches than secondary branches resulting in simplification of the branching pattern. Herein rachis meristem activity will be delayed by the deferral of degenerative point formation that fixes the terminal primary branch of the inflorescence. Therefore the determinate rice panicle may be made relatively indeterminate like barley thanks to the prolonged activity of rachis meristem. The compromise achieved on branch meristem activities would simplify panicle architecture, where floral meristems (spikelets) will be borne directly on inflorescence meristems in place of branch meristems; a situation closure to the indeterminate habit of Arabidopsis inflorescence. This feature will help grain filling because of increased laxity in space and consequential decline in friction-induced ethylene production of closely packed spikelets detrimental to grain filling. Alleles for high grain filling are available in large panicle rice (Meng et al. 2016; Das et al. 2018), but the spatial and temporal distribution of spikelets on the panicle in such type of cultivars has not been scrutinized. Terao et al. (2010) revealed the gene that controls primary branch number also controls vascular bundle formation in rice panicle and hence, has the potential to improve harvest index and grain yield. Designer rice panicle, therefore, should possess more primary branches in place of the secondary branches for a yield advantage.

As discussed previously, branching pattern is determined by spatiotemporal developmental decisions in inflorescence, branch, and spikelet meristems between the maintenance of determinacy and indeterminacy. There are reports that *Arabidopsis* genes *LEAFY (LFY)* and *APETALA1(AP1)*, *CAULIFLOWER (CAL)*, *FRUITFUL (FUL)*, and *Antirrhinum* genes *FLORICAULA (FLO)* and *SQUAMOSA (SQUA)* promote determinacy by terminal flower formation, whereas *TERMINAL FLOWER (TFL1)* gene of the former and *CENTRORADIALIS (CEN)* gene of the latter encourage an indeterminate inflorescence (cited from (Lu et al. 2017). *RCN1* and *RCN2* are rice homologs of *Arabidopsis* genes *TFL1* and *CEN* have the potential for promotion of indeterminacy and production of more panicle branches (Nakagawa et al. 2002). In transgenic rice *RCN1* and *RCN 2* delay phase change of branch shoot to floral meristem state leading to initiation of higher-order panicle branches and formation of a highly branched and denser panicle morphology. Further, Lu et al. (2017) identified that *OsRAMOSA2* (*OsRA2*) gene follows the same pathway of axillary meristem gene *LAX1 (LAX PANICLE 1)* for control of rice panicle architecture. The gene acts downstream *RCN2* to regulate pedicel and branch length while action upstream of *RCN2* controls secondary branch number. Recently Zhao et al. (2016) reported that single nucleotide polymorphism (G/C) largely affects the number of primary and secondary branches and grain number per panicle in dense and erect panicle rice through screening of 72 high yielding japonica rice. They found the genetic diversity of the *DEP1* locus very narrow and suggested modification of the SNP(G/C) at the promoter region to contribute to the flexible application of *DEP1* in rice breeding for greater yield. With technology permissive of indeterminacy panicle architecture, designer rice could be more

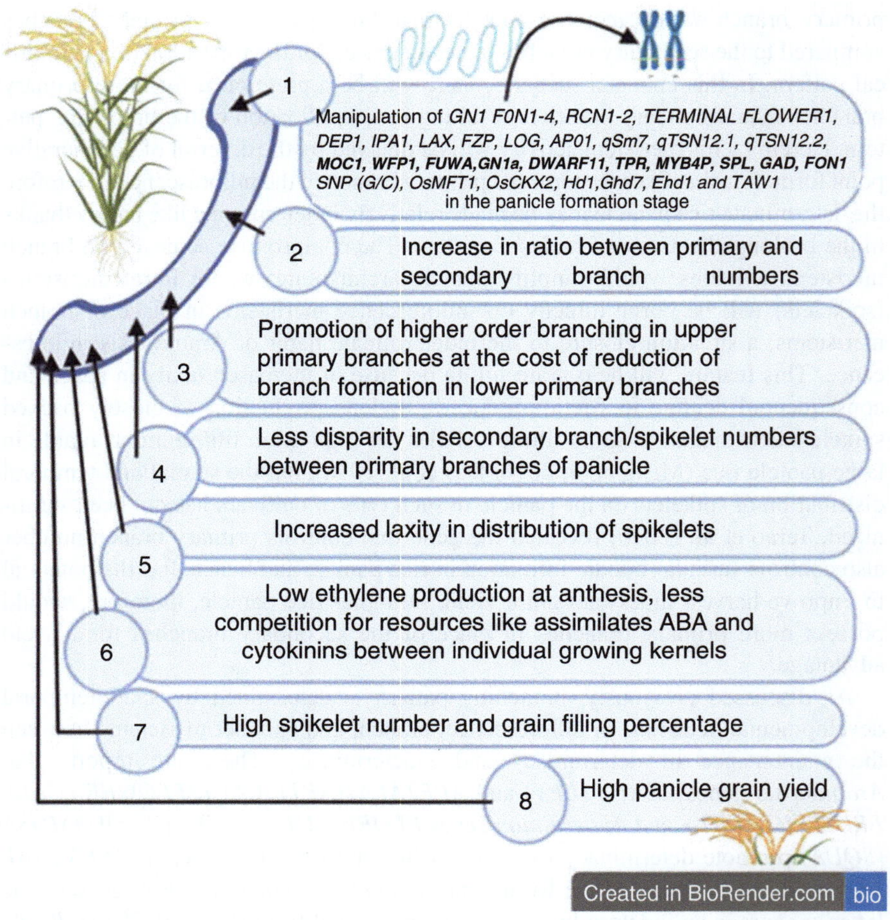

Fig. 13.8 Schematic flow of events for designer rice panicle development

branched, although the intent is not for an increase of secondary branch number, but an increase of primary branch number. Herein we propose a schematic flow of events in pursuit of the development of designer rice panicle. (Fig. 13.8).

Designing Panicle with Multi-Floret Spikelets

As of now most of the genes identified for panicle grain number have influenced panicle architecture by regulating number of primary and secondary branches and spikelet density. An extra-large panicle possessing spikelets in compact orientation is not always an ideal option for high grain filling because of coincidental loss of apical dominance. Apical dominance exists in panicle, when spikelets originate and develop in heterogeneous fashion. Heterogeneous architecture induces co-morbidity of the early formed inferior spikelets at the proximal part of panicle, while promoting growth of the apical spikelets. The asynchronous flowering system has been a hall mark of rice adaptation to diversified and unstable ecological conditions that could not be dispensed with for gain of yield benefits. In pursuance, Mohapatra (2013) proposed modification of panicle architecture to a simple indeterminate inflorescence bearing spikelets more on primary than secondary branches that would be possible once rachis meristem activity is extended beyond degenerative point. Spikelet number higher on primary than secondary branches would allow lax orientation for derailment of ethylene production and promotion of grain filling capacity.

Notwithstanding promotion of spikelets placement on primary than secondary branches, transfer of meristem determinacy to indeterminacy have potential for formation of florets more than one per spikelet, a trait embellished in some indigenous rice mutants (Mishra et al. 1992, Sahu et al. 2017). The typical rice spikelet possesses one fertile floret and one grain. But several workers (Ren et al. 2013, 2018, 2019; Zhang et al. 2017) have shown that depression of some gene functions lead to formation of multi-floret spikelets in the panicle. These mutants have floral organ defects or negative trait genotypes antithetic for effective pyramidization into high yielding rice. Hence, they have not been exploited for use in novel rice breeding technology hitherto in spite of high potential values.

In a recent review, Ren et al. (2020) have described new approaches for altering meristem determinacy and restoration of lateral floret formation for developing a multi-grain spikelet. As noted earlier, rice spikelet meristem produces two rudimentary glumes prior to development of two lateral floral meristems for two sterile lemmas, and one terminal meristem for a fertile floret; all three organs combined together constitute one spikelet, unlike wheat where all florets of spikelet are fertile. Transformation of terminal meristem to indeterminacy or restoration of degraded lateral floral meristem can rejuvenate meristematic tissue for formation of more number of fertile florets. Both the options have been discussed in order to develop multi-floret spikelet.

(a) Transformation of spikelet from determinacy to indeterminacy (Fig. 13.9)

It is proposed that three types of genes are involved in terminating indeterminate fate of spikelet meristem of rice panicle and thereby allowing consumption of the whole meristem for development of one single pistillate spikelet. Instead continuity of indeterminacy can result in formation of 2–6 pistils (Mishra et al. 1992). In determinate spikelet type1 *LHS/OsMADS1* (*LEAFY HULL STERILE1/OsMADS1*)-like genes express in the in the terminal floral meristem that results in specification of terminal meristem. Conversely, *LHS1* genes are expressed in all the floral meristems of indeterminate spikelet. Type 2 comprises *IDS1*(INDETERMINATE *SPIKELET1*)-*like* genes of the AP2/ERF family and they initiate transformation of spikelet meristem into floral meristem. Mutation of these genes prolongs spikelet meristem activity for production of more florets (Ren et al. 2013). Similarly, *MULTI-FLORET SPIKELET1*

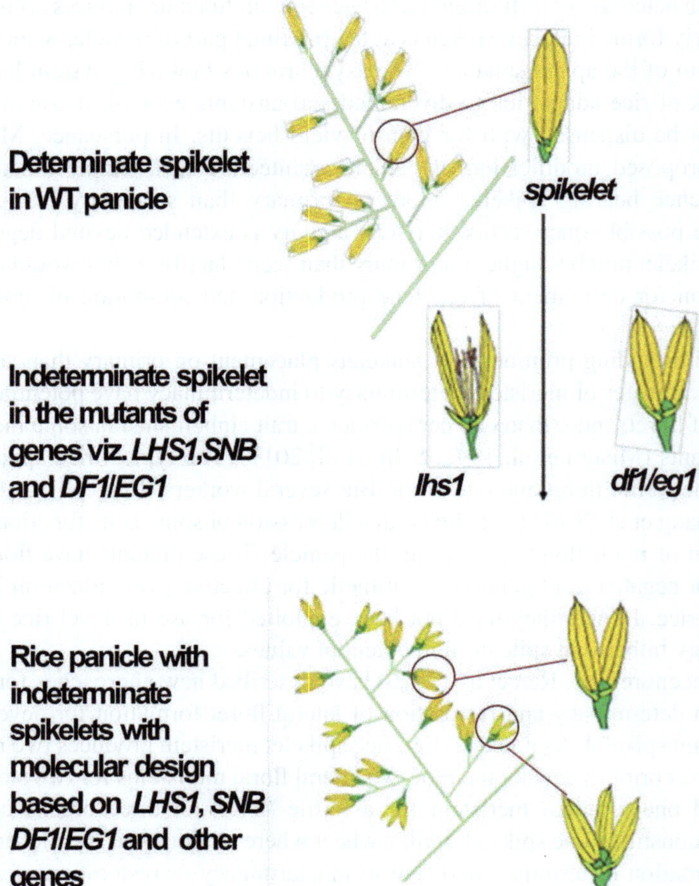

Fig. 13.9 Transformation of determinate rice panicle to indeterminate structure. For this achievement, it is necessary to get rid of negative traits, such as the leafy hull in the *lhs1* mutant, and incomplete penetrance in the *df1* mutant by molecular and/or genetic modification of related genes. For example, in the *lhs1* mutant, the spikelet produces two independent florets, but the lemma and palea are elongated and leafy. In the *df1/eg1* mutant, the spikelet produces two normal florets and grains, but has a low percentage of two-floret spikelet. *WT* wild-type. (Adopted from Ren et al. 2020)

(*MFS1*) encodes an ERF domain protein and loss of function in *mfs1* mutant gives rise to extra lemmas or two florets. Mutation of *IDS1-like* gene (*SUPER NUMERARY BRACT* (*SNB*) in both single as *snb* or double as *snb.ids1*leads to formation of more fertile florets or sterile secondary florets per spikelet (Lee et al. 2007; Lee and An 2012). These workers found that both *SNB* and *OsIDS1* genes synergistically control the meristem establishment of rice. The third type genes, namely *DF1/EG1*, *FON4* and *TOB1*, encode a lipase, a receptor-like kinase and one YABBY protein respectively, for regulation of spikelet determinacy in rice. Suppression of these gene expression induces the formation of two florets or extra lemmas (Li et al. 2009; Tanaka et al. 2012; Ren et al. 2018). Loss of function of the genes in *eg1* and *tob1* mutants leads to formation of an extra lemma-like organ (sterile floret), whereas *dfl* (allelic to *eg1*) mutant induces formation of two fertile florets per spikelet. The mutants of *FON4* have enabled formation of multiple florets in spikelet (Ren et al. 2019).

(b) Restoration of degenerative lateral spikelets (Fig. 13.10).

The other way to design a multi-floret spikelet is to restore floral identity of two sterile lemmas at the base of fertile floret of each spikelet of rice panicle. Presence of sterile lemmas is indicative of degeneration of lemma of lateral florets specific to the position on the pedicel. This composition of two sterile

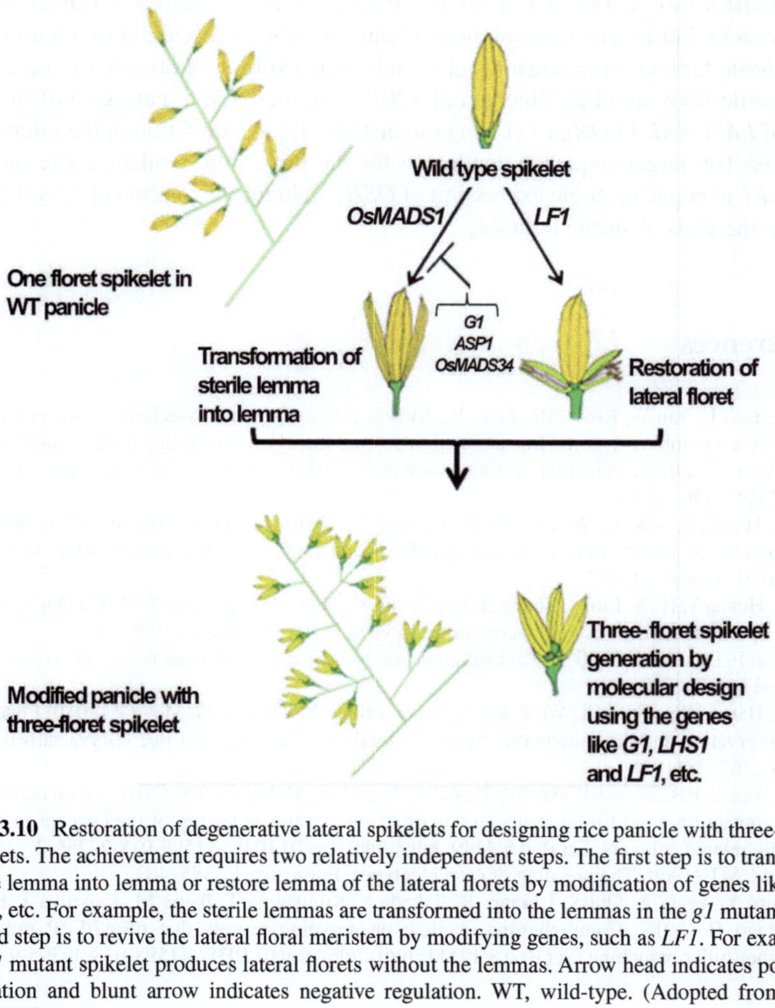

Fig. 13.10 Restoration of degenerative lateral spikelets for designing rice panicle with three-floret spikelets. The achievement requires two relatively independent steps. The first step is to transform sterile lemma into lemma or restore lemma of the lateral florets by modification of genes like *G1*, *LHS1*, etc. For example, the sterile lemmas are transformed into the lemmas in the *g1* mutant. The second step is to revive the lateral floral meristem by modifying genes, such as *LF1*. For example, the *lf1* mutant spikelet produces lateral florets without the lemmas. Arrow head indicates positive regulation and blunt arrow indicates negative regulation. WT, wild-type. (Adopted from Ren et al. 2020)

florets and one fertile floret gravitate support for the three floret spikelet hypothesis of rice. There are several genes amenable for modification to restore fertile floret identity of sterile lemma. The gene *LONG STERILE LEMMA/ ELONGATED EMPTY GLUME1* (*G1/ELE1*) belonging to a plant specific gene family, encodes DUF640 protein, but loss of expression in *g1/ele* mutant encourages transformation of sterile lemma to a lemma-like organ (Yoshida et al. 2009; Hong et al. 2010). Similarly *PANICLE PHYTOMER2* (*PAP2*) encodes *OsMADS34*, responsible for development of the sterile lemma, but interruption of its expression in *pap2/osmads34* mutation leads to formation of an elongated sterile lemma with lemma identity (Gao et al. 2010; Kobayashi et al. 2010). *OsMADS1/LHS1* gene specifies identity of lemma and palea; the gene is not found expressed in sterile lemma, but its ectopic expression induces to an enlarged lemma (Wang et al. 2017). Yoshida et al. (2012) reported that *ABBERANT SPIKELET AND PANICLE1* (*ASP1*) produces a TOPLESS-like protein, but loss of function in *asp1* mutant induces formation of a lemma-like sterile lemma. In contrast to retrograde gene expression prompting increase of fertile floret number, Zhang et al. (2017) reported a dominant gain of function of *LATERAL FLORET1* (*LF1*) mutant embodying a mutation in the microRNA 165/166 target sequence conducive for the three floret spikelet. The mutated *LF1* generates ectopic expression of *OSH1* to induce formation of lateral florets in the axils of sterile lemmas.

References

Bai X, Luo L, Yan W, Kovi MR, Zhan W, Xing Y (2010) Genetic dissection of rice grain shape using a recombinant inbred line population derived from two contrasting parents and fine mapping a pleiotropic quantitative trait locus qGL7. BMC Genet 11:16. https://doi.org/10.1186/1471-2156-11-16

Bai X, Huang Y, Mao D, Wen M, Zhang L, Xing Y (2016) Regulatory role of *FZP* in the determination of panicle branching and spikelet formation in rice. Sci Rep 6:19022. https://doi.org/10.1038/srep19022

Bai X, Huang Y, Hu Y, Liu H, Zhang B, Smaczniak C, Hu G, Han Z, Xing Y (2017) Duplication of an upstream silencer of *FZP* increases grain yield in rice. Nat Plants 3:885–893

Chardon F, Damerval C (2005) Phylogenomic analysis of the PEBP gene family in cereals. J Mol Evol 61:579–590

Cui R, Han J, Zhao S, Su K, Wu F, Du X, Xu Q, Chong K, Theiben G, Meng Z (2010) Functional conservation and diversification of class E floral homeotic genes in rice (Oryza sativa). Plant J 61:767–781

Das K, Panda BB, Shaw BP, Das SR, Dash SK, Kariali E, Mohapatra PK (2018) Grain density and its impact on grain filling characteristic of rice: mechanistic testing of the concept in genetically related cultivars. Sci Rep 8:4149. https://doi.org/10.1038/s41598-018-22256-2

Donald CMT (1968) The breeding of crop ideotypes. Euphytica 17:385–403

Fujishiro Y, Agata A, Ota S, Ishihara R, Takeda Y, Kunishima T, Ikeda M, Kyozuka J, Hob T, Kitano H (2018) Comprehensive panicle phenotyping reveals that qSrn7/FZP influences higher-order branching. Sci Rep 8:12511. https://doi.org/10.1038/s41598-018-30395-9

Gao X, Liang W, Yin C, Ji S, Wang H, Su X, Guo C, Kong H, Xue H, Zhang D (2010) The SEPALLATA-like gene OsMADS34 is required for rice inflorescence and spikelet development. Plant Physiol 153:728–740

Hong L, Qian Q, Zhu KM, Tang D, Huang ZJ, Gao L, Li M, Gu MH, Cheng ZK (2010) ELE restrains empty glumes from developing into lemmas. J Genet Genomics 37:101–115

Ikeda M, Hirose Y, Takashi T, Shibata Y, Yamamura T, Komura T, Doi K, Ashikari M, Matsuoka M, Kitano H (2010) Analysis of rice panicle traits and detection of QTLs using an image analyzing method. Breed Sci 60:55–64

Ikeda-Kawakatsu K, Yasuno N, Oikawa T, Iida S, Ngato Y, Maekawa M, Kyuzuka J (2009) Expression level of *ABERRANT PANICLE ORGANIZATION1* determines rice inflorescence form through control of cell proliferation in the meristem. Plant Physiol 150:736–747

Jing H, Strader LC (2019) Interplay of auxin and cytokinin in lateral root development. Int J Mol Sci 20:486. https://doi.org/10.3390/ijms20030486

Khush GS (1995) Breaking the yield frontier of rice. GeoJournal 35:329–332

Kobayashi K, Maekawa M, Miyao A, Hirochika H, Kyozuka J (2010) PANICLE PHYTOMER2 (PAP2), encoding a SEPALLATA subfamily MADS-box protein, positively controls spikelet meristem identity in rice. Plant Cell Physiol 51:47–57

Komatsu M, Maekawa M, Shimamoto K, Kyozuka J (2001) The *LAX1* and *FRIZZY PANICLE 2* genes determine the inflorescence architecture of rice by controlling rachis-branch and spikelet development. Dev Biol 231:364–373

Komatsu M, Chujo A, Nagato Y, Shimamoto K, Kyozuka J (2003) FRIZZY PANICLE is required to prevent the formation of axillary meristems and to establish floral meristem identity in rice spikelets. Development 130:3841–3850

Kyozuka J, Tokunaga H, Yoshida A (2014) Control of grass inflorescence form by the fine-tuning of meristem phase change. Curr Opin Plant Biol 17:110–115. https://doi.org/10.1016/j.pbi.2013.11.010

Lee DY, Lee J, Moon S, Park SY, An G (2007) The rice heterochronic gene SUPERNUMERARY BRACT regulates the transition meristem to floral meristem. Plant J 49:64–78

Lee DY, An G (2012) Two AP2 family genes, *Supernumerary bract* (*SNB*) and *Osindeterminate spikelet 1* (*OsIDS1*), synergistically control inflorescence architecture and floral meristem establishment in rice. Plant J 69:445–461

Li HG, Xue DW, Gao ZY, Yan MX, Xu WY, Xin Z, Huang DN, Qian Q, Xue YB (2009) A putative lipase gene EXTRA GLUME1 regulates both empty-glume fate and spikelet development in rice. Plant J 57:593–605

Lu Z, Yu H, Xiong G, Wang J, Jiao Y, Liu G, Jing Y, Meng X, Hu X, Qian Q, Fu X, Li J (2013) Genome-wide binding analysis of the transcription activator IDEAL PLANT ARCHITECTURE1 reveals a complex network regulating rice plant architecture. Plant Cell 25:3743–3759. https://doi.org/10.1105/tpc.113.113639

Lu H, Dai Z, Li L, Wang J, Miao X, Shi Z (2017) *OsRAMOSA2* shapes panicle architecture through regulating pedicel length. Front Plant Sci 8:1538

Meng T-Y, Wei H-H, Chao L, Dai Q, Xu K, Huo Z, Wei H, Guo B, Zhang H (2016) Morphological and physiological traits of large-panicle rice varieties with high filled-grain percentage. J Integr Agric 15:1751–1762

Mishra RN, De RN, Bose LK, Panigrahi A (1992) A productive multi-pistilate mutation in rice. Oryza 29:356–360

Mohapatra PK (2013) A physiological approach to design panicle architecture in the quest for high grain yield. In: Muralidharan K, Siddiq EA (eds) International dialogue on perception and prospects of designer rice. Society for Advancement of Rice Research, Directorate of Rice Research, Hyderabad, pp 131–142

Mohapatra PK, Panigrahi R, Turner NC (2011) Physiology of spikelet development on the rice panicle: is manipulation of apical dominance crucial for grain yield improvement? Adv Agron 110:333–359

Mohapatra P, Sarkar R, Kuanar S (2009) Starch synthesizing enzymes and sink strength of grains of contrasting rice cultivars. Plant Sci 176:256–263

Muralidharan K, Prasad G, Rao C, Siddiq E (2019) Genetic gain for yield in rice breeding and rice production in India to meet with the demand from increased human population. Curr Sci 116:544–560

Nakagawa M, Shimamoto K, Kyozuka J (2002) Overexpression of *RCN1* and *RCN2*, rice *TERMINAL FLOWER 1/CENTRORADIALIS* homologs, confers delay of phase transition and altered panicle morphology in rice. Plant J 29:743–750

Panda B, Badoghar A, Sekhar S, Shaw B, Mohapatra PK (2016) 1-MCP treatment enhanced expression of genes controlling endosperm cell division and starch biosynthesis for improvement of grain filling in a dense-panicle rice cultivar. Plant Sci 246:11–25

Panigrahi R, Kariali E, Panda BB, Lafarge T, Mohapatra PK (2019) Controlling the trade-off between spikelet number and grain filling: the hierarchy of starch synthesis in spikelets of rice panicle in relation to hormone dynamics. Funct Plant Biol 46:507–523

Peng S, Cassman KG, Virmani S, Sheehy J, Khush G (1999) Yield potential trends of tropical rice since the release of IR8 and the challenge of increasing rice yield potential. Crop Sci 39:1552–1559

Peng S, Khush GS, Virk P, Tang Q, Zou Y (2008) Progress in ideotype breeding to increase rice yield potential. Field Crop Res 108:32–38

Ren D, Li Y, Zhao F, Sang X, Shi JQ, Wang N, Guo S, Ling Y, Zhang C, Yang Z, He G (2013) MULTI-FLORETSPIKELET1, which encodes an AP2/ERF protein, determines spikelet meristem fate and sterile lemma identity in rice. Plant Physiol 162:872–884

Ren D, Yu H, Rao Y, Xu Q, Zhou T, Hu J, Zhang Y, Zhang G, Zhu L, Gao Z, Chen G, Guo L, Zeng D, Qian Q (2018) Two-floret spikelet as a novel resource has the potential to increase rice yield. Plant Biotechnol J 16:351–353

Ren D, Xu Q, Qiu Z, Cui Y, Zhou T, Zeng D, Guo L, Qian Q (2019) FON4 prevents the multi-floret spikelet in rice. Plant Biotechnol J 17:1007–1009

Ren D, Li L, He G, Qian Q (2020) Multi-floret spikelet improves rice yield. New Phytol 225:2301–2306

Sahu GR, Sarawgi AK, Tiwari JK (2017) New genotypes with multiple pistils and inheritance pattern of this strain in rice. Electron J Plant Breed 8:998–1000

Sasaki K, Fujita D, Koide Y, Lumanglas PD, Gannaban RB, Tagle AG, Obara M, Fukuta Y, Kobayashi N, Ishimaru T (2017) Fine mapping of a quantitative trait locus for spikelet number per panicle in a new plant type rice and evaluation of a nearisogenic line for grain productivity. Jr Exp Bot 68:2693–2702

Sekhar S, Panda BB, Mohapatra T, Das K, Shaw BP, Kariali E, Mohapatra PK (2015) Spikelet-specific variation in ethylene production and constitutive expression of ethylene receptors and signal transducers during grain filling of compact-and lax-panicle rice (Oryza sativa) cultivars. J Plant Physiol 179:21–34

Song S, Wang G, Hu Y, Liu H, Bai X, Qin R, Xing Y (2018) *OsMFT1* increases spikelets per panicle and delays heading date in rice by suppressing *Ehd1*, *FZP* and *SEPALLATA*-like genes. J Exp Bot 69:4283–4293

Tamaki H, Reguera M, Abdel-Tawab YM, Takebayashi Y, Kasahara H, Blumwald E (2015) Targeting hormone-related pathways to improve grain yield in rice: a chemical approach. PLoS One 10:e0131213. https://doi.org/10.1371/journal.pone.0131213

Tanaka W, Toriba T, Ohmori Y, Yoshida A, Kawai A, Mayama-Tsuchida T, Ichikawa H, Mitsuda N, Ohme-Takagi M, Hirano H (2012) The YABBY gene TONGARI-BOUSHI1 is involved in lateral organ development and maintenance of meristem organization in the rice spikelet. Plant Cell 24:80–95

Terao T, Nagata K, Morino K, Hirose T (2010) A gene controlling the number of primary rachis branches also controls the vascular bundle formation and hence is responsible to increase the harvest index and grain yield in rice. Theor Appl Genet 120:875–893

Van Wart J, Kersebaumb KC, Peng S, Milner M, Cassman KG (2013) Estimating crop yield potential at regional to national scales. Field Crop Res 143:34–43

Virk P, Khush G, Peng S (2005) Breeding to enhance yield potential of rice at IRRI: the ideotype approach [2004]. International Rice Research Notes. AGRIS:5–9

Wang Y, Li J (2008) Molecular basis of plant architecture. Annu Rev Plant Biol 59:253–279

Wang L, Zeng X, Zhuang H, Shen Y, Chen H, Wang Z, Long J, Ling Y, He G, Li Y (2017) Ectopic expression of OsMADS1 caused dwarfism and spikelet alteration in rice. Plant Growth Regul 81:433–442

Wang W, Xu X, Zhu C, Gu J, Zhang W, Liu G, Zhu J (2019) Elevated CO 2-induced changes in cytokinin and nitrogen metabolism are associated with different responses in the panicle architecture of two contrasting rice genotypes. Plant Growth Regul 89:119–129

Wang Q, Zhu Y, Zou X, Li F, Zhang J, Kang Z, Li X, Yin C, Lin Y (2020) Nitrogen deficiency-induced decrease in cytokinins content promotes rice seminal root growth by promoting root meristem cell proliferation and cell elongation. Cell 9:916. https://doi.org/10.3390/cells9040916

Xu H, Zhao M, Zhang Q, Xu Z, Xu Q (2016) The *DENSE AND ERECT PANICLE 1* (*DEP1*) gene offering the potential in the breeding of high-yielding rice. Breed Sci 66:659–667

Xu X, Zhang M, Xu Q, Feng Y, Yaun X, Yu H, Wang Y, Wei X, Yang Y (2020) Quantitative trait loci identification and genetic diversity analysis of panicle structure and grain shape in rice. Plant Growth Regul 90:89–100. https://doi.org/10.1007/s10725-019-00549-4

Yamaki S, Miyabayashi T, Eiguchi M, Kitano H, Nonomura K-I, Kurata N (2010) Diversity of panicle branching patterns in wild relatives of rice. Breed Sci 60:586–596

Yamburenko MV, Kieber JJ, Schaller GE (2017) Dynamic patterns of expression for genes regulating cytokinin metabolism and signaling during rice inflorescence development. PLoS One 12(4):e0176060. https://doi.org/10.1371/journal.pone

Yang W, Peng S, Laza RC, Visperas RM, Dionisio-Sese ML (2007) Grain yield and yield attributes of new plant type and hybrid rice. Crop Sci 47:1393–1400

Yoshida S (1981) Fundamentals of rice crop science. International Rice Research Institute, Philippines

Yoshida A, Sasao M, Yasuno N, Takagi K, Daimon Y, Chen R, Yamazaki R, Tokunaga R, Kitaguchi Y, Sato Y (2013) *TAWAWA1*, a regulator of rice inflorescence architecture, functions through the suppression of meristem phase transition. Proc Nat Acad Sci USA 110:767–772

Yoshida A, Suzaki T, Tanaka W, Hirano HY (2009) The homeotic gene long sterile lemma (G1) specifies sterile lemma identity in the rice spikelet. Proc Nat Acad Sci USA 106:20103–20108

Yoshida A, Ohmori Y, Kitano H, Taguchi-Shiobara F, Hirano H (2012) ABERRANT SPIKELET AND PANICLE1, encoding a TOPLESS-related transcriptional co-repressor, is involved in the regulation of meristem fate in rice. Plant J 70:327–339

Yu H, Qiu Z, Xu Q, Wang Z, Zeng D, Hu J, Zhang G, Zhu L, Gao Z, Chen G, Guo L, Qian Q, Ren D (2017) Fine mapping of *LOW TILLER 1*, a gene controlling tillering and panicle branching in rice. Plant Growth Regul 83:93–104

Zhang T, Li Y, Ma L, Sang X, Ling Y, Wang Y, Yu P, Zhuang H, Huang J, Wang N, Zhao F, Zhang C, Yang Z, Fang L, He G (2017) LATERAL FLORET 1 induced the three florets spikelet in rice. Proc Nat Acad Sci USA 114:9984–9989

Zhao M, Sun J, Xiao Z, Chang F, Xu H, Tang L, Chen W, Xu Z, Xu Q (2016) Variations in *DENSE AND ERECT PANICLE 1* (*DEP1*) contribute to the diversity of the panicle trait in high-yielding japonica rice varieties in northern China. Breed Sci 16058. https://doi.org/10.1270/jsbbs.16058

Zhu G, Ren Z, Liu Y, Lu F, Gu L, Shi Y, Liu J, Zhou G, Nimir NTA, Mohapatra PK (2020) Optimization of leaf properties and plant phenotype through yield-based genetic improvement of rice over a period of seventy years in the Yangtze River Basin of China. Food Energy Secur 2020:e223. https://doi.org/10.1002/fes3.223

Zhu Z, Tan L, Fu Y, Liu F, Cai H, Xie D, Wu F, Wu J, Matsumoto T, Son C (2013) Genetic control of inflorescence architecture during rice domestication. Nat Commun 4:2200. https://doi.org/10.1038/ncomms3200

MIX
Papier aus verantwortungsvollen Quellen
Paper from responsible sources
FSC® C105338
FSC
www.fsc.org

If you have any concerns about our products,
you can contact us on
ProductSafety@springernature.com

In case Publisher is established outside the EU,
the EU authorized representative is:
Springer Nature Customer Service Center GmbH
Europaplatz 3, 69115 Heidelberg, Germany

Printed by Libri Plureos GmbH
in Hamburg, Germany